T0135087

Springer Proceedings in Mathematics & Statistics

Volume 262

Springer Proceedings in Mathematics & Statistics

This book series features volumes composed of selected contributions from workshops and conferences in all areas of current research in mathematics and statistics, including operation research and optimization. In addition to an overall evaluation of the interest, scientific quality, and timeliness of each proposal at the hands of the publisher, individual contributions are all refereed to the high quality standards of leading journals in the field. Thus, this series provides the research community with well-edited, authoritative reports on developments in the most exciting areas of mathematical and statistical research today.

More information about this series at http://www.springer.com/series/10533

Luigi G. Rodino · Joachim Toft
Editors

Mathematical Analysis and Applications—Plenary Lectures

ISAAC 2017, Växjö, Sweden

Editors
Luigi G. Rodino
Dipartimento di Matematica
Università di Torino
Turin, Italy

Joachim Toft
Department of Mathematics
Linnaeus University
Växjö, Sweden

ISSN 2194-1009 ISSN 2194-1017 (electronic)
Springer Proceedings in Mathematics & Statistics
ISBN 978-3-030-13151-7 ISBN 978-3-030-00874-1 (eBook)
https://doi.org/10.1007/978-3-030-00874-1

Mathematics Subject Classification (2010): 35-XX, 46-XX, 60-XX, 32-XX, 47-XX, 65-XX

This Springer imprint is published by the registered company Springer Nature Switzerland AG
The registered company address is: Gewerbestrasse 11, 6330 Cham, Switzerland

Preface

This volume is a collection of articles devoted to current research topics in Mathematical Analysis, with emphasis on Fourier analysis and general theory of partial differential equations. It originates from plenary lectures given at the 11th International ISAAC Congress, held during 14–18 August 2017 at the University of Växjö, Sweden.

The papers are authored by six eminent specialists and aim at presenting to a large audience some challenging and attractive themes of the modern Mathematical Analysis, namely as follows:

- The contribution of Nils Dencker is devoted to the local solvability of sub-principal type operators. Dencker proved in 2006 the so-called Nirenberg–Treves conjecture, expressing a necessary and sufficient condition for the solvability of the operators of principal type. In this paper, he addresses to operators with multiple characteristics. Namely, the principal symbol of the operator is assumed to vanish of at least second order on an involutive manifold. Precise necessary conditions are expressed for solvability, in terms of the sub-principal symbol, involving lower order terms.
- In these last years, the fractional-order Laplacian was frequently used in Mathematical Physics, Differential Geometry, Probability and Finance. The paper of Gerd Grubb is devoted to this operator and its generalizations, with attention to homogeneous and non-homogeneous boundary value problems and corresponding heat equations. The paper represents the first general and rigorous approach to non-local problems of this type, in the setting of pseudo-differential operators.
- The paper of Abdelhamid Meziani is devoted to degenerate complex vector fields in the plane, basic example being the Mizohata operator. The corresponding equations share many properties with the Cauchy–Riemann equation. Combining with the theory of the hypo-analytic structures of Treves, Meziani presents an elegant treatment of the subject, including generalizations of the Riemann–Hilbert problem.

- The paper of Alberto Parmeggiani gives a survey on the problem of the lower bounds. Does the positivity of the symbol imply the positivity of the corresponding operator? A precise answer to this question is largely open. Main reference is the classical theorem of Fefferman–Phong, stating positivity of the operator modulo small errors. Here Parmeggiani addresses also to the case of systems of operators; for them, precise lower bounds represent a largely unexplored area.
- Following the original idea of Morrey [1], Peetre and others introduced the function spaces which are nowadays called Morrey spaces. They have a large number of applications in different contexts, and their definition was extended in different directions. Yoshihiro Sawano, in his contribution to the volume, presents a review addressed to non-experts, including the interpolation theory and the weighted version.
- Localization operators were defined by Berezin [2] in the frame of Quantum Mechanics, and later used by Daubechies [3] and others in Signal Theory. The paper of Nenad Teofanov is devoted to this important topic. The definition of localization operator is given here in terms of the Grossmann–Royer transform, which simplifies the proof of several known results. Particular emphasis is given to the action on Gelfand–Shilov and modulation spaces.

Besides plenary talks, about 250 scientific communications were delivered during the Växjö ISAAC Congress. Their texts are published in an independent volume. On the whole, the Congress demonstrated, in particular, the relevant role of the Nordic European countries in several research areas of Mathematical Analysis.

Turin, Italy — Luigi G. Rodino
Växjö, Sweden — Joachim Toft
July 2018

References

1. Morrey Jr., C.B.: On the solutions of quasi-linear elliptic partial differential equations. Trans. Am. Math. Soc. **43**(1), 126–166 (1938)
2. Berezin, F.A.: Wick and anti-wick symbols of operators. Mat. Sb. (N.S.) **86**(128), 578–610 (1971)
3. Daubechies, I.: Time-frequency localization operators: a geometric phase space approach. IEEE Trans. Inform. Theory **34**(4), 605–612 (1988)

Contents

Solvability of Subprincipal Type Operators 1
Nils Dencker

Fractional-Order Operators: Boundary Problems, Heat Equations 51
Gerd Grubb

A Class of Planar Hypocomplex Vector Fields: Solvability and Boundary Value Problems 83
A. Meziani

Almost-Positivity Estimates of Pseudodifferential Operators 109
Alberto Parmeggiani

Morrey Spaces from Various Points of View 139
Yoshihiro Sawano

The Grossmann–Royer Transform, Gelfand–Shilov Spaces, and Continuity Properties of Localization Operators on Modulation Spaces 161
Nenad Teofanov

Contributors

Nils Dencker Centre for Mathematical Sciences, Lund University, Lund, Sweden

Gerd Grubb Department of Mathematical Sciences, Copenhagen University, Copenhagen, Denmark

A. Meziani Department of Mathematics, Florida International University, Miami, FL, USA

Alberto Parmeggiani Department of Mathematics, University of Bologna, Bologna, Italy

Yoshihiro Sawano Department of Mathematics, Tokyo Metropolitan University, Hachioji Tokyo, Japan

Nenad Teofanov Department of Mathematics and Informatics, University of Novi Sad, Novi Sad, Serbia

Solvability of Subprincipal Type Operators

Nils Dencker

Abstract In this paper we consider the solvability of pseudodifferential operators in the case when the principal symbol vanishes of order $k \geq 2$ at a nonradial involutive manifold Σ_2. We shall assume that the operator is of subprincipal type, which means that the kth inhomogeneous blowup at Σ_2 of the refined principal symbol is of principal type with Hamilton vector field parallel to the base Σ_2, but transversal to the symplectic leaves of Σ_2 at the characteristics. When $k = \infty$ this blowup reduces to the subprincipal symbol. We also assume that the blowup is essentially constant on the leaves of Σ_2, and does not satisfying the Nirenberg–Treves condition (Ψ). We also have conditions on the vanishing of the normal gradient and the Hessian of the blowup at the characteristics. Under these conditions, we show that P is not solvable.

1 Introduction

We shall consider the solvability for a classical pseudodifferential operator $P \in \Psi^m_{cl}(X)$ on a C^∞ manifold X of dimension n. This means that P has an expansion $p_m + p_{m-1} + \dots$ where $p_j \in S^j_{hom}$ is homogeneous of degree $j, \forall j$, and $p_m = \sigma(P)$ is the principal symbol of the operator. A pseudodifferential operator is said to be of principal type if the Hamilton vector field H_{p_m} of the principal symbol does not have the radial direction $\xi \cdot \partial_\xi$ on $p_m^{-1}(0)$, in particular $H_{p_m} \neq 0$. We shall consider the case when the principal symbol vanishes of at least second order at an involutive manifold Σ_2, thus P is not of principal type.

P is locally solvable at a compact set $K \subseteq X$ if the equation

$$Pu = v \tag{1.1}$$

N. Dencker (✉)
Centre for Mathematical Sciences, Lund University, Box 118, 221 00 Lund, Sweden
e-mail: dencker@maths.lth.se

L. G. Rodino and J. Toft (eds.), *Mathematical Analysis and Applications—Plenary Lectures*, Springer Proceedings in Mathematics & Statistics 262,
https://doi.org/10.1007/978-3-030-00874-1_1

has a local solution $u \in \mathcal{D}'(X)$ in a neighborhood of K for any $v \in C^\infty(X)$ in a set of finite codimension. We can also define microlocal solvability of P at any compactly based cone $K \subset T^*X$, see Definition 2.6.

For pseudodifferential operators of principal type, local solvability is equivalent to condition (Ψ) on the principal symbol, see [3, 12]. This condition means that

$$\begin{aligned}&\operatorname{Im} ap_m \text{ does not change sign from } - \text{ to } + \\ &\qquad\qquad \text{along the oriented bicharacteristics of } \operatorname{Re} ap_m \end{aligned} \tag{1.2}$$

for any $0 \neq a \in C^\infty(T^*X)$. The oriented bicharacteristics are the positive flow of the Hamilton vector field $H_{\operatorname{Re} ap_m} \neq 0$ on which $\operatorname{Re} ap_m = 0$, these are also called semibicharacteristics of p_m. Condition (1.2) is invariant under multiplication of p_m with nonvanishing factors, and symplectic changes of variables, thus it is invariant under conjugation of P with elliptic Fourier integral operators. Observe that the sign changes in (1.2) are reversed when taking adjoints, and that it suffices to check (1.2) for some $a \neq 0$ for which $H_{\operatorname{Re} ap} \neq 0$ according to [13, Theorem 26.4.12].

For operators which are not of principal type, the situation is more complicated and the solvability may depend on the lower order terms. Then the *refined principal symbol*

$$p_{sub} = p_m + p_{m-1} + \frac{i}{2} \sum_j \partial_{\xi_j} \partial_{x_j} p_m \tag{1.3}$$

is invariantly defined modulo S^{m-2} under changes of coordinates, see Theorem 18.1.33 in [13]. In the Weyl quantization the refined principal symbol is given by $p_m + p_{m-1}$.

When Σ_2 is not involutive, there are examples where the operator is solvable for any lower order terms. For example when P is effectively hyperbolic, then even the Cauchy problem is solvable for any lower order term, see [10, 14, 18]. There are also results in the cases when the principal symbol is a product of principal type symbols not satisfying condition (Ψ), see [1, 8, 9, 20, 25].

In the case where the principal symbol is real and vanishes of at least second order at the involutive manifold there are several results, mostly in the case when the principal symbol is a product of real symbols of principal type. Then the operator is not solvable if the imaginary part of the subprincipal symbol has a sign change of finite order on a bicharacteristic of one the factors of the principal symbol, see [7, 19, 22, 23].

This necessary condition for solvability has been extended to some cases when the principal symbol is real and vanishes of second order at the involutive manifold. The conditions for solvability then involve the sign changes of the imaginary part of the subprincipal symbol on the limits of bicharacteristics from outside the manifold, thus on the leaves of the symplectic foliation of the manifold, see [15–17, 26].

This has been extended to more general limit bicharacteristics of real principal symbols in [4]. There we assumed that the bicharacteristics converge in C^∞ to a limit bicharacteristic. We also assumed that the linearization of the Hamilton vector field is tangent to and has uniform bounds on the tangent spaces of some Lagrangean manifolds at the bicharacteristics. Then P is not solvable if condition Lim(Ψ) is not satisfied on the limit bicharacteristics. This means that the quotient of the imaginary part of the subprincipal symbol with the norm of the Hamilton vector field switches sign from $-$ to $+$ on the bicharacteristics and becomes unbounded when converging to the limit bicharacteristic. This was generalized in [6] to operators with complex principal symbols. There we assumed that the normalized complex Hamilton vector field of the principal symbol converges to a real vector field. Then the limit bicharacteristics are uniquely defined, and one can invariantly define the imaginary part of the subprincipal symbol. Thus condition Lim(Ψ) is well defined and we proved that it is necessary for solvability.

In [5] we considered the case when the principal symbol (not necessarily real valued) vanishes of at least second order at a nonradial involutive manifold Σ_2. We assumed that the operator was of *subprincipal type*, i.e., that the subprincipal symbol on Σ_2 is of principal type with Hamilton vector field tangent to Σ_2 at the characteristics, but transversal to the symplectic leaves of Σ_2. Then we showed that the operator is not solvable if the subprincipal symbol is essentially constant on the symplectic leaves of Σ_2 and does not satisfy condition (Ψ), which we call Sub(Ψ). In the case when the sign change is of infinite order, we also had conditions on the vanishing of both the Hessian of the principal symbol and the complex part of the gradient of the subprincipal symbol.

The difference between [5, 6] is that in the first case the Hamilton vector field of the principal symbol dominates, and in the second the Hamilton vector field of the subprincipal symbol dominates. In this paper, we shall study the case when condition (Ψ) is not satisfied for the refined principal symbol (1.3) which combines both the principal and subprincipal symbols. We shall assume that the principal symbol vanishes of at least order $k \geq 2$ on at a nonradial involutive manifold Σ_2. When $k < \infty$ then the kth jet of the principal symbol is well defined at Σ_2, but since the refined principal symbol is inhomogeneous we make an inhomogeneous blowup, called *reduced subprincipal symbol* by Definition 2.1. We assume that the operator is of *subprincipal type*, i.e., the reduced subprincipal symbol is of principal type, see Definition 2.2. We define condition $\text{Sub}_k(\Psi)$, which is condition (Ψ) on the reduced subprincipal symbol, see Definition 2.3. We assume that the blowup of the refined principal symbol is essentially constant on the symplectic leaves of Σ_2, see (2.27). We also have conditions on the rate of the vanishing of the normal gradient (2.19) and when $k = 2$ of the Hessian of the reduced subprincipal symbol (2.21). When $k = \infty$ all the Taylor terms vanish and condition $\text{Sub}_\infty(\Psi)$ reduces to condition Sub(Ψ) on Σ_2 from [5]. Under these conditions, we show that if condition $\text{Sub}_k(\Psi)$ is not satisfied near a bicharacteristic of the reduced subprincipal symbol then the operator is not solvable near the bicharacteristic, see Theorem 2.1 which is the main result of the paper. In the case when the sign change of $\text{Sub}_k(\Psi)$ is on Σ_2 we get a different result than in [5], since now we localize the pseudomodes with the the phase function instead of the amplitude.

The plan of the paper is as follows. In Sect. 2 we make the definitions of the symbols we are going to use, state the conditions and the main result, Theorem 2.1.

In Sect. 3 we present some examples, and in Sect. 4 we develop normal forms of the operators, which are different in the case when the principal symbol vanishes of finite or infinite order at Σ_2. The approximate solutions, or pseudomodes, are defined in Sect. 5. In Sect. 6 we solve the eikonal equation in the case when the principal symbol vanishes of finite order, in Sect. 7 we solve it in the case when the bicharacteristics are on Σ_2 and in Sect. 8 we solve the transport equations. In order to solve the eikonal and transport equations uniformly we use the estimates of Lemma 6.1, which is proved in Sect. 9. Finally, Theorem 2.1 is proved in Sect. 10.

2 Statement of Results

Let $\sigma(P) = p \in S^m_{\mathrm{hom}}$ be the homogeneous principal symbol of P, we shall assume that

$$\sigma(P) \text{ vanishes of at least second order at } \Sigma_2 \subset T^*X \setminus 0 \tag{2.1}$$

where

$$\Sigma_2 \text{ is a nonradial involutive manifold of codimension } d \tag{2.2}$$

where $0 < d < n-1$ with $n = \dim X$. Here nonradial means that the radial direction $\langle \xi, \partial_\xi \rangle$ is not in the span of the Hamilton vector fields of the manifold, i.e., not equal to H_f on Σ_2 for any $f \in C^1$ vanishing at Σ_2. Then by a change of local homogeneous symplectic coordinates we may assume that locally

$$\Sigma_2 = \{\, \eta = 0 \,\} \qquad (\xi, \eta) \in \mathbf{R}^{n-d} \times \mathbf{R}^d \qquad \xi \neq 0 \tag{2.3}$$

for some $0 < d < n-1$, which can be achieved by a conjugation with elliptic Fourier integral operators.

Now, since p vanishes of at least second order at Σ_2 we can define the *order* of p as

$$2 \le \kappa(w) = \min\{\, |\alpha| : \partial^\alpha p(w) \neq 0 \,\} \qquad w \in \Sigma_2 \tag{2.4}$$

and $\kappa(\omega) = \min_{w \in \omega} \kappa(w)$ for $\omega \subseteq \Sigma_2$, which is equal to ∞ when p vanishes of infinite order. This is an upper semicontinuous function on Σ_2, but since $\kappa(w)$ is has values in $\mathbf{N} \cup \infty$, it attains its minimum $\kappa(\omega)$ on any set $\omega \subseteq \Sigma_2$.

If P is of principal type near Σ_2 then, since solvability is an open property, we find that a necessary condition for P to be solvable at Σ_2 is that condition (Ψ) for the principal symbol is satisfied in some neighborhood of Σ_2. Naturally, this condition is empty on Σ_2 where we instead have conditions on the *refined principal symbol:*

$$p_{sub} = p + p_{m-1} + \frac{i}{2} \sum_j \partial_{x_j} \partial_{\xi_j} p \tag{2.5}$$

(for the Weyl quantization, the refined principal symbol is given by $p + p_{m-1}$). The refined principal symbol is invariantly defined as a function on T^*X modulo S^{m-2} under conjugation with elliptic Fourier integral operators, see [13, Theorem 18.1.33] and [11, Theorem 9.1]. (The latter result is for the Weyl quantization, but the result easily carries over to the Kohn–Nirenberg quantization for classical operators.) The subprincipal symbol

$$p_s = p_{m-1} + \frac{i}{2}\sum_j \partial_{x_j}\partial_{\xi_j} p \tag{2.6}$$

is invariantly defined on Σ_2 under conjugation with elliptic Fourier integral operators.

Remark 2.1 When $\Sigma_2 = \{\, \xi_1 = \xi_2 = \cdots = \xi_j = 0 \,\}$ is involutive, the refined principal symbol is equal to $p_s = p_{m-1}$ at Σ_2.

In fact, this follows since $\partial_\xi p \equiv 0$ on Σ_2. When composing P with an elliptic pseudodifferential operator C, the value of the refined principal symbol of CP is equal to $cp_{sub} + \frac{i}{2}H_p c$ which is equal to cp_s at Σ_2, where $c = \sigma(C)$. Observe that the refined principal symbol is complexly conjugated when taking the adjoint of the operator, see [13, Theorem 18.1.34].

The conormal bundle $N^*\Sigma_2 \subset T^*(T^*X)$ of Σ_2 is the dual of the normal bundle $T_{\Sigma_2}T^*X/T\Sigma_2$. The conormal bundle can be parametrized by first choosing local homogeneous symplectic coordinates so that Σ_2 is given by $\{\, \eta = 0 \,\}$. Then the fiber of $N^*\Sigma_2$ can be parametrized by $\eta \in \mathbf{R}^d$, $d = \operatorname{Codim} \Sigma_2$, so that $N^*\Sigma_2 \cong \Sigma_2 \times \mathbf{R}^d$ and different parametrizations gives linear transformations on the fiber.

We define the kth jet $J^k_w(f)$ of a C^∞ function f at $w \in \Sigma_2$ as the equivalence class of f modulo functions vanishing of order $k+1$ at w. If $k = \kappa(\omega) < \infty$ is given by (2.4) for the open neighborhood $\omega \subset \Sigma_2$ then for $w = (x, y, \xi, 0) \in \Sigma_2$ we find that $J^k_w(p)$ is a well defined homogeneous function on $N^*\Sigma_2$ given by

$$N^*_w\Sigma_2 \ni (w, \eta) \mapsto J^k_w(p)(\eta) = \partial^k_\eta p(w)(\eta) \tag{2.7}$$

since $\partial^j p \equiv 0$ on ω, $j < k$. Here $\partial^k_\eta p(w)$ is the k-form given by the Taylor term of order k of p. If $\kappa(\omega) = \infty$ then of course any jet of p vanishes identically on ω. Here and in the following, the η variables will be treated as parameters.

Definition 2.1 When $k = \kappa(\omega) < \infty$ for some open set $\omega \subset \Sigma_2$ we define the *reduced subprincipal symbol* by

$$N^*\Sigma_2 \ni (w, \eta) \mapsto p_{s,k}(w, \eta) = J^k_w(p)(\eta) + p_s(w) \qquad w \in \omega \tag{2.8}$$

which is a polynomial in η of degree k and is given by the blowup of the refined principal symbol at Σ_2 see Remark 2.4. If $\kappa(\omega) = \infty$ then we define $p_{s,\infty} = p_s\big|_{\Sigma_2}$ so we have (2.8) for any k.

Remark 2.2 The reduced subprincipal symbol is well-defined up to nonvanishing factors under conjugation with elliptic homogeneous Fourier integral operators and under composition with classical elliptic pseudodifferential operator.

In fact, the reduced subprincipal symbol is equal to the refined principal symbol modulo terms homogeneous of degree m vanishing at Σ_2 of order $k+1$ and terms homogeneous of degree $m-1$ vanishing at Σ_2. When composing with an elliptic pseudodifferential operator, both the terms in the refined subprincipal symbol gets multiplied with the same nonvanishing factor, and the terms proportional to ∂p vanish on Σ_2. Observe that if we multiply p_{sub} with c then $p_{s,k}$ gets multiplied with $c\big|_{\Sigma_2}$.

Since p_s is only defined on Σ_2, the Hamilton field $H_{p_{s,k}}$ is only well defined modulo terms that are tangent to the symplectic leaves of Σ_2, which are spanned by the Hamilton vector fields of functions vanishing on Σ_2. Therefore, we shall assume that the reduced principal symbol essentially is constant on the leaves of Σ_2 for fixed η by assuming that

$$\left|dp_{s,k}\big|_{TL}\right| \le C_0 |p_{s,k}| \quad \text{at } \omega \text{ when } |\eta - \eta_0| \ll 1 \tag{2.9}$$

for any leaf L of Σ_2 where $\omega \subset \Sigma_2$. Since $p_{s,k}$ is determined by the Taylor coefficients of the refined principal symbol at Σ_2 we find that (2.9) is determined on Σ_2. When $\eta = 0$ we get condition (2.9) on p_s at Σ_2 which was used in [5]. Condition (2.9) is invariant under multiplication with nonvanishing factors and when $dp_{s,k} \neq 0$ on $p_{s,k}^{-1}(0)$ it is equivalent to the fact that $p_{s,k}$ is constant on the leaves up to nonvanishing factors by the following lemma.

Lemma 2.1 *Assume that $f(x, y, \zeta) \in C^\infty$ is a polynomial in ζ of degree m for $(x, y) \in \Omega$, such that $\partial_x f \neq 0$ when $f = 0$. Assume that Ω is an open bounded C^∞ domain such that $\Omega_{x_0} = \Omega \bigcap \{ x = x_0 \}$ is simply connected for all x_0. Let $\Xi = \{ (x, \zeta) : \exists\, y\ (x, y) \in \Omega$ and $|\zeta - \zeta_0| < c \}$ be the projection on the (x, ζ) variables and assume that there exist $y_0(x) \in C^\infty$ such that $(x, y_0(x), \zeta) \in \Omega \times \{ |\zeta - \zeta_0| < c \}$, $\forall (x, \zeta) \in \Xi$. Then*

$$|\partial_y f| \le C_0 |f| \quad \text{in } \Omega \text{ when } |\zeta - \zeta_0| < c \tag{2.10}$$

for $c > 0$ implies that

$$f(x, y, \zeta) = c(x, y, \zeta) f_0(x, \zeta) \quad \text{in } \Omega \text{ when } |\zeta - \zeta_0| < c \tag{2.11}$$

where $0 < c_0 \le c(x, y, \zeta) \in C^\infty$ and $f_0(x, \zeta) = f(x, y_0(x), \zeta) \in C^\infty$, which implies that $\left|\partial_y^\alpha f\right| \le C_\alpha |f|$, $\forall\, \alpha$, in Ω when $|\zeta - \zeta_0| < c$.

If $\partial_w p_{s,k} \neq 0$ when $p_{s,k} = 0$ and $p_{s,k}$ satisfies (2.9), then we find from Lemma 2.1 after possibly shrinking ω that $p_{s,k}$ is constant on the leaves of Σ_2 in ω when $|\eta - \eta_0| < c_0$ after multiplication with a nonvanishing factor.

Proof Let $\Xi_0 = \{\, (x, \zeta) \in \Xi : f(x, y_0(x), \zeta) = 0 \,\}$. We shall first prove the result when $(x, \zeta) \in \Xi \setminus \Xi_0$. Then $f \neq 0$ at $(x, y_0(x), \zeta)$ and (2.10) gives that $\partial_y \log f$ is uniformly bounded near $(x, y_0(x), \zeta)$, where $\log f$ is a branch of the complex logarithm. Thus, by integrating with respect to y starting at $y = y_0(x)$ in the simply connected $\Omega_x \times \{\, |\zeta - \zeta_0| < c \,\}$ we find that $\log f(x, y, \zeta) - \log f(x, y_0(x), \zeta) \in C^\infty$ is bounded and by exponentiating we obtain

$$f(x, y, \zeta) = c(x, y, \zeta) f_0(x, \zeta) \quad \text{in } \Omega_x \text{ for } |\zeta - \zeta_0| < c \tag{2.12}$$

when $(x, \zeta) \in \Xi \setminus \Xi_0$. Here $f_0(x, \zeta) = f(x, y_0(x), \zeta) \in C^\infty$ and $0 < c_0 \leq c(x, y, \zeta) \in C^\infty$ is uniformly bounded such that $c(x, y_0(x), \zeta) \equiv 1$. This gives that $f^{-1}(0)$ is constant in y when $(x, \zeta) \notin \Xi_0$.

Since $\partial_x f \neq 0$ when $f = 0$ we find that Ξ_0 is nowhere dense. Let $(x_0, \zeta_0) \in \Xi_0$ and choose $z \in \mathbf{C}$ such that $\partial_x \operatorname{Re} z f(x_0, y_0(x_0), \zeta_0) \neq 0$. Let $S_\pm = \{\, \pm \operatorname{Re} z f(x, y_0(x), \zeta) > 0 \,\}$ then

$$f(x, y, \zeta) = c_\pm(x, y, \zeta) f(x, y_0(x), \zeta) \quad \text{in } S_\pm \tag{2.13}$$

where $0 < c_0 \leq c_\pm(x, y, \zeta) \in C^\infty$ is uniformly bounded. By taking the limit of (2.13) at $S = \{\, \operatorname{Re} z f(x, y_0(x), \zeta) = 0 \,\}$ we find that $c_+ = c_-$ when $f \neq 0$ at S. When $f = 0$ at S then by differentiating (2.13) in x we find that $c_+ = c_-$. By repeatedly differentiating (2.13) in x we obtain by recursion that $c_\pm$ extends to $c_0 < c(x, y, \zeta) \in C^\infty$ in Ω when $|\zeta - \zeta_0| < c$ so that (2.11) holds. □

If $p_{s,k}$ is constant in y in a neighborhood of the semibicharacteristic, then the Hamilton field $H_{p_{s,k}}$ will be constant on the leaves and defined modulo tangent vector to the leaves. Therefore we shall introduce a special symplectic structure on $N^*\Sigma_2$. Recall that the symplectic annihilator to a linear space consists of the vectors that are symplectically orthogonal to the space. Let $T\Sigma_2^\sigma$ be the symplectic annihilator to $T\Sigma_2$, which spans the symplectic leaves of Σ_2. If $\Sigma_2 = \{\, \eta = 0 \,\}$, $(x, y) \in \mathbf{R}^{n-d} \times \mathbf{R}^d$, then the leaves are spanned by ∂_y. Let

$$T^\sigma \Sigma_2 = T\Sigma_2 / T\Sigma_2^\sigma \tag{2.14}$$

which is a symplectic space over Σ_2 which in these coordinates is parametrized by

$$T^\sigma \Sigma_2 = \left\{\, ((x_0, y_0; \xi_0, 0); (x, 0; \xi, 0)) \in T\Sigma_2 : \ (x, \xi) \in T^*\mathbf{R}^{n-d} \,\right\} \tag{2.15}$$

This is isomorphic to the symplectic manifold $T^*\mathbf{R}^{n-d}$ with $w \in \Sigma_2$ as parameter.

We define the symplectic structure of $N^*\Sigma_2$ by lifting the structure of Σ_2 to the fibers, so that the leaves of $N^*\Sigma_2$ are given by $L \times \{\, \eta_0 \,\}$ where L is a leaf of Σ_2 for $\eta_0 \in \mathbf{R}^d$. In the chosen coordinates, these leaves are parametrized by $\{\, (x_0, y; \xi_0, 0) \times \{\, \eta_0 \,\} : \ y \in \mathbf{R}^d \,\}$. The radial direction in $N^*\Sigma_2$ will be the radial direction in Σ_2, i.e. $\langle \xi, \partial_\xi \rangle$, lifted to the fibers. Similarly, a vector field $V \in T(N^*\Sigma_2)$ is parallel to the base of $N^*\Sigma_2$ if it is in $T\Sigma_2$, which means that $V\eta = 0$.

If $p_{s,k}$ is constant in y then $H_{p_{s,k}}$ coincides with the Hamilton vector field of $p_{s,k}$ on $p_{s,k}^{-1}(0) \subset N^*\Sigma_2$ with respect the symplectic structure on the symplectic manifold $N^*\Sigma_2$. In fact, in the chosen coordinates we obtain from (2.9) that

$$H_{p_{s,k}} = \partial_\xi p_{s,k}\partial_x - \partial_x p_{s,k}\partial_\xi \tag{2.16}$$

modulo ∂_y, which is nonvanishing if $\partial_{x,\xi} p_{s,k} \neq 0$. Thus $H_{p_{s,k}}$ is well-defined modulo terms containing ∂_y making it well defined on $T^\sigma\Sigma_2 \times \mathbf{R}^d$. Now, if $p_{s,k} = 0$ then by (2.9) we find that $dp_{s,k}\big|_{T\Sigma_2}$ vanishes on $T\Sigma_2^\sigma$ so $dp_{s,k}\big|_{T\Sigma_2}$ is well defined on $T^\sigma\Sigma_2$. We may identify $T(N^*\Sigma_2)$ with $T\Sigma_2 \times \mathbf{R}^d$ since the fiber η is linear.

Definition 2.2 We say that the operator P is of *subprincipal type* on $N^*\Sigma_2$ if the following hold when $p_{s,k} = 0$ on $N^*\Sigma_2$: $H_{p_{s,k}}$ is parallel to the base,

$$dp_{s,k}\big|_{T^\sigma\Sigma_2} \neq 0 \tag{2.17}$$

and the corresponding Hamilton vector field $H_{p_{s,k}}$ of (2.17) does not have the radial direction. The (semi)bicharacteristics of $p_{s,k}$ with respect to the symplectic structure of $N^*\Sigma_2$ are called the subprincipal (semi)bicharacteristics.

Clearly, if coordinates are chosen so that (2.3) holds, then (2.17) gives that $\partial_{x,\xi} p_{s,k} \neq 0$ when $p_{s,k} = 0$ and the condition that the Hamilton vector field does not have the radial direction means that $\partial_\xi p_{s,k} \neq 0$ or $\partial_x p_{s,k} \nparallel \xi$ when $p_{s,k} = 0$. Because of (2.17) we find that $H_{p_{s,k}}$ is transversal to the foliation of $N^*\Sigma_2$ and by (2.9) it is parallel to the base at the characteristics. The semibicharacteristic of $p_{s,k}$ can be written $\Gamma = \Gamma_0 \times \{\eta_0\} \subset T(N^*\Sigma_2)$, where $\Gamma_0 \subset \Sigma_2$ is transversal to the leaves of Σ_2 and η_0 is fixed. The definition can be localized to an open set $\omega \subset N^*\Sigma_2$. It is a generalization of the definition of subprincipal type in [5], which is the special case when $\eta = 0$. When P is of subprincipal type and satisfies (2.9), then we find from Lemma 2.1 that $p_{s,k}$ is constant on the leaves of Σ_2 near a semibicharacteristic after multiplication with a nonvanishing factor. We can now state a condition corresponding to (Ψ) on the reduced subprincipal symbol.

Definition 2.3 If $k = \kappa(\omega)$ for an open set $\omega \subset N^*\Sigma_2$, then we say that P satisfies condition $\mathrm{Sub}_k(\Psi)$ if $\mathrm{Im}\, ap_{s,k}$ does not change sign from $-$ to $+$ when going in the positive direction on the subprincipal bicharacteristics of $\mathrm{Re}\, ap_{s,k}$ in ω for any $0 \neq a \in C^\infty$.

Observe that when $k < \kappa(\omega)$ or $k = \kappa(\omega) = \infty$ then $p_{s,k} = p_s\big|_{\Sigma_2}$ on ω and $\mathrm{Sub}_k(\Psi)$ means that the subprincipal symbol p_s satisfies condition (Ψ) on $T^\sigma\Sigma_2$, which is condition $\mathrm{Sub}(\Psi)$ in [5]. In general, we have that condition $\mathrm{Sub}_k(\Psi)$ is condition (Ψ) given by (1.2) on the reduced subprincipal symbol $p_{s,k}$ with respect to the symplectic structure of $N^*\Sigma_2$. But it is equivalent to the condition (Ψ) on the reduced subprincipal symbol $p_{s,k}$ with respect to the standard symplectic structure. In fact, condition $\mathrm{Sub}_k(\Psi)$ means that condition (Ψ) holds for $p_{s,k}\big|_{\eta=\eta_0}$ for any η_0. By using Lemma 2.1 we may assume that $p_{s,k}$ is independent of y after multiplying with $0 \neq a \in C^\infty$. In that case, the conditions are equivalent and both are invariant under multiplication with nonvanishing smooth factors.

By the invariance of condition (Ψ) given by [13, Theorem 26.4.12] it suffices to check condition $\mathrm{Sub}_k(\Psi)$ for some a such that $H_{\mathrm{Re}\, ap_{s,k}} \neq 0$. We also find that condition $\mathrm{Sub}_k(\Psi)$ is invariant under symplectic changes of variables, thus it is invariant under conjugation of the operator by elliptic homogeneous Fourier integral operators. Observe that the sign change is reversed when taking the adjoint of the operator.

Next, we assume that condition $\mathrm{Sub}_k(\Psi)$ is not satisfied on a semibicharacteristic Γ of $p_{s,k}$, i.e., that $\mathrm{Im}\, ap_{s,k}$ changes sign from $-$ to $+$ on the positive flow of $H_{\mathrm{Re}\, ap_{s,k}} \neq 0$ for some $0 \neq a \in C^\infty$, where η is constant on Γ. Thus, by Lemma 2.1 we may assume that $p_{s,k}$ is constant on the leaves in a neighborhood ω of Γ, and by multiplying with a we may assume that $a \equiv 1$ and that y is constant on the semibicharacteristic.

Definition 2.4 Let p be of subprincipal type on $N^*\Sigma_2$ and Γ a subprincipal semibicharacteristic of p. We say that a C^∞ section of spaces $L \subset T(N^*\Sigma_2)$ is *gliding* for Γ if L is symplectic of maximal dimension $2n - 2(d+1) \geq 2$ so that L is the symplectic annihilator of $T\,\Gamma$ and the foliation of Σ_2, which gives $L \subset T\,\Sigma_2$ since η is constant on L. We say that a C^∞ foliation of $N^*\Sigma_2$ with symplectic leaves M is gliding for Γ if the section of tangent spaces TM is a gliding section for Γ.

Actually, he gliding foliation M for a subprincipal semibicharacteristic Γ is uniquely defined at Γ, since it is determined by the unique annihilator TM and Γ is transversal to the foliation of Σ_2 when $p = 0$ by (2.17). This definition can be localized to a neighborhood of a subprincipal semibicharacteristic.

Example 2.1 Let p be of subprincipal type on $N^*\Sigma_2$. Assume that $\Sigma_2 = \{\, \eta = 0 \,\}$, $\partial_y p = \{\, \eta, p \,\} = 0$ and $\partial_{x,\xi}$ spans TM of the gliding foliation M of $N^*\Sigma_2$ for the bicharacteristics of $H_{\mathrm{Re}\, p} \neq 0$. Then we may complete x, ξ, $\tau = \mathrm{Re}\, p$ and η to a symplectic coordinate system $(t, x, y; \tau, \xi, \eta)$ so that the foliation M is given by intersection of the level sets of τ, t, y and η. In fact, in that case we have $\partial\, \mathrm{Re}\, p \neq 0$ but $\partial_x \mathrm{Re}\, p = \partial_\xi \mathrm{Re}\, p = 0$.

In the case when $\eta_0 \neq 0$ and $k = \kappa(\omega) < \infty$ we will have estimates on the rate of vanishing of $\partial_\eta p_{s,k}$ on the subprincipal semibicharacteristic. Recall that the semibicharacteristic can be written $\Gamma \times \{\, \eta_0 \,\}$. Observe that

$$\partial_\eta p_{s,k} = \mathcal{J}^{k-1}(\partial_\eta p) = \mathcal{J}^{k-1}(\partial p) \tag{2.18}$$

since p vanishes of at least order k at Σ_2 and that the normal derivatives ∂_η is well-defined modulo nonvanishing factors at $\eta = 0$. Let $\omega \subset \Sigma_2$ be a neighborhood of the subprincipal semibicharacteristic Γ and let M be the local C^∞ foliation of $N^*\Sigma_2$ at ω which is gliding for the semibicharacteristics. When $\eta_0 \neq 0$ we shall assume that there exists $\varepsilon > 0$ so that

$$\left| V_1 \cdots V_\ell\, \partial_\eta p_{s,k} \right| \leq C_\ell |p_{s,k}|^{1/k+\varepsilon} \quad \text{on } \omega \text{ when } |\eta - \eta_0| \ll 1 \tag{2.19}$$

for any vector fields $V_j \in TM$, $0 \le j \le \ell$ and any ℓ. Condition (2.19) gives that $V_1 \cdots V_\ell\, \partial_\eta p_{s,k}$ vanishes when $p_{s,k} = 0$. This definition is invariant under symplectic changes of coordinates and multiplication with nonvanishing factors. Observe that $V_1 \cdots V_\ell\, \partial_\eta p_{s,k} = 0$ when $\eta_0 = 0$ since then $p = \partial_\eta p = 0$ and $V_j \in TM \subset T\Sigma_2$. Condition (2.19) with $\ell = 0$ gives that $\eta \mapsto |p_{s,k}(w,\eta)|^{(k-1)/k-\varepsilon}$ is Lipschitz continuous, thus $\eta \mapsto p_{s,k}(w,\eta)$ vanishes at η_0 of order 3 when $k = 2$ and order 2 when $k > 2$.

In the case $k = \kappa(\omega) = 2$ we shall also have a similar condition on the rate of vanishing of $\partial^2_\eta p_{s,k}$ on the subprincipal semibicharacteristic. Then

$$\partial^2_\eta p_{s,k} = \mathcal{J}^0(\partial^2 p) = \operatorname{Hess} p\big|_{\Sigma_2} \tag{2.20}$$

is the Hessian of the principal symbol p at Σ_2, which is well defined on the normal bundle $N\Sigma_2$ since it vanishes on $T\Sigma_2$. Since $p = \partial_\eta p = 0$ on Σ_2, we find that $\operatorname{Hess} p$ is invariant modulo nonvanishing smooth factors under symplectic changes of variables and multiplication of P with elliptic pseudodifferential operators. With the gliding C^∞ foliation M of $N^*\Sigma_2$ for the semibicharacteristics we shall assume that there exists $\varepsilon > 0$ so that

$$\|V_1 \cdots V_\ell \operatorname{Hess} p\| \le C'_\ell |p_{s,k}|^\varepsilon \quad \text{on } \omega \tag{2.21}$$

for any vector fields $V_j \in TM, 0 \le j \le \ell$ and any ℓ. This definition is invariant under symplectic changes of coordinates and multiplication with nonvanishing factors.

Remark 2.3 Conditions (2.19) and (2.21) are well defined and invariant under multiplication with elliptic pseudodifferential operators and conjugation with elliptic Fourier integral operators.

Examples 3.1–3.3 show that conditions (2.19) and (2.21) are essential for the necessity of $\mathrm{Sub}_k(\Psi)$ when $k = 2$.

Example 2.2 If $\operatorname{Re} p_{s,k} = \tau$, $\Sigma_2 = \{\eta = 0\}$, TM is spanned by $\partial_{x,\xi}$ and $t \mapsto \operatorname{Im} p_{s,k}$ vanishes of order $3 \le \ell < \infty$ at $t = t_0(y,\eta) \in C^\infty$ then (2.19) and (2.21) hold. If $t_0(y)$ is independent of η then conditions (2.19) and (2.21) hold for any finite $\ell > 0$.

In fact, if $0 < \ell < \infty$ then we can write $\operatorname{Im} p_{s,k} = a(t - t_0(y,\eta))^\ell$ with $a \ne 0$. If $\ell > \frac{k}{k-1}$ then for any α we find that $\partial^\alpha_{x,\xi}\partial_\eta \operatorname{Im} p_{s,k}$ vanishes of order $\ell - 1 > \ell/k$ at $t = t_0$, and if $\ell > 2$ then $\partial^\alpha_{x,\xi}\partial^2_\eta \operatorname{Im} p_{s,k}$ vanishes of order $\ell - 2 > 0$ at $t = t_0$. If t_0 is independent of η then $\partial^\alpha_{x,\xi}\partial^j_\eta \operatorname{Im} p_{s,k}$ vanishes of order ℓ for any j and α.

Since $\partial_\eta p_{s,k}$ is homogeneous of degree $k-1$ in η, we find from Euler's identity that $\partial_\eta p_{s,k}(w,\eta) = (k-1)\eta \cdot \operatorname{Hess} p(w,\eta)$. Thus (2.21) implies that $|V_1 \cdots V_\ell\, \partial_\eta p_{s,k}| \lesssim |p_{s,k}|^\varepsilon$ when $\eta \ne 0$, but we shall only use condition (2.21) when (2.19) holds, see Theorem 2.1. Here $a \lesssim b$ means $a \le Cb$ for some constant C, and similarly for $a \gtrsim b$.

Now, by (2.9) we have assumed that the reduced subprincipal symbol $p_{s,k}$ is constant on the leaves of Σ_2 near Γ up to multiplication with nonvanishing factors, but when $\kappa < \infty$ we will actually have that condition on the following symbol.

Definition 2.5 If $k = \kappa(\omega) < \infty$ is the order of p on an open set $\omega \subseteq \Sigma_2$ then we define the *extended subprincipal symbol* on $N^*\omega$ by

$$N^*\Sigma_2 \ni (w, \eta) \mapsto q_{s,k}(w, \eta, \lambda) = \lambda J_w^{2k-1}(p)(\eta/\lambda^{1/k}) \\ + J_w^{k-1}(p_s)(\eta/\lambda^{1/k}) \cong p_{s,k}(w, \eta) + \mathcal{O}(\lambda^{-1/k}) \quad (2.22)$$

which is a weighted polynomial in η of degree $2k - 1$. When $\kappa(\omega) = \infty$ we define $q_{s,\infty} \equiv p_s$.

By the invariance of p and p_s, the extended subprincipal symbol transforms as jets under homogeneous symplectic changes of coordinates. It is well defined up to non-vanishing factors and terms proportional to the jet $J_w^{k-1}(\partial_\eta p)(\eta/\lambda^{1/k}) \cong \lambda^{1/k-1}\partial_\eta^k p$ modulo $\mathcal{O}(\lambda^{-1})$ under multiplication with classical elliptic pseudodifferential operators. The extended and the reduced subprincipal symbols are complexly conjugated when taking adjoints.

Remark 2.4 The extended subprincipal symbol (2.22) is given by the blowup of the reduced principal symbol at $\eta = 0$ so that

$$\lambda^{2-m} p_{sub}(x, y; \lambda\xi, \lambda^{1-1/k}\eta) \cong \lambda q_{s,k}(x, y; \xi, \eta, \lambda) \\ = \lambda p_{s,k}(x, y; \xi, \eta) + \mathcal{O}(\lambda^{1-1/k}) \quad \text{modulo } \mathcal{O}(1) \quad (2.23)$$

We also have that

$$\lambda^{2-m}\partial_\eta p_{sub}(x, y; \lambda\xi, \lambda^{1-1/k}\eta) \cong \lambda^{1/k}\partial_\eta q_{s,k}(x, y; \xi, \eta, \lambda) \\ = \lambda^{1/k}\partial_\eta p_{s,k}(x, y; \xi, \eta) + \mathcal{O}(1) \quad (2.24)$$

modulo $\mathcal{O}(\lambda^{1/k-1})$ and

$$\lambda^{2-m}\partial_\eta^2 p_{sub}(x, y; \lambda\xi, \lambda^{1-1/k}\eta) \cong \lambda^{2/k-1}\partial_\eta^2 q_{s,k}(x, y; \xi, \eta, \lambda) \\ = \lambda^{2/k-1}\partial_\eta^2 p_{s,k}(x, y; \xi, \eta) + \mathcal{O}(\lambda^{1/k-1}) \quad (2.25)$$

modulo $\mathcal{O}(\lambda^{2/k-2})$. Observe that if P is of subprincipal type then $dq_{s,k}|_{T^\sigma\Sigma_2} \neq 0$ when $q_{s,k} = 0$ for $\lambda \gg 1$ since this holds for $p_{s,k}$.

In fact, $dq_{s,k} \cong dp_{s,k}$ modulo $\mathcal{O}(\lambda^{-1/k})$ and since $|dp_{s,k}| \neq 0$ the distance between $q_{s,k}^{-1}(0)$ and $p_{s,k}^{-1}(0)$ is $\mathcal{O}(\lambda^{-1/k})$ for $\lambda \gg 1$. Observe that composition of the operator P with elliptic pseudodifferential operators gives factors proportional to $J_w^{k-1}(\partial_\eta p)$ $(\eta/\lambda^{1/k})$ which we shall control with (2.19).

By (2.19) we have that $\partial_\eta p_{s,k} = 0$ when $p_{s,k} = 0$ at ω. We shall also assume this for the next term in the expansion of $q_{s,k}$,

$$\partial_\eta q_{s,k} = \mathcal{O}(\lambda^{-2/k}) \text{ when } p_{s,k} = 0 \text{ at } \omega \text{ for } |\eta - \eta_0| < c_0 \text{ and } \lambda \gg 1 \quad (2.26)$$

Actually, we only need this where $p_{s,k} \wedge d\overline{p}_{s,k}$ vanishes of infinite order at $p_{s,k}^{-1}(0)$ in ω, where $dp_{s,k} \wedge d\overline{p}_{s,k}$ is the complex part of $dp_{s,k}$.

We shall also assume a condition similar to (2.9) on the extended subprincipal symbol

$$\left|dq_{s,k}\big|_{TL}\right| \le C_0|q_{s,k}| \qquad \text{at } \omega \text{ when } |\eta - \eta_0| < c_0 \text{ and } \lambda \gg 1 \tag{2.27}$$

for any leaf L of $N^*\Sigma_2$ where $\omega \subset \Sigma_2$. By letting $\lambda \to \infty$ we obtain that (2.9) holds, since $q_{s,k} \cong p_{s,k}$ modulo $\mathcal{O}(\lambda^{-1/k})$. Also, multiplication of p_{sub} by $a \cong a_0 + a_{-1} + \dots$ with a_j homogeneous of degree j and $a_0 \neq 0$ gives that $q_{s,k}$ gets multiplied by the expansion of $\eta \mapsto a_0(x, y; \xi, \eta/\lambda^{1/k})$ since $ap_{sub} \cong a_0 p_{sub} + a_{-1}p \cong a_0 p_{sub}$ modulo terms in S^{m-1} vanishing of order k at Σ_2. Thus, condition (2.27) is invariant under multiplication of p_{sub} with classical elliptic symbols. Also, (2.27) is invariant under changes of homogeneous symplectic coordinates that preserves $\Sigma_2 = \{\,\eta = 0\,\}$ and TL. Now, we have $\partial_{x,\xi} q_{s,k} \neq 0$ when $q_{s,k} = 0$ and $\lambda \gg 1$ since P is of subprincipal type.

Remark 2.5 Since the semibicharacteristic is transversal to the leaves of Σ_2 and condition (2.27) holds near the semibicharacteristic, Lemma 2.1 gives that

$$q_{s,k}(x, y; \xi, \eta, \lambda) = c(x, y; \xi, \eta, \lambda)\widetilde{q}_{s,k}(x; \xi, \eta, \lambda) \qquad \lambda \gg 1 \tag{2.28}$$

for $|\eta - \eta_0| < c_0$ near the semibicharacteristic. Here $\widetilde{q}_{s,k}(x; \xi, \eta, \lambda)$ is the value of $q_{s,k}$ at the intersection of the semibicharacteristic and the leaf. In fact, the proof of the lemma extends to symbols depending uniformly on the parameter $0 < \lambda^{-1/k} \ll 1$.

Condition (2.27) is *not* invariant under multiplication of P with elliptic pseudodifferential operators or conjugation with elliptic Fourier integral operators. In fact, if A has symbol a then the refined principal symbol of the composition AP is equal to $ap_{sub} + \frac{1}{2i}\{\,a, p\,\}$ which adds $\frac{i}{2}\lambda^{1/k-1}\partial_y a \partial_\eta p_{s,k}$ to $q_{s,k}$. But (2.26) is invariant, since the term containing the factor $\partial_\eta p_{s,k}$ is $\mathcal{O}(\lambda^{-2/k})$ when $k > 2$ and has vanishing η derivative at $p_{s,k}^{-1}(0)$ by (2.21) when $k = 2$.

This is one reason why we have to control the terms with $\partial_\eta p_{s,k}$ with (2.19). When $k < \infty$, $q_{s,k}$ is a polynomial in $\eta/\lambda^{1/k}$ of degree $2k - 1$ and c in (2.28) is an analytic function in $\eta/\lambda^{1/k}$ on ω when $|\eta - \eta_0| < c_0$. Actually, it suffices to expand c in $\eta/\lambda^{1/k}$ up to order k in order to obtain (2.28) modulo $\mathcal{O}(\lambda^{-1})$. If $C(x, y; \xi, \eta/|\xi|^{1/k}) = c(x, y; \xi, \eta, |\xi|)$ in (2.28) then we obtain that Cp_{sub} is constant in y modulo S^{m-2} in ω when $\left|\eta - \eta_0|\xi|^{1-1/k}\right| < c_0|\xi|^{1-1/k}$.

In the case when the principal symbol p is real, a necessary condition for solvability of the operator is that the imaginary part of the subprincipal symbol does not change sign from $-$ to $+$ when going in the positive direction on a C^∞ limit of normalized bicharacteristics of the principal symbol p at Σ_2, see [4]. When p vanishes of exactly order k on $\Sigma_2 = \{\,\eta = 0\,\}$ and the localization

$$\eta \mapsto \sum_{|\alpha|=k} \partial_\eta^\alpha p(x, y; 0, \xi)\eta^\alpha/\alpha!$$

is of principal type when $\eta \neq 0$ such limit bicharacteristics are tangent to the leaves of Σ_2. In fact, then $|\partial_\eta p(x, y; \xi, \eta)| \cong |\eta|^{k-1}$ and $|\partial_{x,y,\xi} p(x, y; \xi, \eta)| = \mathcal{O}(|\eta|^k)$, which gives $H_p = \partial_\eta p \partial_y + \mathcal{O}(|\eta|^k)$. Thus the normalized Hamilton vector field is equal to $A\partial_y$, $A \neq 0$, modulo terms that are $\mathcal{O}(|\eta|)$, so the normalized Hamilton vector fields have limits that are tangent to the leaves. That the η derivatives dominates ∂p can also be seen from Remark 2.4. When the principal symbol is proportional to a real valued symbol, this gives examples of nonsolvability when the subprincipal symbol is not constant on the leaves of Σ_2, see Example 3.4 and [4] in general. Thus condition (2.27) is natural for the the study of the necessity of $\mathrm{Sub}_k(\Psi)$ if there are no other conditions on the principal symbol.

We shall study the microlocal solvability of the operator, which is given by the following definition. Recall that $H^{loc}_{(s)}(X)$ is the set of distributions that are locally in the L^2 Sobolev space $H_{(s)}(X)$.

Definition 2.6 If $K \subset S^*X$ is a compact set, then we say that P is microlocally solvable at K if there exists an integer N so that for every $f \in H^{loc}_{(N)}(X)$ there exists $u \in \mathcal{D}'(X)$ such that $K \bigcap \mathrm{WF}(Pu - f) = \emptyset$.

Observe that solvability at a compact set $K \subset X$ is equivalent to solvability at $S^*X\big|_K$ by [13, Theorem 26.4.2], and that solvability at a set implies solvability at a subset. Also, by [13, Proposition 26.4.4] the microlocal solvability is invariant under conjugation by elliptic Fourier integral operators and multiplication by elliptic pseudodifferential operators. We can now state the main result of the paper.

Theorem 2.1 *Assume that $P \in \Psi^m_{\mathrm{cl}}(X)$ has principal symbol that vanishes of at least second order at a nonradial involutive manifold $\Sigma_2 \subset T^*X \setminus 0$. We assume that P is of subprincipal type, satisfies conditions* (2.26) *and* (2.27) *but does not satisfy condition* $\mathrm{Sub}_k(\Psi)$ *near the subprincipal semibicharacteristic $\Gamma \times \{\eta_0\}$ in $N^*\Sigma_2$ where $\Gamma \subset \omega \subset \Sigma_2$ and $k = \kappa(\omega)$.*

In the case when $\eta_0 \neq 0$ we assume that P satisfies conditions (2.19) *and when $k = 2$ we also assume condition* (2.21) *for a gliding symplectic foliation M of $N^*\Sigma_2$ for the subprincipal semibicharacteristics near Γ.*

In the case $\eta_0 = 0$ and $k = 2$ we assume condition (2.21) *for a gliding symplectic foliation M of $N^*\Sigma_2$ for the subprincipal semibicharacteristics near Γ, and when $k > 2$ we assume no extra condition.*

Under these conditions, P is not locally solvable near $\Gamma \subset \Sigma_2$.

Examples 3.1–3.3 show that conditions (2.19) and (2.21) are essential for the necessity of $\mathrm{Sub}_k(\Psi)$ when $k = 2$. Due to the results of [4], condition (2.27) is natural if there are no other conditions on the principal symbol, see Example 3.4. Observe that for effectively hyperbolic operators, which are always solvable, Σ_2 is not an involutive manifold, see Example 3.7.

Remark 2.6 It follows from the proof that we don't need condition (2.26) in the case when condition (2.27) holds on the leaves of Σ_2 that intersect the semibicharacteristic. In the case when $\eta = 0$ on the subprincipal semibicharacteristics, condition (2.21) only involves Hess p at Σ_2. This gives a different result than Theorem 2.7 in [5], since in that result condition (2.21) is not used, condition (2.27) only involves p_s but we also have conditions on $|dp_s \wedge d\overline{p}_s|$ and Hess p on Σ_2.

Now let $S^*X \subset T^*X$ be the cosphere bundle where $|\xi| = 1$, and let $\|u\|_{(k)}$ be the L^2 Sobolev norm of order k for $u \in C_0^\infty$. In the following, P^* will be the L^2 adjoint of P. To prove Theorem 2.1 we shall use the following result.

Remark 2.7 If P is microlocally solvable at $\Gamma \subset S^*X$, then Lemma 26.4.5 in [13] gives that for any $Y \Subset X$ such that $\Gamma \subset S^*Y$ there exists an integer ν and a pseudodifferential operator A so that $\mathrm{WF}(A) \cap \Gamma = \emptyset$ and

$$\|u\|_{(-N)} \le C(\|P^*u\|_{(\nu)} + \|u\|_{(-N-n)} + \|Au\|_{(0)}) \qquad u \in C_0^\infty(Y) \tag{2.29}$$

where N is given by Definition 2.6.

We shall prove Theorem 2.1 in Sect. 10 by constructing localized approximate solutions to $P^*u \cong 0$ and use (2.29) to show that P is not microlocally solvable at Γ.

3 Examples

Example 3.1 Consider the operator

$$P = D_t + ia(t)\Delta_y \qquad (x, y) \in \mathbf{R}^{n-d} \times \mathbf{R}^d \tag{3.1}$$

where $0 < d < n$, $a(t)$ is real and has a sign change from $-$ to $+$. This operator is equal to the Mizohata operator when $a(t) = t$. We find that P is of subprincipal type, $k = 2$ and $p_{s,2}(t, \tau, \eta) = \tau + ia(t)|\eta|^2$ is constant on the leaves of $\Sigma_2 = \{\eta = 0\}$. Condition (2.27) hold but $\mathrm{Sub}_2(\Psi)$ does not hold since $t \mapsto a(t)|\eta|^2$ changes sign from $-$ to $+$ when $\eta \ne 0$. Since $|\partial_\eta p_{s,2}| \cong \|\,\mathrm{Hess}\ p_{s,2}\| \cong |a(t)|$ when $\eta \ne 0$ and $p_{s,2}$ is independent of (x, ξ) we find that conditions (2.19) and (2.21) hold. Theorem 2.1 gives that P is not locally solvable.

Example 3.2 The operator

$$P = D_t + i(D_{x_1} D_{x_2} + tD_{x_2}^2) \qquad x \in \mathbf{R}^n \quad n \ge 3 \tag{3.2}$$

is solvable, see [2]. We find that P is of subprincipal type, $k = 2$, $\Sigma_2 = \{\xi_1 = \xi_2 = 0\}$ and $p_{s,2}(t, \tau, \xi) = \tau + i(\xi_1\xi_2 + t\xi_2^2)$. Condition $\mathrm{Sub}_2(\Psi)$ does not hold since $t \mapsto \xi_1\xi_2 + t\xi_2^2$ changes sign from $-$ to $+$ when $\xi_1 = -t\xi_2$ and $\xi_2 \ne 0$. Since $|\partial_\xi p_{s,2}| \cong \|\,\mathrm{Hess}\ p_{s,2}\| \cong 1 \gg |p_{s,2}| \cong |t|$ when $\xi_2 \ne 0$ and $\tau = \xi_1 = 0$, we find that conditions (2.19) and (2.21) do not hold.

Example 3.3 Consider the following generalization of Example 3.2 given by

$$P = D_t + i(D_{x_1}D_{x_2} + t^{2j+1}D_{x_2}^2) + i(2j^2 + j)t^{2j-1}x_1^2 \quad x \in \mathbf{R}^n \tag{3.3}$$

for $j > 0$ and $n \geq 3$. We find that P is of subprincipal type, $k = 2$, $\Sigma_2 = \{\,\xi_1 = \xi_2 = 0\,\}$ and $p_{s,2}(t, \tau, \xi) = \tau + i(\xi_1\xi_2 + t^{2j+1}\xi_2^2)$. Thus $\mathrm{Sub}_2(\Psi)$ does not hold since $t \mapsto \xi_1\xi_2 + t^{2j+1}\xi_2^2$ changes sign from $-$ to $+$ when $\xi_1 = -t^{2j+1}\xi_2$ and $\xi_2 \neq 0$. Since $|\partial_\xi p_{s,2}| \cong \|\,\mathrm{Hess}\, p_{s,2}\| \cong 1$ and $|p_{s,2}| \cong |t|^{2j+1}$ when $\tau = \xi_1 = 0$ and $\xi_2 \neq 0$, we find that conditions (2.19) and (2.21) do not hold. By choosing $x_2 - t^{2j+1}x_1$ as new x_2 coordinate we obtain that

$$P = D_t + i\big(D_{x_1} + i(2j+1)t^{2j}x_1\big)D_{x_2} + i(2j^2 + j)t^{2j-1}x_1^2 \tag{3.4}$$

Then by conjugating P with $e^{(2j+1)t^{2j}x_1^2/2}$ we obtain $P = D_t + iD_{x_1}D_{x_2}$ which has constant coefficients and is solvable.

Example 3.4 Consider the operator

$$P = D_t + f(t, y, D_x) + i\,\Box_y \quad (x, y) \in \mathbf{R}^{n-2} \times \mathbf{R}^2 \tag{3.5}$$

where $f(t, y, \xi) \in S^1_{\mathrm{hom}}$ is real and $\Box_y = \partial_{y_1}\partial_{y_2}$ is the wave operator in $y \in \mathbf{R}^2$. We find that P is of subprincipal type, $k = 2$, $\Sigma_2 = \{\,\eta = 0\,\}$ and $p_{s,2}(t, y, \tau, \xi, \eta) = \tau + f(t, y, \xi) - i\eta_1\eta_2$ so (2.27) is not satisfied if $\partial_y f \neq 0$. Since $-iP = \Box_y - iD_t - if(t, y, D_x)$ it follows from Theorem 1.2 in [16] that P is not solvable if $\partial_y f \neq 0$.

Example 3.5 Consider the operator

$$P = D_t + if(t, x, D_x) + B(t, x, D_y) \quad (x, y) \in \mathbf{R}^{n-d} \times \mathbf{R}^d \tag{3.6}$$

where $0 < d < n$, $f(t, x, \xi) \in S^1_{\mathrm{hom}}$ is real and $B(t, x, \eta) \in S^2_{\mathrm{hom}}$ vanishes of degree $k \geq 2$ at $\Sigma_2 = \{\,\eta = 0\,\}$. Then $p_{s,k} = \tau + if(t, x, \xi) + B_k(t, x, \eta)$ where B_k is the kth Taylor term at Σ_2 of the principal symbol of B, so (2.27) is satisfied everywhere.

Assume that $B(t, \eta)$ is independent of x and the sign change in $t \mapsto f(t, x, \xi) + \mathrm{Im}\, B_k(t, \eta)$ is from $-$ to $+$ of order $\ell < \infty$ at $t = t_0$. If $t \mapsto \partial_\eta B_k(t, \eta)$ vanishes of order greater than ℓ/k at $t = t_0$ then (2.19) holds. If $k = 2$ and $t \mapsto \partial_\eta^2 B_k(t, \eta)$ vanishes at $t = t_0$ then (2.21) holds. Then P is not solvable by Theorem 2.1 and Remark 2.6.

If $\mathrm{Im}\, B(x, \eta) \neq 0$ is constant in t and k is odd with $\mathrm{Im}\, B_k(x, \eta) \gtrless 0, \forall\, x$, then condition $\mathrm{Sub}_k(\Psi)$ implies that $t \mapsto f(t, x, \xi)$ is nonincreasing. In fact, Sard's theorem gives for almost all values f_0 of f that there exists (t, x, ξ) so that $f(t, x, \xi) = f_0$ and $\partial_t f(t, x, \xi) \neq 0$. Then one can choose η so that $f(t, x, \xi) + \mathrm{Im}\, B_k(x, \eta) = 0$ so $\mathrm{Sub}_k(\Psi)$ gives $\partial_t f(t, x, \xi) \leq 0$.

If $t \mapsto f(t, x, \xi)$ is nonincreasing, $B(x, \eta)$ is constant in t and $\mathrm{Re}\, B \equiv 0$, then P is solvable. In fact, then $[P^*, P] = 2i[\mathrm{Re}\, P, \mathrm{Im}\, P] = 2\partial_t f \leq 0$ so $\|\,\mathrm{Re}\, Pu\|^2 \lesssim \|Pu\|^2 + \|P^*u\|^2 \lesssim \|P^*u\|^2 + \|u\|^2$ and $\|u\| \ll \|\,\mathrm{Re}\, Pu\|$ if $|t| \ll 1$ in the support of $u \in C_0^\infty$.

Example 3.6 The linearized Navier–Stokes equation

$$\partial_t u + \sum_j a_j(t,x)\partial_{x_j} u + \Delta_x u = f \qquad a_j(x) \in C^\infty \tag{3.7}$$

is of subprincipal type. The symbol is

$$i\tau + i\sum_j a_j(t,x)\xi_j - |\xi|^2 \tag{3.8}$$

so P is of subprincipal type, $k = 2$, $\Sigma_2 = \{\,\xi = 0\,\}$ and $p_{s,2}(\tau,\xi) = i\tau - |\xi|^2$. Thus $\mathrm{Sub}_2(\Psi)$ holds since $-|\xi|^2$ does not change sign when t changes.

Example 3.7 Effectively hyperbolic operators P are weakly hyperbolic operators for which the fundamental matrix F has two real eigenvalues, here $F = \mathcal{J}\,\mathrm{Hess}\, p\big|_{\Sigma_2}$ with $p = \sigma(P)$ and $\mathcal{J}(x,\xi) = (\xi, -x)$ is the symplectic involution. Then P is solvable for any subprincipal symbol by (see [14, 18]) but in this case Σ_2 is *not* an involutive manifold.

4 The Normal Form

We are going to prepare the operator microlocally near the semibicharacteristic. We have assumed that P^* has the symbol expansion $p_m + p_{m-1} + \ldots$ where $p_j \in S^j_{\mathrm{hom}}$ is homogeneous of degree j. By multiplying P^* with an elliptic classical pseudodifferential operator, we may assume that $m = 2$ and $p = p_2$. By choosing local homogeneous symplectic coordinates $(x, y; \xi, \eta)$ we may assume that $X = \mathbf{R}^n$ and $\Sigma_2 = \{\,\eta = 0\,\} \subset T^*\mathbf{R}^n \setminus 0$ with the symplectic foliation by leaves spanned by ∂_y. If p vanishes of order $k < \infty$ at $\omega \subset \Sigma_2$ we find that

$$p(x,y;\xi,\eta) = \sum_{|\alpha|=k} B_\alpha(x,y;\xi,\eta)\eta^\alpha/\alpha! \qquad (x,y,\xi) \in \omega \tag{4.1}$$

where B_α is homogeneous of degree $2-k$, and $B_\alpha(x,y;\xi,0) \not\equiv 0$ for some $|\alpha| = k$ and some $(x,y,\xi,0) \in \omega$. When p vanishes of infinite order we get (4.1) for any k.

We shall first consider the case when $k = \kappa(\omega) < \infty$. Recall the reduced subprincipal symbol $p_{s,k}(w,\eta) = J^k_w(p)(\eta) + p_s(w)$, $w \in \Sigma_2$, by Definition 2.1, and the extended subprincipal symbol $q_{s,k}(w,\eta,\lambda) = \lambda J^{2k-1}_w(p)(\eta/\lambda^{1/k}) + J^{k-1}_w(p_s)(\eta/\lambda^{1/k})$ by Definition 2.5. Observe that these are invariantly defined and are the complex conjugates of the corresponding symbols of P by Remark 2.4. We also find from Remark 2.4 that

$$\begin{aligned} p_{sub}(x,y;\xi,\eta/|\xi|^{1/k}) &\cong |\xi| q_{s,k}(x,y;\xi_0,\eta_0,|\xi|) \\ &= |\xi| p_{s,k}(x,y;\xi_0,\eta_0) + \mathcal{O}(|\xi|^{1-1/k}) \quad \text{modulo } \mathcal{O}(1) \end{aligned} \tag{4.2}$$

where $(\xi_0, \eta_0) = |\xi|^{-1}(\xi, \eta)$. We also have

$$\partial_\eta p_{sub}(x,y;\xi,\eta/|\xi|^{1/k}) \cong |\xi|^{1/k} \partial_\eta p_{s,k}(x,y;\xi_0,\eta_0) \quad \text{modulo } \mathcal{O}(1) \tag{4.3}$$

and

$$\begin{aligned} &\partial_\eta^2 p_{sub}(x,y;\xi,\eta/|\xi|^{1/k}) \\ &\qquad \cong |\xi|^{2/k-1} \partial_\eta^2 p_{s,k}(x,y;\xi_0,\eta_0) \quad \text{modulo } \mathcal{O}(|\xi|^{1/k-1}) \end{aligned} \tag{4.4}$$

When $k < \infty$ we shall localize with respect to the metric

$$g_k(dx,dy;d\xi,d\eta) = |dx|^2 + |dy|^2 + |d\xi|^2/\Lambda^2 + |d\eta|^2/\Lambda^{2-2/k} \tag{4.5}$$

where $\Lambda = (|\xi|^2 + 1)^{1/2}$. If $g_{\varrho,\delta}$ is the metric corresponding to the symbol classes $S^m_{\varrho,\delta}$ we find that

$$g_{1,0} \le g_k \le g_{1-1/k,0}$$

When $k = \infty$ we shall let $g_\infty = g_{1,0}$ which is the limit metric when $k \to \infty$.

We shall use the Weyl calculus symbol notation $S(m, g_k)$ where m is a weight for g_k, one example is $\Lambda^m = (|\xi|^2 + 1)^{m/2}$. Observe that we have the usual asymptotic expansion when composing $S(\Lambda^m, g_k)$ with $S^j_{\varrho,\delta}$ when $\varrho > 0$ and $\delta < 1 - \frac{1}{k}$.

Remark 4.1 If $k < \infty$, f is homogeneous of degree m and vanishes of order j at Σ_2 then $f \in S(\Lambda^{m-j/k}, g_k)$ when $|\eta| \lesssim |\xi|^{1-1/k}$.

One example is $p = \sigma(P^*) \in S(\Lambda, g_k)$ in ω when $|\eta| \lesssim |\xi|^{1-1/k}$ for $k = \kappa(\omega)$. In fact, when $|\eta| \lesssim |\xi|^{1-1/k}$ we have $|f| \lesssim |\xi|^{m-j}|\eta|^j \lesssim |\xi|^{m-j/k}$. Differentiation in x or y does not change this estimate, differentiation in ξ lowers the homogeneity by one and when taking derivatives in η we may lose a factor $\eta_j = \mathcal{O}(|\xi|^{1-1/k})$. We shall prepare the symbol in domains of the type

$$\widetilde{\Omega} = \left\{(x,y,\lambda\xi,\lambda^{1-1/k}\eta) : (x,y,\xi,\eta) \in \Omega \subset S^*\mathbf{R}^n, \quad \lambda > 0\right\} \tag{4.6}$$

which is a g_k neighborhood consisting of the inhomogeneous rays through Ω.

Now for $k < \infty$ we shall use the blowup mapping

$$\chi : (x,y;\xi,\eta) \mapsto (x,y;\xi,\eta/|\xi|^{1/k}) \tag{4.7}$$

which is a bijection when $|\xi| \ne 0$. The pullback by χ maps symbols in $S(\Lambda^m, g_k)$ where $|\eta| \lesssim |\xi|^{1-1/k}$ to symbols in $S^m_{1,0}$ where $|\eta| \lesssim |\xi|$, see for example (4.2). Also Taylor expansions in η where $|\eta| \lesssim |\xi|^{1-1/k}$ get mapped by χ^* to polyhomogeneous expansions, and a conical neighborhood Ω is mapped by χ to the g_k neighborhood $\widetilde{\Omega}$.

The blowup

$$p_{sub} \circ \chi(x, y; \xi, \eta) = q(x, y; \xi, \eta) = |\xi| q_{s,k}(x, y; \xi/|\xi|, \eta/|\xi|, |\xi|)$$
$$\cong |\xi| p_{s,k}(x, y; \xi/|\xi|, \eta/|\xi|) \in S^1_{1,0} \quad \text{modulo } S^{1-1/k}_{1,0} \tag{4.8}$$

is a sum of terms homogeneous of degree $1 - j/k$ for $j \geq 0$ by Definition 2.5. We shall prepare the blowup $q_{s,k}$ and get it on a normal form after multiplication with pseudodifferential operators and conjugation with elliptic Fourier integral operators.

We have assumed that P is of subprincipal type and does not satisfy condition $\text{Sub}_k(\Psi)$ near a subprincipal semicharacteristic $\Gamma \times \{\eta_0\} \subset N^*\Sigma_2$, which is transversal to the leaves of $N^*\Sigma_2$. By changing Γ and η_0 we may obtain that $\operatorname{Im} ap_{s,k}$ changes sign from $+$ to $-$ on the bicharacteristic $\Gamma \times \{\eta_0\}$ of $\operatorname{Re} ap_{s,k}$ for some $0 \neq a \in C^\infty$. The differential inequality (2.27) in these coordinates means that

$$|\partial_y q_{s,k}| \leq C|q_{s,k}| \quad \text{when } |\xi| \gg 1 \tag{4.9}$$

in a conical neighborhood ω in $N^*\Sigma_2$ containing $\Gamma \times \{\eta_0\}$. By shrinking ω we may obtain that the intersections of ω and the leaves of Σ_2 are simply connected. Then by putting $|\xi| = \lambda$ we obtain from Remark 2.5 that

$$\widetilde{q}_{s,k}(x, \xi, \eta) \cong c(x, y, \xi, \eta) q_{s,k}(x, y, \xi, \eta) \quad \text{at } \omega \text{ when } |\xi| \gg 1 \tag{4.10}$$

modulo $S^0_{1,0}$. Here $\widetilde{q}_{s,k}$ is the value of $q_{s,k}$ at the intersection of the semibicharacteristic and the leaf. Here $0 \neq c \in S^0_{1,0}$ is a sum of terms homogeneous of degree $-j/k$ for $j \geq 0$ such that $|c| > 0$ when $|\xi| \gg 1$. In fact, c has an expansion in $\eta/|\xi|^{1/k}$ and it suffices to take terms up to order k in c to get (4.10) modulo $S^0_{1,0}$. Thus the term homogeneous of degree 0 in c is nonvanishing in the conical neighborhood ω. By cutting off the coefficients of the lower order terms of c where $|\xi| \gg 1$, we may assume that $c \neq 0$ in ω.

By multiplying P with a pseudodifferential operator with symbol $C = c \circ \chi^{-1} \in S(1, g_k)$ when $|\eta| \lesssim |\xi|^{1-1/k}$, we obtain by Remark 2.5 the refined principal symbol

$$\widetilde{p}_{sub} + \frac{1}{2i}\partial_y C \partial_\eta p_{sub} \quad \text{modulo } S(1, g_k) \text{ in } \chi(\omega) \tag{4.11}$$

for $\widetilde{P} = CP$. Here $\widetilde{p}_{sub} = Cp_{sub}$ is constant on the leaves of $N^*\Sigma_2$ modulo $S(1, g_k)$ in $\chi(\omega)$. We have that

$$\partial_\eta p_{sub} = C^{-1}\partial_\eta \widetilde{p}_{sub} + \partial_\eta C^{-1} \widetilde{p}_{sub}$$

where $\partial_\eta C^{-1} \in S(\Lambda^{1/k-1}, g_k)$ when $|\eta| \lesssim |\xi|^{1-1/k}$. Thus by multiplying $\widetilde{P}$ with a pseudodifferential operator with symbol $1 - \frac{1}{2i}\partial_\eta C^{-1}\partial_y C \in S(1, g_k)$, we obtain the refined principal symbol

$$\widetilde{p}_{sub} + c_0 \partial_\eta \widetilde{p}_{sub} \qquad \text{modulo } S(1, g_k) \text{ in } \chi(\omega) \tag{4.12}$$

for some $c_0 \in S(1, g_k)$ which may depend on y. Then we find that

$$\partial_y(\widetilde{p}_{sub} + c_0 \partial_\eta \widetilde{p}_{sub}) \cong \partial_y c_0 \partial_\eta \widetilde{p}_{sub} \qquad \text{modulo } S(1, g_k)$$

By putting $q = \widetilde{p}_{sub} \circ \chi$ obtain that $\partial_y q = 0$ in ω when $|\xi| \gg 1$. We shall control the term proportional to $\partial_\eta \widetilde{p}_{sub} \circ \chi = |\xi|^{1/k} \partial_\eta q \in S^{1/k}$ by using condition (2.19), see Lemma 6.1. Observe that $q \cong q_1 = p_{s,k} \circ \chi$ modulo $S^{1-1/k}$, which is homogeneous and independent of y near ω. By the invariance of the condition, we may assume that a is independent of y. Then the semibicharacteristics are constant in η so we may choose a independent of (y, η).

Observe that changing a changes Γ and η_0 by the invariance, but we may assume that $\Gamma \times \{ \eta_0 \}$ is arbitrarily close to the original semibicharacteristic by [13, Theorem 26.4.12]. Since $\operatorname{Im} aq_1$ changes sign on $\Gamma \times \{ \eta_0 \}$ there is a maximal semibicharacteristic $\Gamma' \times \{ \eta_0 \}$ on which $\operatorname{Im} aq_1 = 0$ and because of the sign change we may shrink Γ so that it is not a closed curve. Here Γ' could be a point, which is always the case if the sign change is of finite order. By continuity, $\partial_{x,\xi} \operatorname{Re} aq_1 \neq 0$ near $\Gamma' \times \{ \eta \}$ for η close to η_0 and we may extend a to a nonvanishing symbol that is homogeneous of degree 0 near Γ. Multiplying P with an elliptic pseudodifferential operator with symbol $a = a \circ \chi^{-1}$ we may assume that $a \equiv 1$.

Recall that conditions (2.19) (and (2.21) when $k = 2$) holds in some neighborhood of $\Gamma \times \{ \eta_0 \}$ with the gliding foliation M of $N^*\Sigma_2$. By using Darboux' theorem we can choose local coordinate functions (x, ξ) such that TM is spanned by ∂_x and ∂_ξ for the leaves of M. Now $0 \neq H_{\operatorname{Re} q_1}$ is tangent to $\Gamma' \times \{ \eta_0 \}$, transversal to the symplectic foliation of Σ_2, constant in y and in the symplectic annihilator of TM. Since TM is symplectic, this gives that $\operatorname{Re} q_1$ and η are constant on the leaves M. Now take $\tau = \operatorname{Re} q_1$ when $\eta = \eta_0$ and extend it is so that τ is independent of η. Then we can complete τ, y and η to a homogeneous symplectic coordinate system $(t, x, y; \tau, \xi, \eta)$ in a conical neighborhood ω of Γ' in Σ_2 so that $(x, \xi)\big|_{\eta=\eta_0}$ is preserved. Since the change of variables preserves the (y, η) variables, it preserves $\Sigma_2 = \{ \eta = 0 \}$ and its symplectic foliation and the fact that $\partial_y q = \{ \eta, q \} = 0$. When $\eta = \eta_0$ we have that $\operatorname{Re} q_1 = \tau$ and the leaves TM of the foliation M are spanned by $\partial_x = H_\xi$ and $\partial_\xi = -H_x$ modulo ∂_y when $\eta = \eta_0$. Since q is independent of y we may assume that V_ℓ is in the span of $\partial_{x,\xi}$ in (2.19) and (2.21). Since the η variables are preserved, the blowup map χ and the inhomogeneous rays are preserved and the coordinate change is an isometry with respect to the metric g_k.

By conjugating with elliptic Fourier integral operators in the variables (t, x) independently of y microlocally near $\Gamma' \subset \Sigma_2$, we obtain that $\operatorname{Re} q_1 = \tau$ in a conical neighborhood of Γ when $\eta = \eta_0$. This gives $q_1 = \tau + \varrho(t, x, \tau, \xi, \eta)$ in a neighborhood of $\Gamma' \times \{ \eta_0 \}$, where ϱ is homogeneous and $\operatorname{Re} \varrho \equiv 0$ when $\eta = \eta_0$. Since this is a change of symplectic variables $(t, x; \tau, \xi)$ for fixed (y, η) we find by the invariance that $t \mapsto \operatorname{Im} \varrho(t, x, 0, \xi, \eta_0)$ changes sign from $+$ to $-$ near Γ'. Observe that the reduced principal symbol is invariant under the conjugation by Remark 2.2 so the condition that $\partial_y q = \{\eta, q\} = 0$ is preserved, but we may also have a term $c\partial_\eta q \in S^{1/k}$ where c could depend on y.

Next, we shall use the Malgrange preparation theorem on q_1. Since $\partial_\tau q_1 \neq 0$ near $\Gamma' \times \{\eta_0\}$ we obtain that

$$\tau = c(t, x, \tau, \xi, \eta) q_1(t, x, \tau, \xi, \eta) + r(t, x, \xi, \eta) \tag{4.13}$$

locally for η close to η_0 when $|\xi| = 1$, and by a partition of unity near $\Gamma' \times \{\eta_0\}$, which can be extended by homogeneity so that c is homogeneous of degree 0 and r is homogeneous of degree 1. Observe that this gives that $\partial_\eta q_1 = \partial_\eta c^{-1}(\tau - r) - c^{-1}\partial_\eta r$ by (4.13). By taking the τ derivative of (4.13) using that $q_1 = 0$ and $\partial_\tau q_1 = 1$ at $\Gamma' \times \{\eta_0\}$ we obtain that $c = 1$ on $\Gamma' \times \{\eta_0\}$. Multiplying the operator P^* with a pseudodifferential operator with symbol $c \circ \chi^{-1} \in S(1, g_k)$ when $|\eta| \lesssim |\xi|^{1-1/k}$ we obtain that $q_1(t, x, \tau, \xi, \eta) = \tau - r(t, x, \xi, \eta)$ in a conical neighborhood of $\Gamma' \times \{\eta_0\}$.

Writing $r = r_1 + i r_2$ with r_j real, we may complete $\tau - r_1(t, x, \xi, \eta_0)$, t, y and η to a homogeneous symplectic coordinate system $(t, x, y; \tau, \xi, \eta)$ near $\Gamma' \times \{\eta_0\}$ so that $(x, \xi)\big|_{t=0}$ is preserved. This is a change of coordinates in $(t, x; \tau, \xi)$ which as before is independent of the variables (y, η). We find that $\partial_\tau r = \{r, t\} = 0$ and $\partial_y r = \{\eta, r\} = 0$ are preserved and $\operatorname{Re} r\big|_{\eta=\eta_0} \equiv 0$. By the invariance we find that $t \mapsto r_2(t, x, \xi, \eta_0)$ changes sign from $+$ to $-$ near Γ'. As before, the blowup map χ and the inhomogeneous rays are preserved and the coordinate change is an approximate isometry with respect to the metric g_k.

By conjugating with elliptic Fourier integral operators in (t, x) which are constant in y microlocally near $\Gamma' \subset \Sigma_2$, the calculus gives as before that

$$q_1(t, x, \tau, \xi, \eta) = \tau + f(t, x, \xi, \eta) \tag{4.14}$$

where $f = -r$. When $\eta = \eta_0$ we have $\operatorname{Re} f \equiv 0$ and $\operatorname{Im} f$ has a sign change from $+$ to $-$ as t increases near Γ' by the invariance of condition (Ψ). In fact, the reduced principal symbol is invariant under the conjugation so the condition that $\partial_y q = \{\eta, q\} = 0$ is preserved, but we may also have a term $c\partial_\eta q \in S^{1/k}$ where c could depend on y. Observe that τ is invariant under the blowup mapping χ given by (4.7). By the invariance, we find that (2.19) (and (2.21) if $k = 2$) holds for q_1 with TM spanned by ∂_x and ∂_ξ when $\eta = \eta_0$. In fact, since y, η and t are independent of (x, ξ) we find that the span of $\partial_{x,\xi}$ is invariant modulo terms proportional to ∂_τ. Since $\partial_\eta q_1$ is independent of τ by (4.14) we may take V_j in the span of $\partial_{x,\xi}$ in (2.19) (and (2.21) if $k = 2$) when $\eta = \eta_0$. Thus, by putting $\tau = -\operatorname{Re} f$ in (2.19) we obtain that there exists $\varepsilon > 0$ so that

$$|\partial_{x,\xi}^\alpha \partial_\eta f| \lesssim |\operatorname{Im} f|^{1/k+\varepsilon} \quad \forall \alpha \quad |\eta - \eta_0| < c \tag{4.15}$$

near Γ' in Σ_2. Observe that on Γ', where $\operatorname{Im} f$ vanishes, (4.15) gives that $\partial^\alpha_{x,\xi}\partial_\eta f$ vanishes $\forall\,\alpha$. Similarly, it follows from (2.21) that there exist $\varepsilon > 0$ so that

$$|\partial^\alpha_{x,\xi}\partial^2_\eta f| \lesssim |\operatorname{Im} f|^\varepsilon \qquad \forall\,\alpha \tag{4.16}$$

near Γ' in Σ_2. As before, (4.16) gives that $\partial^\alpha_{x,\xi}\partial^2_\eta f$ vanishes $\forall\,\alpha$, when $\operatorname{Im} f$ vanishes.

We shall next consider on the lower order terms in the expansion $q = q_1 + q_2 + \dots$ where $q_j \in S^{1-(j-1)/k}$. Observe that $\partial_\eta q_2 = 0$ when $q_1 = 0$ by condition (2.26). (Actually, since $dq_1 \wedge d\overline{q}_1 = 2idf \wedge d\tau$ it suffices that this holds when f vanishes of infinite order.) We shall use the Malgrange preparation theorem on q_j, $j \ge 2$, in a conical neighborhood of $\Gamma' \times \{\eta_0\}$. Since $\partial_\tau q_1 \ne 0$ we obtain

$$q_j(t,x,\tau,\xi,\eta) = c_j(t,x,\tau,\xi,\eta)q_1(t,x,\tau,\xi,\eta) + r_j(t,x,\xi,\eta) \qquad j \ge 2 \tag{4.17}$$

locally for η close to η_0 when $|\xi| = 1$, and by a partition of unity near $\Gamma' \times \{\eta_0\}$. This can be extended to a conical neighborhood of $\Gamma' \times \{\eta_0\}$ so that $r_j \in S^{1-(j-1)/k}$ is independent of τ and $c_j \in S^{-(j-1)/k}$. Multiplying the operator P^* with a pseudodifferential operator with symbol $1 - c_j \circ \chi^{-1} \in S(1, g_k)$ where $c_j \circ \chi^{-1} \in S(\Lambda^{-(j-1)/k}, g_k)$ when $|\eta| \lesssim |\xi|^{1-1/k}$ we obtain that $q_j(t,x,\tau,\xi,\eta) = r_j(t,x,\xi,\eta)$. Since q_2 is now independent of τ we find by putting $\tau = -\operatorname{Re} f$ that $\partial_\eta q_2 = 0$ when $\operatorname{Im} f = 0$ (of infinite order) by condition (2.26).

When $j = k$ then we find from (4.12) and (4.13) that $q_k \in S^{1/k}$ also contains the term $c_0\partial_\eta c^{-1}(\tau + f) + c_0 c^{-1}\partial_\eta f$ modulo S^0, where $c^{-1} \in S^0$ and $c_0 \in S^{1/k}$ may depend on y. By using (4.17) and multiplying the operator P^* with a pseudodifferential operator with symbol $1 - (c_k + c_0\partial_\eta c^{-1}) \circ \chi^{-1} \in S(1, g_k)$ when $|\eta| \lesssim |\xi|^{1-1/k}$ we obtain that $q_k = r_k + c_0 c^{-1}\partial_\eta f$, where $c_0 \in S^{1/k}$ may depend on y.

This preparation can be done for all lower order terms of pullback of the full symbol of P^* given by $\sigma(P^*) \circ \chi$, where the terms are in $S^{-j/k}$ for $j \ge 0$. These terms may depend on y, but that does not change the already prepared terms since $c_j \in S^{-1-j/k}$ in (4.17). We shall cut off in a g_k neighborhood of the bicharacteristic and then we have to measure the error terms of the preparation.

Definition 4.1 In the case $k < \infty$ and $R \in \Psi^\mu_{\varrho,\delta}$ where $\varrho + \delta \ge 1$, $\varrho > 0$ and $\delta < 1 - \frac{1}{k}$ we say that $T^*X \ni (t_0, x_0, y_0; \tau_0, \xi_0, \eta_0) \notin \mathrm{WF}_{g_k}(R)$ if the symbol of R is $\mathcal{O}(|\xi|^{-N})$, $\forall\,N$, when the g_k distance to the inhomogeneous ray $\{\,(t_0, x_0, y_0; \varrho\tau_0, \varrho\xi_0, \varrho^{1-1/k}\eta_0) : \varrho \in \mathbf{R}_+\,\}$ is less than $c > 0$. If $k = \infty$ and $R \in \Psi^\mu_{\varrho,\delta}$, for $\varrho > 0$ and $\delta < 1$, then $\mathrm{WF}_{g_k}(R) = \mathrm{WF}(R)$.

For example, $(t_0, x_0, y_0; \tau_0, \xi_0, \eta_0) \notin \mathrm{WF}_{g_k}\big(R(D)\big)$ if R is the cutoff function

$$R(\tau,\xi,\eta) = 1 - \chi\big(c|(\tau,\xi)|^{1/k-1}(\eta - \eta_0)\big) \in S(1, g_k)$$

with $\chi \in C_0^\infty$ such that $0 \notin \operatorname{supp}(1 - \chi)$ and $c > 0$. By the calculus, Definition 4.1 means that there exists $A \in S(1, g_k)$ so that $A \ge c > 0$ in a g_k neighborhood of the inhomogeneous ray such that $AR \in \Psi^{-N}$ for any N. By the conditions on ϱ and δ, it follows from the calculus that Definition 4.1 is invariant under composition with classical elliptic pseudodifferential operators and under conjugation with elliptic homogeneous Fourier integral operators preserving the fiber and $\Sigma_2 = \{\eta = 0\}$. We also have that $\mathrm{WF}_{g_k}(R)$ grows when k increases and $\mathrm{WF}_{g_k}(R) \subseteq \mathrm{WF}(R)$, with equality when $k = \infty$.

Cutting off where $|\eta - \eta_0| \lesssim |\xi|^{1-1/k}$ we obtain that

$$P^* = D_t + F_1(t,x,y,D_x,D_y) + F_0(t,x,y,D_x,D_y) + R(t,x,y,D_x,D_y) \tag{4.18}$$

where $R \in S(\Lambda^2, g_k)$ such that $\Gamma' \times \{\eta_0\} \bigcap \mathrm{WF}_{g_k}(R) = \emptyset$, $F_j \in S(\Lambda^j, g_k)$ such that

$$F_1 \circ \chi(t,x,y,\xi,\eta) \cong f(t,x,\xi,\eta) + r(t,x,\xi,\eta) \in S^1 \tag{4.19}$$

modulo $S^{1-2/k}$, where $r \in S^{1-1/k}$ and $\partial_\eta r = 0$ when $\mathrm{Im}\, f$ vanishes (of infinite order). Also there exists $c \in S(1, g_k)$ so that $F_1 - c\partial_\eta F_1$ is constant in y modulo $S(1, g_k)$.

Next, we study the case when $k = \infty$. We have $q_{s,\infty} = p_s$, $p_{s,\infty} = p_s\big|_{\Sigma_2}$ and $g_\infty = g_{1,0}$. Then we shall not prepare the principal symbol, which vanishes of infinite order at Σ_2. Instead, we shall prepare the lower order terms starting with p_1, which is homogeneous of degree 1. We shall prepare p_1 in a similar way as q near the subprincipal semicharacteristic $\Gamma \subset \Sigma_2$. First we may as before use the differential inequality (2.27) to obtain that p_s is constant in y near Γ after multiplication with an nonvanishing homogeneous $c \in S^0_{1,0}$. By multiplication with an elliptic pseudodifferential operator with symbol c we obtain that p_1 is constant in y modulo terms vanishing of infinite order at Σ_2. In fact, the composition of P with a classical elliptic pseudodifferential operator can only give terms in p_{sub} vanishing of infinite order at Σ_2.

By assumption condition $\mathrm{Sub}_\infty(\Psi)$ is not satisfied, so there exists $0 \neq a \in S^0_{1,0}$ so that $\mathrm{Im}\, ap_1\big|_{\Sigma_2}$ changes sign from $+$ to $-$ on the bicharacteristic $\Gamma \subset \Sigma_2$ of $\mathrm{Re}\, ap_1$ for some $0 \neq a \in C^\infty$ which can be assumed to be homogeneous and constant in y and η. By multiplication with an elliptic pseudodifferential operator with symbol a we may assume that $a \equiv 1$. Let $\Gamma' \subset \Gamma$ be the subset on which p_1 vanishes. Since $0 \neq H_{\mathrm{Re}\, p_1}$ is tangent to $\Gamma \subset \Sigma_2$ and $\partial_y p_1 = \{p_1, \eta\} = 0$ we can complete $\tau = \mathrm{Re}\, p_1$ and η to a homogeneous symplectic coordinate system $(t,x,y;\tau,\xi,\eta)$ in a conical neighborhood ω of Γ', which preserves the foliation of Σ_2. Then conjugating with an elliptic Fourier integral operators, we obtain $p_1 = \tau + i\,\mathrm{Im}\, p_1$ modulo terms vanishing of infinite order at Σ_2. The conjugation also gives terms proportional to $\partial_\eta p \in S^1$ which vanish of infinite order at Σ_2. As before, we find that condition $\mathrm{Sub}_\infty(\Psi)$ is not satisfied in any neighborhood of Γ' in Σ_2 by the invariance.

Since $\partial_\tau p_1 \neq 0$ we can use the Malgrange preparation theorem as before to obtain

$$\tau = c(t,x,\tau,\xi,\eta)p_1(t,x,\tau,\xi,\eta) + r(t,x,\xi,\eta) \tag{4.20}$$

locally, and by a partition of unity near $\Gamma' \subset \Sigma_2$. This may be extended by homogeneity to a conical neighborhood of Γ', thus for η close to 0. Then $cp_1 = \tau - r$

where $r \in S^1$ is constant in τ and $0 \neq c \in S^0$ near Γ'. In fact, this follows by taking the τ derivative of (4.20) and using that $p_1 = 0$ and $\partial_\tau p_1 \neq 0$ at Γ'.

Multiplying the operator P^* with an elliptic pseudodifferential operator with symbol c we obtain that $p_1(t, x, \tau, \xi, \eta) = \tau - r(t, x, \xi, \eta)$ in a conical neighborhood of Γ'. By writing $r = r_1 + ir_2$ with r_j real, we may complete $\tau - r_1(t, x, \xi, \eta)$, η and t to a homogeneous symplectic coordinate system $(t, x, y; \tau, \xi, \eta)$ in a conical neighborhood of Γ'. By conjugating with elliptic Fourier integral operators we obtain that

$$p_1(t, x, \tau, \xi, \eta) = \tau + f(t, x, \xi, \eta) \tag{4.21}$$

near $\Gamma' \subset \Sigma_2$, where $f = -ir_2$ modulo terms vanishing of infinite order at Σ_2. By the invariance we find that $t \mapsto \operatorname{Im} f(t, x, \xi, 0)$ changes sign from $+$ to $-$ near Γ'.

We shall next consider on the lower order terms in the expansion $p + p_1 + p_0 + \cdots$ where $p_j \in S^j$ near Γ' may depend on y when $j \leq 0$. Observe that $\partial_\eta p_0 = 0$ when $p_1 = 0$ by condition (2.26). (Actually, since $dp_1 \wedge d\overline{p}_1 = 2idf \wedge d\tau$ it suffices that this holds when f vanishes of infinite order.) We shall use the Malgrange preparation theorem on p_j, $j \leq 0$, in a conical neighborhood of Γ'. Since $\partial_\tau p_1 \neq 0$ we obtain for $j \leq 0$ that

$$p_j(t, x, y, \tau, \xi, \eta) = c_j(t, x, y, \tau, \xi, \eta) p_1(t, x, \tau, \xi, \eta) + r_j(t, x, y, \xi, \eta) \tag{4.22}$$

locally and by a partition of unity near Γ'. Extending by homogenity we obtain that $r_j \in S^j$ and $c_j \in S^{j-1} \subset S^{-1}$ near Γ' for η close to 0. After multiplication with an elliptic pseudodifferential operator with symbol $1 - c_j$ we obtain that $p_j \cong r_j$ is independent of τ modulo terms vanishing of infinite order at Σ_2. Since p_0 is now independent of τ we find by putting $\tau = -\operatorname{Re} f$ that $\partial_\eta p_0 = 0$ when $\operatorname{Im} f = 0$ (of infinite order) by condition (2.26). Continuing in this way, we can make any lower order term in the expansion of P independent of τ modulo terms vanishing of infinite order at Σ_2.

Since condition $\operatorname{Sub}_k(\Psi)$, $k \leq \infty$, is *not* satisfied, we find that $t \mapsto \operatorname{Im} f(t, x_0, \xi_0, \eta_0)$ changes sign from $+$ to $-$ as $t \in I$ increases and we assume that $\operatorname{Im} f(t, x_0, \xi_0, \eta_0) = 0$ when $t \in I' \subset I$. Observe that we shall keep η_0 fixed and when $k = \infty$ we have $\eta_0 = 0$. If (4.15) holds then we find that $\partial^\alpha_{x,\xi} \partial_\eta f = 0$ on $\Gamma' \times \{\eta_0\}$, $\forall\, \alpha\, \beta$, and if (4.16) holds then we find that $\partial^\alpha_{x,\xi} \partial^2_\eta f = 0$ on $\Gamma' \times \{\eta_0\}$, $\forall\, \alpha\, \beta$. Observe that we have $\partial^\alpha_x \partial^\beta_\xi \operatorname{Re} f = 0$, $\forall\, \alpha\, \beta$, when $\eta = \eta_0$.

Now if $|I'| \neq 0$, then by reducing to *minimal bicharacteristics* near which $\operatorname{Im} f$ changes sign as in [12, p. 75], we may assume that $\partial^\alpha_x \partial^\beta_\xi \operatorname{Im} f$ vanishes on a bicharacteristic $\Gamma' \times \{\eta_0\}$, $\forall\, \alpha\, \beta$, which is arbitrarily close to the original bicharacteristic (see [24, Sect. 2] for a more refined analysis).

In fact, if $\operatorname{Im} f(a, x, \xi, \eta_0) > 0 > \operatorname{Im} f(b, x, \xi, \eta_0)$ for some (x, ξ) near (x_0, ξ_0) and $a < b$, then we can define

$$L(x,\xi) = \inf\{\, t - s : a < s < t < b \text{ and } \operatorname{Im} f(s,x,\xi,\eta_0) > 0 > \operatorname{Im} f(t,x,\xi,\eta_0) \,\}$$

when (x,ξ) is close to (x_0,ξ_0), and we put $L_0 = \liminf_{(x,\xi)\to(x_0,\xi_0)} L(x,\xi)$. If $L_0 > 0$ then for every $\varepsilon > 0$ there exists an open neighborhood V_ε of (x_0,ξ_0) such that the diameter of V_ε is less than ε and $L(x,\xi) > L_0 - \varepsilon/2$ when $(x,\xi) \in V_\varepsilon$. By definition, there exists $(x_\varepsilon,\xi_\varepsilon) \in V_\varepsilon$ and $a < s_\varepsilon < t_\varepsilon < b$ so that $t_\varepsilon - s_\varepsilon < L_0 + \varepsilon/2$ and $\operatorname{Im} f(s_\varepsilon,x_\varepsilon,\xi_\varepsilon,\eta_0) > 0 > \operatorname{Im} f(t_\varepsilon,x_\varepsilon,\xi_\varepsilon,\eta_0)$. Then it is easy to see that

$$\partial_x^\alpha \partial_\xi^\beta \operatorname{Im} f(t,x_\varepsilon,\xi_\varepsilon,\eta_0) = 0 \quad \forall\, \alpha\,\beta \quad \text{when} \quad s_\varepsilon + \varepsilon < t < t_\varepsilon - \varepsilon \tag{4.23}$$

since else we would have a sign change in an interval of length less than $L_0 - \varepsilon/2$ in V_ε. We may then choose a sequence $\varepsilon_j \to 0$ so that $s_{\varepsilon_j} \to s_0$ and $t_{\varepsilon_j} \to t_0$, then $L_0 = t_0 - s_0$ and (4.23) holds at (x_0,ξ_0,η_0) for $s_0 < t < t_0$.

Proposition 4.1 *Assume that P satisfies the conditions in Theorem 2.1 with $k = \kappa(\omega)$. Then by conjugating with elliptic Fourier integral operators and multiplication with an elliptic pseudodifferential operator we may assume that*

$$P^* = D_t + F(t,x,y,D_x,D_y) + R(t,x,y,D_t,D_x,D_y) \tag{4.24}$$

microlocally near $\Gamma = \{\,(t,x_0,y_0;0,\xi_0,0) : t \in I\,\} \subset \Sigma_2$. In the case $k < \infty$ we have $R \in S(\Lambda^2, g_k) \subset S^2_{1-1/k,0}$ such that $\Gamma \times \{\,\eta_0\,\} \bigcap \mathrm{WF}_{g_k}(R) = \emptyset$, and $F = F_1 + F_0$ with $F_1 \in S(\Lambda, g_k)$ and $F_0 \in S(1, g_k)$. Here

$$F_1 \circ \chi(t,x,y,\xi,\eta) \cong f(t,x,\xi,\eta) + r(t,x,\xi,\eta) \ \text{ modulo } S^{1-2/k} \tag{4.25}$$

where χ is the blowup map (4.7), $r \in S^{1-1/k}$, $\operatorname{Re} f(t,x,\xi,\eta_0) \equiv 0$ and $\operatorname{Im} f = \operatorname{Im} p_{s,k} \in S^1$ is given by (2.8) such that $t \mapsto \operatorname{Im} f(t,x_0,\xi_0,\eta_0)$ changes sign from $+$ to $-$ when $t \in I$ increases. Also, $\partial_\eta r = 0$ when f vanishes (of infinite order), and there exists $c \in S(1,g_k)$ so that $F_1 - c\partial_\eta F_1$ is constant in y modulo $S(1,g_k)$, where $\partial_\eta F_1 \in S(\Lambda^{1/k}, g_k)$.

If $\eta_0 \neq 0$, then condition (4.15) holds near $\Gamma \times \{\,\eta_0\,\}$. If $k = 2$ then condition (4.16) also holds near $\Gamma \times \{\,\eta_0\,\}$ if $\eta_0 \neq 0$ and near Γ' in Σ_2 if $\eta_0 = 0$. If $f = 0$ on $\Gamma' \times \{\,\eta_0\,\}$ where $\Gamma' \subset \Gamma$ and $|\Gamma'| \neq 0$ we may assume that $\partial_x^\alpha \partial_\xi^\beta \partial_\eta^\gamma f = 0$ on $\Gamma' \times \{\,\eta_0\,\}$ for any α, β and $|\gamma| \le 1$, and when $k = 2$ that $\partial_x^\alpha \partial_\xi^\beta \partial_\eta^\gamma f = 0$ on $\Gamma' \times \{\,\eta_0\,\}$ for any α, β and $|\gamma| \le 2$.

In the case when $k = \infty$ we obtain (4.24) with $R \in S^2_{1,0}$ vanishing of infinite order on Σ_2, $F = F_1 + F_0$ where $F_0(t,x,y;\xi,\eta) \in S^0_{1,0}$ and

$$F_1(t,x,\xi,\eta) = f(t,x;\xi,\eta) \in S^1_{1,0} \tag{4.26}$$

where $t \mapsto \operatorname{Im} f(t,x_0,\xi_0,0)$ changes sign from $+$ to $-$ when t increases and $\operatorname{Re} f(t,x,\xi,0) \equiv 0$. If $\operatorname{Im} f = 0$ on $\Gamma' \times \{\,0\,\}$ with $|\Gamma'| \neq 0$ we may assume that $\partial_x^\alpha \partial_\xi^\beta f = 0$ on $\Gamma' \times \{\,0\,\}$ for any α, β.

5 The Pseudomodes

For the proof of Theorem 2.1 we shall modify the Moyer–Hörmander construction of approximate solutions (or pseudomodes) of the type

$$u_\lambda(t, x, y) = e^{i\lambda\omega_\lambda(t,x,y)} \sum_{j\ge 0} \phi_j(t, x, y)\lambda^{-j\kappa} \qquad \lambda \ge 1 \tag{5.1}$$

with $\kappa > 0$, phase function ω_λ and amplitudes ϕ_j. Here the phase function $\omega_\lambda(t, x, y)$ will be uniformly bounded in C^∞ and complex valued, such that $\operatorname{Im}\omega_\lambda \ge 0$ and $\partial \operatorname{Re}\omega_\lambda \ne 0$ when $\operatorname{Im}\omega_\lambda = 0$. The amplitude functions $\phi_j \in C^\infty$ may depend uniformly on λ. Letting $z = (t, x, y)$ we have the formal expansion

$$\begin{aligned} &p(z, D_z)(\exp(i\lambda\omega_\lambda)\phi) \\ &\qquad \sim \exp(i\lambda\omega_\lambda)\sum_\alpha \partial_\zeta^\alpha p(z, \lambda\partial_z\omega_\lambda(z))\mathcal{R}_\alpha(\omega_\lambda, \lambda, D_z)\phi(z)/\alpha! \end{aligned} \tag{5.2}$$

where $\mathcal{R}_\alpha(\omega_\lambda, \lambda, D_z)\phi(z) = D_w^\alpha(\exp(i\lambda\widetilde{\omega}_\lambda(z, w))\phi(w))\big|_{w=z}$ and

$$\widetilde{\omega}_\lambda(z, w) = \omega_\lambda(w) - \omega_\lambda(z) + (z - w)\partial\omega_\lambda(z)$$

The error term in (5.2) is of the same order in λ as the last term in the expansion. Observe that since the phase is complex valued, the values of the symbol are given by an almost analytic extension at the real parts, see Theorem 3.1 in Chapter VI and Chapter X:4 in [21]. If $P^* = D_t + F(t, x, y, D_{x,y})$ we find from (5.2) that

$$\begin{aligned} &e^{-i\lambda\omega_\lambda} P^* e^{i\lambda\omega_\lambda}\phi \\ &= \left(\lambda\partial_t\omega_\lambda + F(t, x, y, \lambda\partial_{x,y}\omega_\lambda) - i\lambda\partial^2_{\xi,\eta}F(t, x, y, \lambda\partial_{x,y}\omega_\lambda)\partial^2_{x,y}\omega_\lambda/2\right)\phi \\ &\quad + D_t\phi + \partial_{\xi,\eta}F(t, x, y, \lambda\partial_{x,y}\omega_\lambda)D_{x,y}\phi + \partial^2_{\xi,\eta}F(t, x, y, \lambda\partial_{x,y}\omega_\lambda)D^2_{x,y}\phi/2 \\ &\qquad\qquad + \sum_{j\ge 0}\lambda^{-j}R_j(t, x, y, D_{x,y})\phi \end{aligned} \tag{5.3}$$

Here the values of the symbols at $(t, x, y, \lambda\partial_{t,x,y}\omega_\lambda)$ will be replaced by finite Taylor expansions at $(t, x, y, \lambda \operatorname{Re}\partial_{t,x,y}\omega_\lambda)$, which determine the almost analytic extensions.

Now assume that $P^* = D_t + F + R$ is given by Proposition 4.1. In the case $k = \kappa(\omega) < \infty$ in a open neighborhood ω of the bicharacteristic Γ and $\eta_0 \ne 0$ we have $F = F_1 + F_0$ with $F_j \in S(\Lambda^j, g_k)$ and $R \in S(\Lambda^2, g_k)$ with $\Gamma \notin \mathrm{WF}_{g_k}(R)$. In this case, we shall use a nonhomogeneous phase function given by (6.3):

$$\begin{aligned} \omega_\lambda(t, x, y) = w_0(t) &+ \langle x - x_0(t), \xi_0\rangle + \lambda^{-1/k}\langle y - y_0(t), \eta_0\rangle \\ &+ \mathcal{O}(|x - x_0(t)|^2) + \lambda^{\varrho-1}\mathcal{O}(|y - y_0(t)|^2) \end{aligned} \tag{5.4}$$

such that $\partial_y \omega_\lambda = \lambda^{-1/k}\eta_0 + \mathcal{O}(\lambda^{\varrho-1})$ with some $0 < \varrho < 1/2$. We find by Remark 2.4 that $F_1(t, x, y, \lambda\partial_x\omega_\lambda, \lambda\partial_y\omega_\lambda) \cong F_1(t, x, y, \lambda\xi_0, \lambda^{1-1/k}\eta_0)$ gives an approximate blowup of $F_1 \in S(\Lambda, g_k)$. Since $\partial^\alpha\partial_{t,x}\omega_\lambda = \mathcal{O}(1)$ and $\partial^\alpha\partial_y\omega_\lambda = \mathcal{O}(\lambda^{-1/k})$ for any α we obtain the following result from the chain rule.

Remark 5.1 If $0 < \varrho \le 1/2$, $\omega_\lambda(t, x, y)$ is given by (6.3) and $a(t, x, y, \tau, \xi, \eta) \in S(\Lambda^m, g_k)$ then $\lambda^{-m} a(t, x, y, \lambda\partial\omega_\lambda) \in C^\infty$ uniformly.

This gives that $R_0(t, x, y) = F_0(t, x, y, \lambda\partial_{x,y}\omega_\lambda)$ is bounded in (5.3) and $R_m(t, x, y, D_{t,x,y})$ are bounded differential operators of order i in t, order j in x and order ℓ in y, where $i + j + \ell \le m + 2$ for $m > 0$. In fact, derivatives in τ and ξ of $F_1 \in S(\Lambda, g_k)$ lowers the order of λ by one, but derivatives in η lowers the order only by $1 - 1/k$ until we have taken k derivatives, thereafter by 1. Thus for R_m, which is the coefficient for λ^{-m}, we find that $-m \le 1 - i - j - \ell(1 - 1/k)$ so that $i + j + \ell \le m + 1 + \ell/k \le m + 2$ for $\ell \le k$, else $-m \le 1 - i - j - k(1 - 1/k) - (\ell - k) = 2 - i - j - \ell$ which also gives $i + j + \ell \le m + 2$. For the term R we shall use the following result when $k < \infty$.

Remark 5.2 If $R \in S(\Lambda^m, g_k) \subset S^m_{1-1/k,0}$, u_λ is given by (5.1) with phase function ω_λ in (6.3) and

$$\left\{(t, x, y, \lambda\partial_{t,x,y}\omega_\lambda) : (t, x, y) \in \bigcup_j \operatorname{supp}\phi_j\right\} \bigcap \mathrm{WF}_{g_k}(R) = \emptyset \qquad \lambda \gg 1 \tag{5.5}$$

then $Ru_\lambda = \mathcal{O}(\lambda^{-N})$, $\forall N$.

In fact, by using the expansion (5.2) we find that $\partial^\alpha R(t, x, y, \lambda\partial_{t,x,y}\omega_\lambda) = \mathcal{O}(\lambda^{-N})$ for any α and N in a neighborhood of the support of ϕ_j for any j when $\lambda \gg 1$.

In the case $k = \kappa(\omega) = \infty$ or $\eta_0 = 0$ we shall use the phase function given by (7.2), then

$$\begin{aligned}\omega_\lambda(t, x, y) = w_0(t) + \langle x - x_0(t), \xi_0\rangle + \lambda^{\varrho-1}\langle y - y_0(t), \eta_0\rangle \\ + \mathcal{O}(|x - x_0(t)|^2) + \lambda^{\varrho-1}\mathcal{O}(|y - y_0(t)|^2)\end{aligned} \tag{5.6}$$

such that $\partial_y\omega_\lambda = \lambda^{\varrho-1}\big(\eta_0 + \mathcal{O}(|y - y_0(t)|)\big)$ with some $0 < \varrho < 1$. If $R \in S^2$ vanishes of infinite order at $\eta = 0$ then $\partial^\alpha R(t, x, y, \lambda\partial_{t,x,y}\omega_\lambda) = \mathcal{O}(\lambda^{-N})$ for any α and N. Thus, we get the expansion (5.3) with bounded $R_0 = F_0(t, x, y, \lambda\partial_{x,y}\omega_\lambda)$ and bounded differential operators $R_m(t, x, y, D_{t,x,y})$ of order i in t, order j in x and order ℓ in y, where $i + j + \ell \le m + 2$ for $m > 0$. When $k < \infty$ this follows as before, and in the case $k = \infty$ we have that derivatives in τ, ξ and η of $F_1 \in S^1_{1,0}$ lowers the order of λ by one. In that case, we find for R_m that $-m \le 1 - i - j - \ell$ so that $i + j + \ell \le m + 1$.

Remark 5.3 If $0 < \varrho \le 1$, $\omega_\lambda(t, x, y)$ is given by (7.2) and $a(t, x, y, \tau, \xi, \eta) \in S^m_{1,0}$ then $\lambda^{-m} a(t, x, y, \lambda\partial\omega_\lambda) \in C^\infty$ uniformly.

This follows from the chain rule since $\partial^\alpha\partial\omega_\lambda = \mathcal{O}(1)$ for any α.

6 The Eikonal Equation

We shall solve the eikonal equation approximately, first in the case when $k = \kappa(\omega) < \infty$ and $\eta_0 \neq 0$. This equation is given by the highest order terms of (5.3):

$$\lambda\partial_t\omega_\lambda + F_1(t, x, y, \lambda\partial_{x,y}\omega_\lambda) - i\lambda\partial_\eta^2 F_1(t, x, y, \lambda\partial_{x,y}\omega_\lambda)\partial_y^2\omega_\lambda = 0 \qquad (6.1)$$

modulo $\mathcal{O}(1)$. Here $F_1 \in S(\Lambda, g_k)$ satisfies $F_1 \circ \chi = f \in S^1$ modulo $S^{1-1/k}$ when $|\eta| \lesssim |\xi|^{1-1/k}$ by (4.25) in Proposition 4.1. Thus if $\partial_y\omega_\lambda = \mathcal{O}(\lambda^{-1/k})$ we obtain the blowup

$$F_1(t, x, y, \lambda\partial_x\omega_\lambda, \lambda\partial_y\omega_\lambda) \cong \lambda f(t, x, \partial_x\omega_\lambda, \lambda^{1/k}\partial_y\omega_\lambda) \qquad (6.2)$$

modulo terms that are $\mathcal{O}(\lambda^{1-1/k})$. Now Re $f \equiv 0$ when $\eta = \eta_0$, f vanishes on $\Gamma' = \{ (t, x_0, \xi_0, \eta_0) : t \in I' \}$ and $t \mapsto \operatorname{Im} f(t, x_0, \xi_0, \eta_0) \in S^1$ changes sign from $+$ to $-$ as t increases in a neighborhood of I'. We may choose coordinates so that $0 \in I'$ thus $f(0, x_0, \xi_0, \eta_0) = 0$. Observe that (4.15) (and (4.16) if $k = 2$) holds near Γ'. If $|I'| \neq 0$ then by Proposition 4.1 we may assume that $\partial_x^\alpha\partial_\xi^\beta\partial_\eta^\gamma f$ vanishes at Γ', $\forall\, \alpha\, \beta$ and $|\gamma| \leq 1$ when $k > 2$ and for $\forall\, \alpha\, \beta$ and $|\gamma| \leq 2$ when $k = 2$. We also have that $F_1 - c\partial_\eta F_1$ is constant in y modulo $S(1, g_k)$ when $|\eta| \lesssim |\xi|^{1-1/k}$, where $c \in S(1, g_k)$ may depend on y. The case when $\eta_0 = 0$, for example when $k = \infty$, will be treated in Sect. 7.

We shall choose the phase function so that $\operatorname{Im}\omega_\lambda \geq 0$, $\partial_x \operatorname{Re}\omega_\lambda \neq 0$ and $\partial_{x,y}^2 \operatorname{Im}\omega_\lambda > 0$ near the interval. We shall adapt the method by Hörmander [12] to inhomogeneous phase functions. The phase function $\omega_\lambda(t, x, y)$ is given by the expansion

$$\begin{aligned}\omega_\lambda(t, x, y) &= w_0(t) + \langle\xi_0(t), x - x_0(t)\rangle \\ &\quad + \lambda^{-1/k}\langle\eta_0(t), y - y_0(t)\rangle + \sum_{2\leq i\leq K} w_{i,0}(t)(x - x_0(t))^i/i! \\ &\quad + \lambda^{\varrho-1}\sum_{\substack{2\leq i+j\leq K\\ j\neq 0}} w_{i,j}(t)(x - x_0(t))^i(y - y_0(t))^j/i!j! \end{aligned} \qquad (6.3)$$

for sufficiently large K, where we will choose $0 < \varrho < 1/2$, $\xi_0(0) = \xi_0 \neq 0$, Im $w_{2,0}(0) > 0$, Im $w_{1,1}(0) = 0$ and Im $w_{0,2}(0) > 0$. This gives $\partial_{x,y}^2 \operatorname{Im}\omega_\lambda > 0$ when $t = 0$ and $|x - x_0(0)| + |y - y_0(0)| \ll 1$ which then holds in a neighborhood. Here we use the multilinear forms $w_{i,j} = \{ w_{\alpha,\beta}i!j!/\alpha!\beta! \}_{|\alpha|=i,|\beta|=j}$, $(x - x_0(t))^j = \{ (x - x_0(t))^\alpha \}_{|\alpha|=j}$ and $(y - y_0(t))^j = \{ (y - y_0(t))^\alpha \}_{|\alpha|=j}$ to simplify the notation. Observe that $x_0(t)$, $y_0(t)$, $\xi_0(t)$, $\eta_0(t)$ and $w_{j,k}(t)$ will depend uniformly on λ.

Putting $\Delta x = x - x_0(t)$ and $\Delta y = y - y_0(t)$ we find that

$$
\begin{aligned}
\partial_t \omega_\lambda(t,x,y) = w_0'(t) - \langle x_0'(t), \xi_0(t)\rangle - \lambda^{-1/k}\langle y_0'(t), \eta_0(t)\rangle \\
+ \langle \xi_0'(t) - w_{2,0}(t)x_0'(t) - \lambda^{\varrho-1} w_{1,1}(t) y_0'(t), \Delta x\rangle \\
+ \langle \lambda^{-1/k}\eta_0'(t) - \lambda^{\varrho-1} w_{1,1}(t)x_0'(t) - \lambda^{\varrho-1} w_{0,2}(t) y_0'(t), \Delta y\rangle \\
+ \sum_{2\le i\le K} (w_{i,0}'(t) - w_{i+1,0}(t)x_0'(t) - \lambda^{\varrho-1} w_{i,1}(t)y_0'(t))(\Delta x)^i/i! \\
+ \lambda^{\varrho-1} \sum_{\substack{2\le i+j\le K\\ j\ne 0}} (w_{i,j}'(t) - w_{i+1,j}(t)x_0'(t) - w_{i,j+1}(t)y_0'(t))(\Delta x)^i(\Delta y)^j/i!j!
\end{aligned}
\tag{6.4}
$$

where the terms $w_{i,j}(t) \equiv 0$ for $i + j > K$. We have

$$
\begin{aligned}
\partial_x \omega_\lambda(t,x,y) = \xi_0(t) + \sum_{1\le i\le K-1} w_{i+1,0}(t)(\Delta x)^i/i! \\
+ \lambda^{\varrho-1} \sum_{\substack{1\le i+j\le K-1\\ j\ne 0}} w_{i+1,j}(t)(\Delta x)^i(\Delta y)^j/i!j! \\
= \xi_0(t) + \sigma_0(t,x) + \lambda^{\varrho-1}\sigma_1(t,x,y)
\end{aligned}
\tag{6.5}
$$

Here σ_0 is a finite expansion in powers of Δx and σ_1 is a finite expansion in powers of Δx and Δy. Also

$$
\begin{aligned}
\partial_y \omega_\lambda(t,x,y) = \lambda^{-1/k}\eta_0 + \lambda^{\varrho-1} \sum_{1\le i+j\le K-1} w_{i,j+1}(t)(\Delta x)^i(\Delta y)^j/i!j! \\
= \lambda^{-1/k}\left(\eta_0(t) + \lambda^{1/k+\varrho-1}\sigma_2(t,x,y)\right)
\end{aligned}
\tag{6.6}
$$

where σ_2 is a finite expansion in powers of Δx and Δy.

Since the phase function is complex valued, the values of the symbol will be given by a formal Taylor expansion at the real values. Recall that $F_1 \circ \chi \cong f$ modulo $S^{1-1/k}$ so by the expansion in Remark 2.4 we find

$$
\begin{aligned}
F_1(t,x,y,\lambda\partial_x\omega_\lambda, \lambda\partial_y\omega_\lambda) \cong \lambda f(t,x,\xi_0+\sigma_0+\lambda^{\varrho-1}\sigma_1, \eta_0 + \lambda^{1/k+\varrho-1}\sigma_2) \\
= \lambda \sum_{\alpha,\beta} \partial_\xi^\alpha \partial_\eta^\beta f(t,x,\xi_0+\sigma_0,\eta_0)(\lambda^{\varrho-1}\sigma_1)^\alpha(\lambda^{1/k+\varrho-1}\sigma_2)^\beta/\alpha!\beta!
\end{aligned}
\tag{6.7}
$$

modulo $\mathcal{O}(\lambda^{1-1/k})$, which can then be expanded in Δx and Δy. This expansion can be done for any derivative of F_1, see for example (6.14). Observe that $F_1 - c\partial_\eta F_1$ is constant in y modulo $S(1, g_k)$, where $\partial_\eta F_1 \in S(\Lambda^{1/k}, g_k)$ when $|\eta| \lesssim |\xi|^{1-1/k}$ and $c \in S(1, g_k)$ may depend on y. Remark 2.4 also gives that $\partial_\eta^2 F_1(t,x,y,\lambda\partial_{x,y}\omega_\lambda)$ is bounded and by (6.6) we find that $\partial_y^2\omega_\lambda = \mathcal{O}(\lambda^{\varrho-1})$ so the last term in (6.1) is $\mathcal{O}(\lambda^\varrho)$.

When $x = x_0$ and $y = y_0$ we obtain that $F_1(t, x_0, y_0, \lambda\partial_x\omega_\lambda, \lambda\partial_y\omega_\lambda) \cong \lambda f(t, x_0, \xi_0, \eta_0)$ modulo $\mathcal{O}(\lambda^{1-1/k})$. Thus, taking the value of (6.1) and dividing by λ, we obtain the equation

$$w_0'(t) - \langle x_0'(t), \xi_0(t)\rangle + f(t, x_0(t), \xi_0(t), \eta_0(t)) = 0 \tag{6.8}$$

modulo $\mathcal{O}(\lambda^{-1/k}) + \mathcal{O}(\lambda^{1/k+\varrho-1})$. By taking real and imaginary parts we obtain the equations

$$\begin{cases} \operatorname{Re} w_0'(t) = \langle x_0'(t), \xi_0(t)\rangle - \operatorname{Re} f(t, x_0(t), \xi_0(t), \eta_0(t)) \\ \operatorname{Im} w_0'(t) = -\operatorname{Im} f(t, x_0(t), \xi_0(t), \eta_0(t)) \end{cases} \tag{6.9}$$

modulo $\mathcal{O}(\lambda^{-1/k}) + \mathcal{O}(\lambda^{1/k+\varrho-1}) = \mathcal{O}(\lambda^{-\kappa})$ for some $\kappa > 0$ since $\varrho < 1/2$. After choosing $w_0(0)$ this will determine w_0 when we have determined $(x_0(t), \xi_0(t), \eta_0(t))$. In the following, we shall solve equations like (6.9) modulo $\mathcal{O}(\lambda^{-\kappa})$ for some $\kappa > 0$, which will give the asymptotic solutions when $\lambda \to \infty$.

Using (6.7), we find that the first order terms in Δx of (6.1) will similarly be zero if

$$\begin{aligned} \xi_0'(t) - w_{2,0}(t)x_0'(t) + \partial_x f(t, x_0(t), \xi_0(t), \eta_0(t)) \\ + \partial_\xi f(t, x_0(t), \xi_0(t), \eta_0(t)) w_{2,0}(t) = 0 \end{aligned} \tag{6.10}$$

modulo $\mathcal{O}(\lambda^{-\kappa})$ for some $\kappa > 0$. By taking real and imaginary parts we find that (6.10) gives that

$$\begin{cases} \xi_0' = \operatorname{Re} w_{2,0}x_0' - \operatorname{Re} \partial_x f + \operatorname{Im} \partial_\xi f \operatorname{Im} w_{2,0} - \operatorname{Re} \partial_\xi f \operatorname{Re} w_{2,0} \\ \operatorname{Im} w_{2,0}x_0' = \operatorname{Im} \partial_x f + \operatorname{Im} \partial_\xi f \operatorname{Re} w_{2,0} + \operatorname{Re} \partial_\xi f \operatorname{Im} w_{2,0} \end{cases} \tag{6.11}$$

modulo $\mathcal{O}(\lambda^{-\kappa})$. Here and in what follows, the values of the symbols are taken at $(t, x_0(t), y_0(t), \xi_0(t), \eta_0(t))$. We shall put $(x_0(0), \xi_0(0)) = (x_0, \xi_0)$, which will determine $x_0(t)$ and $\xi_0(t)$ if $|\operatorname{Im} w_2(t)| \neq 0$.

Similarly, the second order terms in Δx of (6.1) vanish if we solve

$$w_{2,0}'/2 - w_{3,0}x_0'/2 + \partial_\xi f w_{3,0}/2 + \partial_x^2 f/2 + \Re\left(\partial_x \partial_\xi f w_{2,0}\right) + w_{2,0}\partial_\xi^2 f w_{2,0}/2 = 0$$

modulo $\mathcal{O}(\lambda^{-\kappa})$ for some $\kappa > 0$ where $\Re A = \frac{1}{2}(A + A^t)$ is the symmetric part of A. This gives

$$w_{2,0}' = w_{3,0}x_0' - \left(\partial_\xi f w_{3,0} + \partial_x^2 f + 2\Re\left(\partial_x \partial_\xi f w_{2,0}\right) + w_{2,0}\partial_\xi^2 f w_{2,0}\right) \tag{6.12}$$

with initial data $w_{2,0}(0)$ such that $\operatorname{Im} w_{2,0}(0) > 0$, which then holds in a neighborhood. Similarly, for $j > 2$ we obtain

$$w_{j,0}'(t) = w_{j+1,0}(t)x_0'(t) - \Big(f\big(t, x, \xi_0(t) + \sigma_0(t,x), \eta_0(t)\big)\Big)_j \tag{6.13}$$

modulo $\mathcal{O}(\lambda^{-\kappa})$, where we have taken the jth term of the expansion in Δx. Observe that (6.11)–(6.13) only involve x_0, ξ_0 and $w_{j,0}$ with $j \le K$.

Next, we will study the y dependent terms. Then, we have to expand $F_1 \circ \chi \cong f + r$ modulo $S^{1-2/k}$ where $r \in S^{1-1/k}$ is independent of y and $\partial_\eta r = 0$ when f vanishes (of infinite order). By expanding r, we get (6.7) with f replaced by r and λ replaced by $\lambda^{1-1/k}$. Since $\partial_\eta F_1 \circ \chi \cong \lambda^{1/k}\partial_\eta f$ modulo S^0 we obtain modulo bounded terms that

$$\begin{aligned}\partial_\eta F_1(t,x,y,\lambda\partial_{x,y}\omega_\lambda) &\cong \lambda^{1/k}\partial_\eta f(t,x_0,\xi_0+\sigma_0+\lambda^{\varrho-1}\sigma_1,\eta_0+\lambda^{1/k+\varrho-1}\sigma_2)\\ &\cong \lambda^{1/k}\partial_\eta f(t,x_0,\xi_0+\sigma_0,\eta_0)+\lambda^{2/k+\varrho-1}\partial_\eta^2 f(t,x_0,\xi_0+\sigma_0,\eta_0)\sigma_2 \qquad (6.14)\end{aligned}$$

modulo $\mathcal{O}(\lambda^{1/k+2\varrho-1})$. Similarly we obtain that

$$\partial_\eta^2 F_1(t,x,y,\lambda\partial_{x,y}\omega_\lambda) \cong \lambda^{2/k-1}\partial_\eta^2 f(t,x_0,\xi_0+\sigma_0,\eta_0)$$

modulo $\mathcal{O}(\lambda^{1/k+\varrho-1})$. Recall that $\partial_y F_1 \cong \partial_y c\partial_\eta F_1 \in S(\Lambda^{1/k}, g_k)$ modulo $S(1,g_k)$ when $|\eta| \lesssim |\xi|^{1-1/k}$, which gives that $\partial_y^\alpha F_1 \cong \partial_y^\alpha c\partial_\eta F_1$ modulo $S(1,g_k)$ for any α, see (7.13). Using (6.7) and (6.14) we find that the coefficients of the first order terms in Δy of (6.1) are

$$\begin{aligned}\lambda^{1-1/k}\eta_0' - \lambda^\varrho w_{0,2}y_0' - \lambda^\varrho w_{1,1}x_0' + \lambda^\varrho\partial_\eta r w_{0,2} + \lambda^\varrho\partial_\xi f w_{1,1}\\ +\lambda^{1/k+\varrho}\partial_\eta f w_{0,2} + \lambda^{1/k}\partial_y c\partial_\eta f - i\lambda^{2/k+\varrho-1}\partial_\eta^2 f w_{0,3} \qquad (6.15)\end{aligned}$$

modulo $\mathcal{O}(1)$ for some $c \in S^0$. Here and in what follows, the values of the symbols are taken at $(t, x_0(t), y_0(t), \xi_0(t)+\sigma_0(t,x), \eta_0(t))$ By taking the real and imaginary parts we obtain the equations

$$\eta_0' = -\lambda^{2/k+\varrho-1}\operatorname{Re}\partial_\eta f w_{0,2} - \lambda^{2/k-1}\operatorname{Re}\partial_y c\partial_\eta f \qquad (6.16)$$

modulo $\mathcal{O}(\lambda^{1/k+\varrho-1})$, and

$$\begin{aligned}\operatorname{Im} w_{0,2}y_0' = -\operatorname{Im} w_{1,1}x_0' + \operatorname{Im}\partial_\eta r w_{0,2} + \operatorname{Im}\partial_\xi f w_{1,1}\\ +\lambda^{1/k}\operatorname{Im}\partial_\eta f w_{0,2} + \lambda^{1/k-\varrho}\operatorname{Im}\partial_y c\partial_\eta f - \lambda^{2/k-1}\operatorname{Re}\partial_\eta^2 f w_{0,3} \qquad (6.17)\end{aligned}$$

modulo $\mathcal{O}(\lambda^{-\varrho})$. This gives that $\eta_0'(0) = \mathcal{O}(\lambda^{-\kappa})$ for some $\kappa > 0$, since (4.15) gives $\partial_\eta f(0,x_0,\xi_0,\eta_0) = 0$. We will choose initial data $\eta_0(0) = \eta_0$, $y_0(0) = y_0$ and $\operatorname{Im} w_{0,2}(0) > 0$, then y_0' is well defined in a neighborhood. In order to control the unbounded terms in (6.16) and (6.17), we shall use scaling and (4.15). Therefore we let

$$\zeta_0(t) = \lambda^{1-1/k-\varrho}(\eta_0(t) - \eta_0) \qquad (6.18)$$

where $\eta_0 = \eta_0(0)$. Then we get from (6.16) that

$$\zeta_0'(t) = -\lambda^{1/k} \operatorname{Re} \partial_\eta f w_{0,2} - \lambda^{1/k-\varrho} \operatorname{Re} \partial_y c \partial_\eta f \tag{6.19}$$

modulo bounded terms. Expanding (6.19) in $\eta_0(t) = \lambda^{1/k+\varrho-1}\zeta_0(t) + \eta_0$ we find

$$\zeta_0'(t) = -\lambda^{1/k} \operatorname{Re} \partial_\eta f_0 w_{0,2} - \lambda^{1/k-\varrho} \operatorname{Re} \partial_y c \partial_\eta f_0 - \lambda^{2/k+\varrho-1} \operatorname{Re} \zeta_0 \partial_\eta^2 f_0 w_{0,2} \\ - \lambda^{2/k-1} \operatorname{Re} \left(\zeta_0 \partial_\eta \partial_y c \partial_\eta f_0 + \partial_y c \partial_\eta^2 f_0 \zeta_0\right) \tag{6.20}$$

modulo $\mathcal{O}(\lambda^{-\kappa})$ for some $\kappa > 0$ where $\partial_\eta^j f_0 = \partial_\eta^j f\big|_{\eta=\eta_0}$ for $j \ge 0$. We shall use the following result.

Lemma 6.1 *Assume $k = \kappa(\omega) < \infty$, ε is given by* (4.15)–(4.16) *and* $\operatorname{Im} w_0(t) \ge 0$ *is the solution to* $\operatorname{Im} w_0'(t) = -\operatorname{Im} f(t, x_0(t), \xi_0(t), \eta_0)$ *with* $\operatorname{Im} w_0(0) = 0$.

If (4.15) *holds, then for any $\delta < \min\left(\varepsilon, 1 - \frac{1}{k}\right)$, α and β, there exists $\kappa > 0$ and $C \ge 1$ with the property that if*

$$\left| \int_0^t \left|\lambda^{1/k} \partial_x^\alpha \partial_\xi^\beta \partial_\eta f(s, x_0(s), \xi_0(s), \eta_0)\right| ds \right| \ge C\lambda^{-\delta} \tag{6.21}$$

with $\lambda \ge C$, then $\lambda \operatorname{Im} w_0(s) \ge \lambda^\kappa / C$ for some s in the interval connecting 0 and t.

If $k = 2$ and (4.16) *holds, then for any $\delta < \varepsilon$, α and β, there exists $\kappa > 0$ and $C \ge 1$ with the property that if*

$$\left| \int_0^t \left|\lambda^{2/k-1} \partial_x^\alpha \partial_\xi^\beta \partial_\eta^2 f(s, x_0(s), \xi_0(s), \eta_0)\right| ds \right| \ge C\lambda^{-\delta} \qquad \forall\, \alpha\, \beta \tag{6.22}$$

with $\lambda \ge C$, then $\lambda \operatorname{Im} w_0(s) \ge \lambda^\kappa / C$ for some s in the interval connecting 0 and t.

Observe that η_0 is constant in Lemma 6.1, but we have to show that $\operatorname{Im} w_0(t)$ has the minimum 0 at $t = 0$. Lemma 6.1 will be proved in Sect. 9. Clearly, (6.22) cannot hold if $k > 2$ and $\delta < 1/3$. We shall use Lemma 6.1 for a fixed $\delta > 0$ and for $|\alpha| + |\beta| \le N$. Since we only have to integrate the eikonal equations in the interval where $\lambda \operatorname{Im} w_0 \lesssim \lambda^\kappa$ for some $\kappa > 0$ and sufficiently large λ, we may assume the integrals in (6.21) and (6.22) are $\mathcal{O}(\lambda^{-\delta})$ when $\lambda \to \infty$. Using this and integrating (6.20) we find that $\zeta_0 = \mathcal{O}(\lambda^{-\kappa})$ for some $\kappa > 0$ if $\varrho \ll 1$, $\operatorname{Im} w_0(t) \ge 0$ and the coefficients $w_{i,j}$ are bounded. This gives a constant asymptotic solution η_0.

Remark 6.1 If $\zeta_0(t) = \lambda^{1-1/k-\varrho}\left(\eta_0(t) - \eta_0\right)$ then we have

$$\partial_x^\alpha \partial_\xi^\beta f(t, x_0(t), \xi_0(t), \eta_0(t)) \cong \partial_x^\alpha \partial_\xi^\beta f(t, x_0(t), \xi_0(t), \eta_0) \\ + \lambda^{1/k+\varrho-1} \partial_x^\alpha \partial_\xi^\beta \partial_\eta f(t, x_0(t), \xi_0(t), \eta_0)\zeta_0(t) \\ + \lambda^{2/k+2\varrho-2} \partial_x^\alpha \partial_\xi^\beta \partial_\eta^2 f(t, x_0(t), \xi_0(t), \eta_0)\zeta_0^2(t)/2 \tag{6.23}$$

for $\varrho \ll 1$ and $t \in I$ modulo $\mathcal{O}\left(\lambda^{-1-\kappa}|\zeta_0(t)|^3\right)$ for some $\kappa > 0$. We also obtain that

$$\begin{aligned}\partial_x^\alpha\partial_\xi^\beta\partial_\eta f(t,x_0(t),\xi_0(t),\eta_0(t)) &\cong \partial_x^\alpha\partial_\xi^\beta\partial_\eta f(t,x_0(t),\xi_0(t),\eta_0)\\ &\quad + \lambda^{1/k+\varrho-1}\partial_x^\alpha\partial_\xi^\beta\partial_\eta^2 f(t,x_0(t),\xi_0(t),\eta_0)\zeta_0(t)\end{aligned} \tag{6.24}$$

for $\varrho \ll 1$ and $t \in I$ modulo $\mathcal{O}(\lambda^{2\varrho-1}|\zeta_0(t)|^2)$ and

$$\partial_x^\alpha\partial_\xi^\beta\partial_\eta^2 f(t,x_0(t),\xi_0(t),\eta_0(t)) \cong \partial_x^\alpha\partial_\xi^\beta\partial_\eta^2 f(t,x_0(t),\xi_0(t),\eta_0) \tag{6.25}$$

when $\varrho \ll 1$ and $t \in I$ modulo $\mathcal{O}(\lambda^{\varrho-1/2}|\zeta_0(t)|)$.

If ζ_0 is bounded and the integrals in (6.21) and (6.22) are $\mathcal{O}(\lambda^{-\delta})$ then for δ and ϱ small enough we may replace η_0 with $\eta_0(t)$ in these integrals with a smaller $\delta > 0$. In fact, then the change in (6.22) is $\mathcal{O}(\lambda^{\varrho-1/2})$ and the change in (6.21) is $\mathcal{O}(\lambda^{\varrho-\delta} + \lambda^{2\varrho-1/2})$ using (6.22). Observe that we only need that $\delta > 0$ for the proof, but we have to show that $\operatorname{Im} w_0(t)$ has the minimum 0 at $t = 0$ which will be done later.

Using (6.7) and (6.14) we find that the second order terms in Δy of (6.1) vanish if

$$\begin{aligned}&\lambda^\varrho(w_{0,2}' - w_{1,2}x_0' - w_{0,3}y_0' + \partial_\eta r w_{0,3} + \partial_\xi f w_{1,2})\\ &\quad + \lambda^{1/k+\varrho}\partial_\eta f w_{0,3} + \lambda^{2/k+2\varrho-1}w_{0,2}\partial_\eta^2 f w_{0,2} + \lambda^{1/k}\partial_y^2 c\partial_\eta f\\ &\qquad + 2\lambda^{2/k+\varrho-1}\partial_y c\partial_\eta^2 f w_{0,2} - i\lambda^{2/k+\varrho-1}\partial_\eta^2 f w_{0,4} = 0\end{aligned} \tag{6.26}$$

modulo $\mathcal{O}(1)$ if $\varrho \ll 1$. This gives

$$\begin{aligned}w_{0,2}' &= w_{1,2}x_0' + w_{0,3}y_0' - \partial_\eta r w_{0,3} - \partial_\xi f w_{1,2}\\ &\quad - \lambda^{1/k}\partial_\eta f w_{0,3} - \lambda^{2/k+\varrho-1}w_{0,2}\partial_\eta^2 f w_{0,2} - \lambda^{1/k-\varrho}\partial_y^2 c\partial_\eta f\\ &\quad - 2\lambda^{2/k-1}\partial_y c\partial_\eta^2 f w_{0,2} + i\lambda^{2/k-1}\partial_\eta^2 f w_{0,4}\end{aligned} \tag{6.27}$$

modulo $\mathcal{O}(\lambda^{-\varrho})$ if $\varrho \ll 1$. By using Lemma 6.1 we may assume that the coefficients in the right hand side are uniformly integrable when ϱ is small enough. We choose the initial value so that $\operatorname{Im} w_{0,2}(0) > 0$ which then holds in a neighborhood.

Similarly, the coefficients for the term $\Delta x^j \Delta y^\ell$ in (6.1) can be found from the expansion in Δx and Δy of

$$\begin{aligned}&\lambda^\varrho\left(\sum_{j,\ell\neq 0}(w_{j,\ell}' - w_{j+1,\ell}x_0' - w_{j,\ell+1}y_0')\Delta x^j\Delta y^\ell/j!\ell! + \partial_\eta r\sigma_2 + \partial_\xi f\sigma_1\right)\\ &\quad + \lambda^{1/k+\varrho}\partial_\eta f\sigma_2 + \lambda^{2/k+2\varrho-1}\sigma_2\partial_\eta^2 f\sigma_2/2 + \lambda^{1/k}\partial_y c\Delta y\partial_\eta f\\ &\qquad + \lambda^{2/k+\varrho-1}\partial_y c\Delta y\partial_\eta^2 f\sigma_2 - i\lambda^{2/k+\varrho-1}\partial_\eta^2 f\partial_y\sigma_2\end{aligned} \tag{6.28}$$

modulo $\mathcal{O}(1)$ if $\varrho \ll 1$. Here the values of the symbols are taken at $(t, x_0(t), y_0(t), \xi_0(t) + \sigma_0(t, x), \eta_0(t))$, so the last terms can be expanded in Δx and Δy which also involves the ξ derivatives. Taking the coefficient for $\Delta x^j \Delta y^\ell$ and dividing by λ^ϱ we obtain that these terms vanish if

$$w'_{j,\ell} = w_{j+1,\ell}x'_0 + w_{j,\ell+1}y'_0 - j!\,\ell!\Big(\partial_\eta r\sigma_2 + \partial_\xi f\sigma_1 + \lambda^{1/k}\partial_\eta f\sigma_2 \\ + \lambda^{2/k+\varrho-1}\sigma_2\partial_\eta^2 f\sigma_2/2 + \lambda^{1/k-\varrho}\partial_y c\Delta y\partial_\eta f \\ + \lambda^{2/k-1}\partial_y c\Delta y\partial_\eta^2 f\sigma_2 - i\lambda^{2/k-1}\partial_\eta^2 f\partial_y\sigma_2\Big)_{j,\ell} \tag{6.29}$$

modulo $\mathcal{O}(\lambda^{-\varrho})$, where in the right hand side we have taken the coefficient of $\Delta x^j \Delta y^\ell$, expanding $\partial_y c$, $\partial_\eta r$, $\partial_\xi f$, $\partial_\eta f$ and $\partial_\eta^2 f$ in Δx and Δy which also involves the ξ derivatives. We choose initial values $w_{j,\ell}(0) = 0$ for $j = \ell = 1$ and $j + \ell > 2$. Observe that the Lagrange error term in the Taylor expansion of (6.7) is $\mathcal{O}(\lambda(|x - x_0(t)| + |y - y_0(t)|)^{K+1})$.

Now assume that $\operatorname{Im} w_0(t) \geq 0$ is a solution to the equation $\operatorname{Im} w'_0(t) = -\operatorname{Im} f(t, x_0(t), \xi_0(t), \eta_0)$ with $\operatorname{Im} w_0(0) = 0$, thus $\operatorname{Im} w_0(t)$ has a minimum at $t = 0$. The Eqs. (6.9), (6.11)–(6.13), (6.17), (6.20), (6.27) and (6.29) form a system of nonlinear ODE for $x_0(t)$, $y_0(t)$, $\xi_0(t)$, $\zeta_0(t)$ and $w_{j,\ell}(t)$ for $j + \ell \leq K$. By Remark 6.1 we can then replace $\eta_0(t)$ in f by $\eta_0 = \eta_0(0)$ when $\varrho \ll 1$. Since we only have to integrate where $\operatorname{Im} w_0(t) \lesssim \lambda^{\kappa-1}$ for some $\kappa > 0$, this system has uniformly integrable coefficients by Lemma 6.1 for $\varrho \ll 1$, which gives a local solution near $(0, x_0, y_0, \xi_0, \eta_0)$.

In the case when $\Gamma' = \big\{ (t, x_0, y_0, \xi_0, \eta_0) : t \in I' \big\}$ for $|I'| \neq 0$, we shall use the following definition.

Definition 6.1 For $a(t) \in L^\infty(\mathbf{R})$ and $\kappa \in \mathbf{R}$ we say that $a(t) \in I(\lambda^\kappa)$ if $\int_0^t a(s)\,ds = \mathcal{O}(\lambda^\kappa)$ uniformly for all $t \in I$ when $\lambda \gg 1$.

We have assumed that $\partial_x^\alpha \partial_\xi^\beta f(t, x_0, \xi_0, \eta_0) = 0$, $\forall\, \alpha\, \beta$, for $t \in I'$. Let I be the interval containing I' such that $\partial_x^\alpha \partial_\xi^\beta \partial_\eta f \in I(\lambda^{-1/k-\delta})$ and $\partial_x^\alpha \partial_\xi^\beta \partial_\eta^2 f \in I(\lambda^{1-2/k-\delta})$ for some $\delta > 0$ by Lemma 6.1. We obtain that $x'_0 = \xi'_0 = 0$ on I' by (6.11) which gives $w'_0 = 0$ on I' by (6.9). We also obtain $\eta'_0 \cong 0$ modulo $I(\lambda^{-\kappa})$ for some $\kappa > 0$ by (6.16), since all the coefficients are in $I(\lambda^{-\kappa})$. If $\varrho \ll 1$ we find from Eqs. (6.12) and (6.13) that $w'_{j,0} = 0$ on I' for $j \geq 2$. By (6.29) we find when $\ell > 0$ that

$$w'_{j,\ell} \cong w_{j,\ell+1}\big(y'_0 - j!\,\ell!\,\partial_\eta r\big) \qquad \text{on } I' \text{modulo } I(\lambda^{-\kappa}) \tag{6.30}$$

where $y'_0 \in I(1)$. Since $w_{j,\ell} \equiv 0$ when $j + \ell > K$ and $w_{j,\ell}(0) = 0$ when $j + \ell > 2$ we recursively find that $w_{j,\ell} \cong 0$ when $j + \ell > 2$ and $w'_{j,\ell} \cong 0$ when $j + \ell = 2$ modulo $I(\lambda^{-\kappa})$. By (6.17) we find that

$$y'_0 \cong (\operatorname{Im} w_{0,2}(0))^{-1} \operatorname{Im} \partial_\eta r w_{0,2}(0) \qquad \text{on } I' \text{modulo } I(\lambda^{-\kappa}) \tag{6.31}$$

which gives $y_0' = o(1)$ on I. In fact, we assume that $\partial_\eta r = 0$ when $\operatorname{Im} f$ vanishes of infinite order. We may choose I' as the largest interval containing 0 such that w_0 vanishes on I'. Then in any neighborhood of an endpoint of I' there exists points where $w_0 > c \ge \lambda^{\kappa-1}$ for $\lambda \gg 1$.

Now f and r are independent of y near the semibicharacteristic, so the coefficients of the system of equations are independent of $y_0(t)$ modulo $I(\lambda^{-\kappa})$. (If the symbols are independent of y in an arbitrarily large y neighborhood we don't need the vanishing condition on $\partial_\eta r$.) Thus, we obtain the solution ω_λ in a neighborhood of $\gamma' = \{ (t, x_0(t), y_0(t)) : t \in I' \}$ for $\lambda \gg 1$. In fact, by scaling we see that the $I(\lambda^{-\kappa})$ perturbations do not change the local solvability of the ordinary differential equation for large enough λ.

But we also have to show that $t \mapsto \operatorname{Im} f(t, x_0(t), \xi_0(t), \eta_0) = f_0(t)$ changes sign from + to − as t increases for some choice of initial data (x_0, ξ_0) and $w_{2,0}(0)$. Then we obtain that $\operatorname{Im} w_0(t) \ge 0$ for the solution to $\operatorname{Im} w_0'(t) = -\operatorname{Im} f(t, x_0(t), \xi_0(t), \eta_0)$ with suitable initial data. Observe that (6.11)–(6.13) only involve x_0, ξ_0 and $w_{j,0}$ with $j \le K$ and are uniformly integrable.

First we shall consider the case when $t \mapsto \operatorname{Im} f(t, x_0, \xi_0, \eta_0)$ changes sign from + to − of first order. Then

$$f_0'(0) = \operatorname{Im} \partial_t f(0, x_0, \xi_0, \eta_0) + \operatorname{Im} \partial_x f(0, x_0, \xi_0, \eta_0) x_0'(0) + \operatorname{Im} \partial_\xi f(0, x_0, \xi_0, \eta_0) \xi_0'(0) \qquad (6.32)$$

where $\operatorname{Im} \partial_t f(0, x_0, \xi_0, \eta_0) < 0$.

Remark 6.2 By (6.11) we may choose $w_{2,0}(0)$ so that $\operatorname{Im} w_{2,0}(0) > 0$ and $|(x_0'(0), \xi_0'(0))| \ll 1$.

In fact, if $\operatorname{Im} \partial_\xi f \ne 0$ then we may choose $\operatorname{Re} w_{2,0}(0)$ so that $\operatorname{Im} \partial_x f + \operatorname{Im} \partial_\xi f \operatorname{Re} w_{2,0} = 0$ at $(0, x_0, \xi_0, \eta_0)$. Since $\operatorname{Re} f \equiv 0$ when $\eta = \eta_0$ we find from (6.11) that $x_0'(0) = 0$ and

$$\xi_0'(0) = \operatorname{Im} \partial_\xi f(0, x_0, \xi_0, \eta_0) \operatorname{Im} w_{2,0}(0) = o(1)$$

if $\operatorname{Im} w_{2,0}(0) \ll 1$. On the other hand, if $\operatorname{Im} \partial_\xi f(t, x_0, \xi_0, \eta_0) = 0$ then by putting $\operatorname{Re} w_{2,0}(0) = 0$ we find from (6.11) that $\xi_0'(0) = 0$ and if $\operatorname{Im} w_{2,0}(0) \gg 1$ we obtain

$$x_0'(0) = (\operatorname{Im} w_{2,0}(0))^{-1} \operatorname{Im} \partial_x f(0, x_0, \xi_0, \eta_0) = o(1)$$

If $(x_0'(0), \xi_0'(0)) = o(1)$ then $f_0'(0) < 0$ by (6.32) so $t \mapsto f_0(t)$ has a sign change from + to − of first order at $t = 0$. By (6.9) we obtain that the asymptotic solution $t \mapsto \operatorname{Im} w_0(t)$ has a local minimum on I, which is also true when $\lambda \gg 1$, and the minima can be made equal to 0 by subtracting a constant depending on λ.

We also have to consider the general case when $t \mapsto \operatorname{Im} f(t, x_0, \xi_0, \eta_0)$ changes sign from + to − of higher order as t increases near I'. If there exist points in any (x, ξ) neighborhood of Γ' for $\eta = \eta_0$ where $\operatorname{Im} f = 0$ and $\partial_t \operatorname{Im} f < 0$, then by changing the initial data we can as before construct approximate solutions for which $t \mapsto \operatorname{Im} w_0(t)$ has a local minimum equal to 0 on I when $\lambda \gg 1$. Otherwise, $\partial_t \operatorname{Im} f \geq 0$ when $\operatorname{Im} f = 0$ in some (x, ξ) neighborhood of Γ' for $\eta = \eta_0$. Then we take the asymptotic solution $(w(t), w_{j,0}(t)) = (x_0(t), \xi_0(t), w_{j,0}(t))$ to (6.11)–(6.13) when $\lambda \to \infty$ with $\eta_0(t) \equiv \eta_0$ and initial data $w(0) = (x, \xi)$ but fixed $w_{j,0}(0)$. This gives a change of coordinates $(t, w) \mapsto (t, w(t))$ near $\gamma' = \{ (t, x_0, \xi_0) : t \in I' \}$ if $\partial_x^\alpha \partial_\xi^\beta f(t, x_0, \xi_0, \eta_0) = 0$ when $t \in I'$. In fact, the solution is constant on Γ' since all the coefficients of (6.11)–(6.13) vanish there. By the invariance of condition (Ψ) there will still exist a change of sign from $+$ to $-$ of $t \mapsto \operatorname{Im} f(t, w(t), \eta_0)$ in any neighborhood of Γ' after the change of coordinates, see [13, Theorem 26.4.12]. (Recall that conditions (2.19), (2.21) and (2.27) hold in some neighborhood of Γ'.) By choosing suitable initial values (x_0, ξ_0) at $t = t_0$ we obtain that $\operatorname{Im} w_0'(t) = -\operatorname{Im} f(t, w(t), \eta_0)$ has a sign change from $-$ to $+$ and $t \mapsto \operatorname{Im} w_0(t)$ has a local minimum on I for $\lambda \gg 1$, which can be assumed to be equal to 0 after subtraction. Thus, we obtain that

$$e^{i\lambda\omega_\lambda(t,x)} \leq e^{-(\lambda(\operatorname{Im} w_0(t)+c|\Delta x|^2)+c\lambda^\varrho|\Delta y|^2)} \qquad |\Delta x + |\Delta y| \ll 1 \tag{6.33}$$

where $\min_I \operatorname{Im} w_0(t) = 0$ with $\operatorname{Im} w_0(t) > 0$ for $t \in \partial I$. This gives an approximate solution to (6.1), and summing up, we have proved the following result.

Proposition 6.1 *Let $\Gamma' = \{ (t, x_0, y_0; 0, \xi_0, \eta_0) ; t \in I' \}$ and assume that $\partial_x^\alpha \partial_\xi^\beta f(t, x_0, \xi_0, \eta_0) = 0$, $\forall\, \alpha\, \beta$, for $t \in I'$ if $|I'| \neq 0$. Then for $\varrho > 0$ small enough and $\lambda \gg 1$ we may solve (6.1) modulo terms that are $\mathcal{O}(1)$ with $\omega_\lambda(t, x)$ given by (6.3) in a neighborhood of $\gamma' = \{ (t, x_0(t), y_0(t)) : t \in I' \}$ modulo $\mathcal{O}(\lambda(|x - x_0(t)| + |y - y_0(t)|)^M)$, $\forall\, M$, such that when $t \in I'$ we have that $(x_0(t), \xi_0(t), \eta_0(t)) = (x_0, \xi_0, \eta_0)$, $w_0(t) = 0$, $w_{1,1}(t) \cong 0$ and $w_{j,k}(t) \cong 0$ for $j + k > 2$ modulo $\mathcal{O}(\lambda^{-\kappa})$ for some $\kappa > 0$, $\operatorname{Im} w_{2,0}(t) > 0$ and $\operatorname{Im} w_{0,2}(t) > 0$.*

Assume that $t \mapsto f(t, x_0, \xi_0, \eta_0)$ changes sign from $+$ to $-$ as t increases near I'. Then by changing the initial values we may obtain that the curve $t \mapsto (t, x_0(t), y_0(t); 0, \xi_0(t), \eta_0(t))$, $t \in I$, is arbitrarily close to Γ, $\min_{t \in I} \operatorname{Im} w_0(t) = 0$ and $\operatorname{Im} w_0(t) > 0$ for $t \in \partial I$.

Since (6.33) holds near Γ' the errors in the eikonal equation will give terms that are bounded by $C_M \lambda^{1-M\varrho/2}$, $\forall\, M$. Observe that cutting off where $\operatorname{Im} w_0 > 0$ will give errors that are $\mathcal{O}(\lambda^{-M})$, $\forall\, M$.

7 The Bicharacteristics on Σ_2

We shall also consider the case when $\eta_0 = 0$ on the bicharacteristics, including the case $k = \infty$. As before, the eikonal equation is given by

$$\lambda \partial_t \omega_\lambda + F_1(t, x, y, \lambda \partial_{x,y} \omega_\lambda) - i\lambda \partial_\eta^2 F_1(t, x, y, \lambda \partial_{x,y} \omega_\lambda) \partial_y^2 \omega_\lambda = 0 \tag{7.1}$$

modulo bounded terms. By Proposition 4.1 we have $F_1 \in S(\Lambda, g_k)$, $F_1 \circ \chi = f \in S^1$ modulo $S^{1-1/k}$ when $|\eta| \lesssim |\xi|^{1-1/k}$ for $k < \infty$, and $F_1 \cong p_1 = f \in S^1$ modulo terms in S^2 vanishing of infinite order at Σ_2 when $k = \infty$. We have assumed that f is independent of y, $\operatorname{Re} f(t,x,\xi,0) \equiv 0$, $t \mapsto \operatorname{Im} f(t,x,\xi,0)$ changes sign from $+$ to $-$ as t increases in a neighborhood of I' and $f = 0$ at $\Gamma' = \{(t, x_0, \xi_0, 0) : \xi_0 \neq 0 \ t \in I'\}$. If $|I'| \neq 0$ then by Proposition 4.1 we may assume that $f\big|_{\Sigma_2}$ vanishes of infinite order at Γ'. When $k < \infty$ there exists $c \in S(1, g_k)$ so that $F_1 - c\partial_\eta F_1$ is independent of y modulo $S(1, g_k)$ when $|\eta| \lesssim |\xi|^{1-1/k}$ and when $k = \infty$ we may assume that f is independent of y.

We will use the phase function

$$
\begin{aligned}
\omega_\lambda(t,x,y) &= w_0(t) + \langle x - x_0(t), \xi_0(t)\rangle + \sum_{2 \le i \le K} w_{i,0}(t)(x - x_0(t))^i / i! \\
&+ \lambda^{\varrho-1}\Big(\langle y - y_0(t), \eta_0(t)\rangle + \sum_{\substack{2 \le i+j \le K \\ j \neq 0}} w_{i,j}(t)(x - x_0(t))^i (y - y_0(t))^j / i! j!\Big)
\end{aligned}
\tag{7.2}
$$

for sufficiently large K, where we will choose $0 < \varrho < 1/2$, $\xi_0(0) = \xi_0 \neq 0$, $\eta_0(0) = 0$, $\operatorname{Im} w_{2,0}(0) > 0$, $\operatorname{Im} w_{1,1}(0) = 0$ and $\operatorname{Im} w_{0,2}(0) > 0$, which will give $\partial^2_{x,y} \operatorname{Im} \omega_\lambda > 0$ when $t = 0$ and $|x - x_0(0)| + |y - y_0(0)| \ll 1$ which then holds in a neighborhood. Here as before we use the multilinear forms $w_{i,j} = \{ w_{\alpha,\beta} i! j! / \alpha! \beta! \}_{|\alpha|=i, |\beta|=j}$, $(x - x_0(t))^j = \{ (x - x_0(t))^\alpha \}_{|\alpha|=j}$ and $(y - y_0(t))^j = \{ (y - y_0(t))^\alpha \}_{|\alpha|=j}$ to simplify the notation. Observe that $x_0(t)$, $y_0(t)$, $\xi_0(t)$, $\eta_0(t)$ and $w_{j,k}(t)$ will depend uniformly on λ.

Putting $\Delta x = x - x_0(t)$ and $\Delta y = y - y_0(t)$ we find that

$$
\begin{aligned}
\partial_t \omega_\lambda(t,x,y) = w_0'(t) &+ \langle x_0'(t), \xi_0(t)\rangle - \lambda^{\varrho-1}\langle y_0'(t), \eta_0(t)\rangle \\
&+ \langle \xi_0'(t) - w_{2,0}(t) x_0'(t) - \lambda^{\varrho-1} w_{1,1}(t) y_0'(t), \Delta x\rangle \\
&+ \lambda^{\varrho-1}\langle \eta_0'(t) - w_{1,1}(t) x_0'(t) - w_{0,2}(t) y_0'(t), \Delta y\rangle \\
+ \sum_{2 \le i \le K} & (w_{i,0}'(t) - w_{i+1,0}(t) x_0'(t) - \lambda^{\varrho-1} w_{i,1}(t) y_0'(t))(\Delta x)^i / i! \\
+ \lambda^{\varrho-1} \sum_{\substack{2 \le i+j \le K \\ j \neq 0}} & (w_{i,j}'(t) - w_{i+1,j}(t) x_0'(t) - w_{i,j+1}(t) y_0'(t))(\Delta x)^i (\Delta y)^j / i! j!
\end{aligned}
\tag{7.3}
$$

where the terms $w_{i,j}(t) \equiv 0$ for $i + j > K$. We have

$$\partial_x \omega_\lambda(t,x,y) = \xi_0(t) + \sum_{1\le i\le K-1} w_{i+1,0}(t)(\Delta x)^i/i! \\ + \lambda^{\varrho-1} \sum_{\substack{1\le i+j\le K \\ j\ne 0}} w_{i+1,j}(t)(\Delta x)^i(\Delta y)^j/i!j! \\ = \xi_0(t) + \sigma_0(t,x) + \lambda^{\varrho-1}\sigma_1(t,x,y) \quad (7.4)$$

Here σ_0 is a finite expansion in powers of Δx and σ_1 is a finite expansion in powers of Δx and Δy. We also find

$$\partial_y \omega_\lambda(t,x,y) = \lambda^{\varrho-1}\Big(\eta_0 + \sum_{1\le i+j\le K-1} w_{i,j+1}(t)(\Delta x)^i(\Delta y)^j/i!j!\Big) \\ = \lambda^{\varrho-1}\big(\eta_0(t) + \sigma_2(t,x,y)\big) \quad (7.5)$$

where σ_2 is a finite expansion in powers of Δx and Δy. The main change from Sect. 6 is that $\lambda^{1/k}\eta_0$ get replaced by $\lambda^{\varrho}\eta_0$ in (6.6). Since the phase function is complex valued, the values will be given by a formal Taylor expansion of the symbol at the real values.

In the case $k < \infty$, the blowup $F_1 \circ \chi$ gives the Taylor expansion of F_1 at $\eta = 0$, for example f is the kth Taylor term of p with the constant term of p_1. By expanding, we find

$$F_1(t,x,y,\lambda\partial_{x,y}\omega_\lambda) \cong \lambda F_1(t,x,y,\xi_0+\sigma_0,0) + \lambda^{\varrho}\partial_\xi F_1(t,x,y,\xi_0+\sigma_0,0)\sigma_1 \\ + \lambda^{\varrho}\partial_\eta F_1(t,x,y,\xi_0+\sigma_0,0)(\eta_0+\sigma_2) \\ + \lambda^{2\varrho}(\eta_0+\sigma_2)\partial_\eta^2 F_1(t,x,y,\xi_0+\sigma_0,0)(\eta_0+\sigma_2)/2 \quad (7.6)$$

modulo $\mathcal{O}(\lambda^{-\kappa})$ for some $\kappa > 0$ if $\varrho \ll 1$. In fact, on Σ_2 we have $p = \partial p = \partial_\xi^2 p = 0$ so $\partial_{\xi,\eta} F_1 \in S^0$. The last term of (7.6) is $\mathcal{O}(\lambda^{2\varrho-1})$ if $k > 2$ since then $\partial_\eta^2 F_1 \in S^{-1}$ on Σ_2. Similarly, we find

$$\partial_\eta F_1(t,x,y,\lambda\partial_{x,y}\omega_\lambda) \cong \partial_\eta F_1(t,x,y,\xi_0+\sigma_0,0) \\ + \lambda^{\varrho}\partial_\eta^2 F_1(t,x,y,\xi_0+\sigma_0,0)(\eta_0+\sigma_2) \quad (7.7)$$

modulo $\mathcal{O}(\lambda^{-\kappa})$, where the last term vanish if $k > 2$.

In the case $k = \infty$ we have that the principal symbol $p \in S^2$ vanishes of infinite order when $\eta = 0$, which gives $p(t,x,y,\lambda\partial_{x,y}\omega_\lambda) = \mathcal{O}(\lambda^{2-j(1-\varrho)})$ for any j. Thus, we may assume that $F_1 \cong f$ modulo S^0 when $k = \infty$ which gives $\partial_y F_1 \cong 0$ modulo S^0. Since $\partial_\eta^2 F_1$ is bounded and $\partial_y^2 \omega_\lambda = \mathcal{O}(\lambda^{\varrho-1})$ by (7.5) we obtain that last term in (7.1) is $\mathcal{O}(\lambda^{\varrho})$. Thus we find from (7.1) and (7.6) that

$$\partial_t \omega_\lambda + f(t,x,\xi_0+\sigma_0,0) \cong 0 \quad (7.8)$$

modulo $\mathcal{O}(\lambda^{-\kappa})$. Observe we shall solve (7.8) modulo terms that are $\mathcal{O}(\lambda^{-1})$. When $x = x_0$ we obtain from (7.6) and (7.8) that

$$w_0' - \langle x_0'(t), \xi_0(t)\rangle + f(t, x_0, \xi_0, 0) = 0 \tag{7.9}$$

modulo $\mathcal{O}(\lambda^{-\kappa})$, which gives the Eq. (6.9) with $\operatorname{Re} f \equiv \eta_0 \equiv 0$.

Similarly, since $F_1 = f$ when $\eta = 0$ the first order terms in Δx of (7.1) vanish if

$$\begin{aligned} \xi_0'(t) - w_{2,0}(t)x_0'(t) + \partial_x f(t, x_0(t), \xi_0(t), 0)& \\ + \partial_\xi f(t, x_0(t), \xi_0(t), 0) w_{2,0}(t) &= 0 \end{aligned} \tag{7.10}$$

modulo $\mathcal{O}(\lambda^{-\kappa})$. By taking real and imaginary parts we find from (7.10) that (6.11) holds with $\eta_0 \equiv 0$. We put $(x_0(0), \xi_0(0)) = (x_0, \xi_0)$, which will determine $x_0(t)$ and $\xi_0(t)$ if $\operatorname{Im} w_{2,0}(t) \neq 0$. The second order terms in Δx vanish if

$$w_{2,0}' - w_{3,0}x_0' + \partial_\xi f w_{3,0} + \partial_x^2 f + 2\Re\left(\partial_x\partial_\xi f w_{2,0}\right) + w_{2,0}\partial_\xi^2 f w_{2,0} = 0 \tag{7.11}$$

modulo $\mathcal{O}(\lambda^{-\kappa})$, where $\Re A = (A + A^t)/2$ is the symmetric part of A. Here and in what follows, the values of the symbols are taken at $(t, x_0(t), y_0(t), \xi_0(t), 0)$. This gives the Eq. (6.12) modulo $\mathcal{O}(\lambda^{-\kappa})$ with $\eta_0 \equiv 0$ and we choose initial data $w_{2,0}(0)$ such that $\operatorname{Im} w_{2,0}(0) > 0$ which then holds in a neighborhood. Similarly, for $j > 2$ we obtain

$$w_{j,0}'(t) = w_{j+1,0}(t)x_0'(t) - \Big(f\big(t, x, \xi_0(t) + \sigma_0(t, x), 0\big)\Big)_j \tag{7.12}$$

modulo $\mathcal{O}(\lambda^{-\kappa})$, where we have taken the jth term of the expansion in Δx. Observe that (7.10)–(7.12) only involve x_0, ξ_0 and $w_{j,0}$ with $j \le K$.

When $k < \infty$, we expand $F_1 \circ \chi \cong f + r$ modulo $S^{1-2/k}$ when $|\eta| \lesssim |\xi|^{1-1/k}$, where $r \in S^{1-1/k}$ is homogeneous, independent of y and $\partial_\eta r = 0$ when f vanishes (of infinite order). Since $\partial_\eta f = 0$ at Σ_2, we find that $\partial_\eta F_1 \circ \chi = |\xi|^{1/k}\partial_\eta r \in S^0$ at Σ_2. Observe that r consists of the Taylor terms of p of order $k+1$ and the first order Taylor terms of p_1 at Σ_2. Now at Σ_2 we find $\partial_\xi F_1 \cong \partial_\xi f$ and $\partial_\eta^2 F_1 \cong |\xi|^{2/k}\partial_\eta^2 f \in S^0$ modulo S^{-1}. We also have $\partial_y F_1 \cong \partial_y c \partial_\eta F_1 \in S(\Lambda^{1/k}, g_k)$ modulo $S(1, g_k)$ so we find

$$\partial_y\partial_\eta F_1 \cong \partial_\eta(\partial_y c\partial_\eta F_1) \cong 0 \tag{7.13}$$

modulo $S(1, g_k)$ when $|\eta| \lesssim |\xi|^{1-1/k}$, which gives $\partial_y^\alpha F_1 \cong \partial_y^\alpha c \partial_\eta F_1 \in S(\Lambda^{1/k}, g_k)$ modulo $S(1, g_k)$. In the case $k = \infty$, we have $r \equiv 0$, $\partial_y F_1 \cong \partial_y f = 0$ modulo S^0 and we may formally put $1/k = 0$ in the formulas.

By (7.13), (7.6) and (7.7) the first order terms in Δy of (7.1) are equal to

$$\begin{aligned} &\lambda^\varrho\eta_0' - \lambda^\varrho w_{0,2}y_0' - \lambda^\varrho w_{1,1}x_0' + \lambda^\varrho\partial_\xi f w_{1,1} + \lambda^\varrho\partial_\eta r w_{0,2} \\ &\quad + 2\lambda^{2/k+2\varrho-1}\eta_0\partial_\eta^2 f w_{0,2} + \lambda^{2/k+\varrho-1}\partial_y c\partial_\eta^2 f\eta_0 - i\lambda^{2/k+\varrho-1}\partial_\eta^2 f w_{0,3} \end{aligned} \tag{7.14}$$

modulo $\mathcal{O}(1)$ since $\partial_y F_1 \cong \partial_y c \partial_\eta F_1 \cong |\xi|^{1/k} \partial_y c \partial_\eta r \cong 0$ on Σ_2 modulo bounded terms. In the case $k = \infty$, we put $r \equiv 0$ and $\partial_\eta^2 f \equiv 0$. The terms in (7.14) vanish if

$$\eta_0' - w_{0,2} y_0' - w_{1,1} x_0' + \partial_\xi f w_{1,1} + \partial_\eta r w_{0,2} + 2\lambda^{2/k+\varrho-1} \eta_0 \partial_\eta^2 f w_{0,2} \\ + \lambda^{2/k-1} \partial_y c \partial_\eta^2 f \eta_0 - i\lambda^{2/k-1} \partial_\eta^2 f w_{0,3} = 0 \qquad (7.15)$$

modulo $\mathcal{O}(\lambda^{-\varrho})$. The real part of (7.15) gives

$$\eta_0' = \operatorname{Re} w_{0,2} y_0' + \operatorname{Re} w_{1,1} x_0' - \operatorname{Re} \partial_\xi f w_{1,1} \\ - \operatorname{Re} \partial_\eta r w_{0,2} - 2\lambda^{2/k+\varrho-1} \operatorname{Re} \eta_0 \partial_\eta^2 f w_{0,2} \\ - \lambda^{2/k-1} \operatorname{Re} \partial_y c \partial_\eta^2 f \eta_0 + \lambda^{2/k-1} \operatorname{Im} \partial_\eta^2 f w_{0,3} \qquad (7.16)$$

modulo $\mathcal{O}(\lambda^{-\kappa})$, and we will choose initial data $\eta_0(0) = 0$.

By taking the imaginary part of (7.15) we find

$$\operatorname{Im} w_{0,2} y_0' = - \operatorname{Im} w_{1,1} x_0' + \operatorname{Im} \partial_\xi f w_{1,1} \\ + \operatorname{Im} \partial_\eta r w_{0,2} + 2\lambda^{2/k+\varrho-1} \operatorname{Im} \eta_0 \partial_\eta^2 f w_{0,2} \\ + \lambda^{2/k-1} \operatorname{Im} \partial_y c \partial_\eta^2 f \eta_0 - \lambda^{2/k-1} \operatorname{Re} \partial_\eta^2 f w_{0,3} \qquad (7.17)$$

modulo $\mathcal{O}(\lambda^{-\kappa})$. When $k > 2$ we have $\partial_\eta^2 f \equiv 0$ on Σ_2 and when $k = 2$ we shall use Lemma 6.1 with $\eta_0 = 0$ to obtain that (7.16) and (7.17) are uniformly integrable if $\varrho \ll 1$.

By using the expansions (7.6) and (7.7), we can obtain the coefficients for the term $\Delta x^j \Delta y^\ell$ in (7.1) from the expansion of

$$\lambda^\varrho \sum_{j,\ell \neq 0} (w_{j,\ell}' - w_{j+1,\ell} x_0' - w_{j,\ell+1} y_0') \Delta x^j \Delta y^\ell / j!\ell! \\ + \lambda^\varrho \partial_\xi f \sigma_1 + \lambda^\varrho \partial_\eta r \sigma_2 + \lambda^{2/k+2\varrho-1} (\eta_0 + \sigma_2) \partial_\eta^2 f (\eta_0 + \sigma_2)/2 \\ + \lambda^{2/k+\varrho-1} \partial_y c \Delta y \partial_\eta^2 f (\eta_0 + \sigma_2) - i\lambda^{2/k+\varrho-1} \partial_\eta^2 f \partial_y \sigma_2 \qquad (7.18)$$

modulo $\mathcal{O}(1)$. In the case $k = \infty$ we put $r \equiv 0$ and $\partial_\eta^2 f \equiv 0$. Here the last terms can be expanded in Δx and Δy which also involves the ξ derivatives. Taking the coefficient for $\Delta x^j \Delta y^\ell$ and dividing by λ^ϱ we obtain that these terms vanish if

$$w_{j,\ell}' = w_{j+1,\ell} x_0' + w_{j,\ell+1} y_0' \\ - j!\ell! \Big(\partial_\xi f \sigma_1 + \partial_\eta r \sigma_2 + \lambda^{2/k+\varrho-1} (\eta_0 + \sigma_2) \partial_\eta^2 f (\eta_0 + \sigma_2)/2 \\ + \lambda^{2/k-1} \partial_y c \Delta y \partial_\eta^2 f (\eta_0 + \sigma_2) - i\lambda^{2/k-1} \partial_\eta^2 f \partial_y \sigma_2 \Big)_{j,\ell} \qquad (7.19)$$

modulo $\mathcal{O}(\lambda^{-\kappa})$, for some $\kappa > 0$.

When $k = \infty$ we find that these equations form a uniformly integrable system of nonlinear ODE. When $k < \infty$ and $\operatorname{Im} w_0(t) \geq 0$ then by using Lemma 6.1 with $\eta_0 = 0$, $\lambda \gg 1$ and $\varrho \ll 1$ we obtain a uniformly integrable system, which gives a local solution near $(0, x_0, y_0, \xi_0, 0)$. When $f(t, x_0, \xi_0, 0) = 0$ for $t \in I'$ when $|I'| \neq 0$, we have assumed that $\partial_x^\alpha \partial_\xi^\beta f(t, x_0, \xi_0, 0) = 0, \forall\, \alpha\, \beta$, for $t \in I'$. When $k = 2$ we use Lemma 6.1 to obtain that $\partial_x^\alpha \partial_\xi^\beta \partial_\eta^2 f(t, x_0, \xi_0, 0) \in I(\lambda^{-\delta})$ for $t \in I'$, $\forall\, \alpha\, \beta$, where $I(\lambda^{-\delta})$ is given by Definition 6.1. Then (7.10) gives that $x_0' = \xi_0' = 0$ on I' and (7.9) gives that $w_0' = 0$ on I'. Equations (7.11) and (7.12) give that $w_{j,0}' = 0$ on I' for $j \geq 2$. By (7.19) we find when $\ell > 0$ that

$$w_{j,\ell}' \cong w_{j,\ell+1}\big(y_0' - j!\,\ell!\,\partial_\eta r\big) \qquad \text{on } I' \text{modulo } I(\lambda^{-\kappa})$$

for some $\kappa > 0$ where $y_0' = I(1)$. Since $w_{j,\ell} \equiv 0$ when $j + \ell > K$ and $w_{j,\ell}(0) = 0$ when $j + \ell > 2$ we find by recursion that $w_{j,\ell}(t) \cong 0$ when $j + \ell > 2$ and $w_{j,\ell}'(t) \cong 0$ for $t \in I'$ modulo $I(\lambda^{-\kappa})$ when $j + \ell = 2$. By (7.17) we find that

$$y_0' \cong (\operatorname{Im} w_{0,2}(0))^{-1} \operatorname{Im} \partial_\eta r w_{0,2}(0) \qquad \text{on } I' \text{ modulo } I(\lambda^{-\kappa}) \tag{7.20}$$

which gives $y_0' = o(1)$ in I, and (7.16) gives $\eta_0' \cong \operatorname{Re}(y_0' - \partial_\eta r) w_{0,2}(0) = o(1)$ modulo $I(\lambda^{-\kappa})$. In fact, we assume that $\partial_\eta r = 0$ when $\operatorname{Im} f$ vanishes of infinite order. We may choose I' as the largest interval containing 0 such that w_0 vanish on I'. Then in any neighborhood of an endpoint of I' there exists points where $w_0 > c \geq \lambda^{\kappa-1}$ for $\lambda \gg 1$.

Now f and r are independent of y near the semibicharacteristic, so the coefficients of the system of equations are independent of $y_0(t)$ modulo $I(\lambda^{-\kappa})$. (If the symbols are independent of y in an arbitrarily large y neighborhood we don't need the vanishing condition on $\partial_\eta r$.) Since we restrict f and r to $\eta = 0$ and these functions are independent of y, the coefficients of the system of equations are independent of $(y_0(t), \eta_0(t))$ modulo $I(\lambda^{-\kappa})$. Thus for $\lambda \gg 1$ the system has a solution ω_λ in a neighborhood of $\gamma' = \big\{\, (t, x_0(t), y_0(t)) : t \in I' \,\big\}$. As before, the Lagrange error term in the Taylor expansion of (7.1) is $\mathcal{O}(\lambda(|x - x_0(t)| + |y - y_0(t)|)^{K+1})$.

But we have to show that $t \mapsto \operatorname{Im} f(t, x_0(t), \xi_0(t), 0) = f_0(t)$ changes sign from $+$ to $-$ as t increases for some choice of initial values (t_0, x_0, ξ_0) and $w_{2,0}(0)$. Then we obtain that $\operatorname{Im} w_0(t) \geq 0$ for the solution to $\operatorname{Im} w_0'(t) = -\operatorname{Im} f(t, x_0(t), \xi_0(t), 0)$ with suitable initial data. We shall use the same argument as in Sect. 6. Observe that (7.10)–(7.12) only involve x_0, ξ_0 and $w_{j,0}$ with $j \leq K$ and are uniformly integrable. When the sign change is of first order we can use Remark 6.2 to choose $w_{2,0}(0)$ so that $|(x_0'(0), \xi_0'(0))| \ll 1$ and $\operatorname{Im} w_{2,0}(0) > 0$. We have

$$\begin{aligned} f_0'(0) = \operatorname{Im} \partial_t f(0, x_0, \xi_0, 0) + \operatorname{Im} \partial_x f(0, x_0, \xi_0, 0) x_0'(0) \\ + \operatorname{Im} \partial_\xi f(0, x_0, \xi_0, 0) \xi_0'(0) \end{aligned} \tag{7.21}$$

and since $\operatorname{Im} \partial_t f(0, x_0, \xi_0, 0) < 0$ we obtain that $t \mapsto f_0(t)$ has a sign change from $+$ to $-$ of first order as t increases if $|(x_0'(0), \xi_0'(0))| \ll 1$.

We also have to consider the general case when $t \mapsto \operatorname{Im} f(t, x_0, \xi_0, 0)$ changes sign from $+$ to $-$ of higher order as t increases near I'. If there exist points in any (x, ξ) neighborhood of Γ' for $\eta = 0$ where $\operatorname{Im} f = 0$ and $\partial_t \operatorname{Im} f < 0$, then by changing the initial data we can as before construct approximate solutions for which $t \mapsto \operatorname{Im} w_0(t)$ has a local minimum equal to 0 on I when $\lambda \gg 1$. Otherwise we have $\operatorname{Im} \partial_t f \geq 0$ when $\operatorname{Im} f = 0$ in some (x, ξ) neighborhood of Γ'. Then we take the asymptotic solution $w(t) = (x_0(t), \xi_0(t), w_{j,0}(t))$ to (7.10)–(7.12) when $\lambda \to \infty$ with $\eta_0(t) \equiv 0$ and initial data $w = (x, \xi)$ but fixed $w_{2,0}(0)$ and $w_{j,0}(0)$. This gives a change of coordinates $(t, x, \xi) \mapsto (t, w(t))$ near Γ'. In fact, the solution is constant on Γ' when $|I'| \neq 0$ since all the coefficients of (7.10)–(7.12) vanish there. By the invariance of condition (Ψ) there would then exist a change of sign of $t \mapsto \operatorname{Im} f(t, w(t), 0)$ from $+$ to $-$ in any neighborhood of Γ'. Thus by choosing suitable initial values (t_0, x_0, ξ_0) arbitrarily close to Γ' we obtain that $t \mapsto f_0(t)$ changes sign from $+$ to $-$ as t increases.

Since $\operatorname{Im} w_0'(t) = -\operatorname{Im} f(t, x_0(t), \xi_0(t), 0)$ we obtain that

$$e^{i\lambda\omega_\lambda(t,x)} \leq e^{-(\lambda(\operatorname{Im} w_0(t)+c|\Delta x|^2)+c\lambda^\varrho|\Delta y|^2)} \qquad |\Delta x| + |\Delta y| \ll 1 \tag{7.22}$$

where $\min_I \operatorname{Im} w_0(t) = 0$ with $\operatorname{Im} w_0(t) > 0$ for $t \in \partial I$. This gives the following result.

Proposition 7.1 *Let $\Gamma' = \{(t, x_0, y_0; 0, \xi_0, 0) : t \in I'\}$ and assume that $\partial_x^\alpha \partial_\xi^\beta f(t, x_0, \xi_0, 0) = 0$, $\forall\, \alpha\, \beta$, for all $t \in I'$ in the case $|I'| \neq 0$. Then for $\varrho \ll 1$ and $\lambda \gg 1$ we may solve (7.1) modulo $\mathcal{O}(\lambda(|x - x_0(t)| + |y - y_0(t)|)^M)$, $\forall\, M$, with $\omega_\lambda(t, x)$ given by (7.2) in a neighborhood of $\gamma' = \{(t, x_0(t), y_0(t)) : t \in I'\}$. When $t \in I'$ we find that $(x_0(t), \xi_0(t)) = (x_0, \xi_0)$, $w_0(t) = 0$, $w_{1,1}(t) \cong 0$ and $w_{j,k}(t) \cong 0$ for $j + k > 2$ modulo $\mathcal{O}(\lambda^{-\kappa})$ for some $\kappa > 0$, $\operatorname{Im} w_{2,0}(t) > 0$ and $\operatorname{Im} w_{0,2}(t) > 0$.*

If $t \mapsto f(t, x_0, \xi_0, 0)$ changes sign from $+$ to $-$ as t increases near I' then by choosing initial values we may obtain that $\{(t, x_0(t), y_0(t); 0, \xi_0(t), 0) : t \in I\}$ is arbitrarily close to Γ, $\min_{t \in I} \operatorname{Im} w_0(t) = 0$ and $\operatorname{Im} w_0(t) > 0$ for $t \in \partial I_0$.

8 The Transport Equations

Next, we shall solve the transport equations for the amplitudes $\phi \in C^\infty$, first in the case when $k < \infty$ and $\eta_0 \neq 0$ as in Sect. 6. Then we use the phase function (6.3), by expanding the transport equation using (6.5)–(6.7) and (6.14) we find that it is given by the following terms in (5.3):

$$\begin{aligned} D_t\phi + \left(\lambda^{1/k}\partial_\eta f(t, x, \xi_0 + \sigma_0, \eta_0) + \lambda^{2/k+\varrho-1}\partial_\eta^2 f(t, x, \xi_0 + \sigma_0, \eta_0)\sigma_2\right) D_y\phi \\ + \lambda^{2/k-1}\partial_\eta^2 f(t, x, \xi_0 + \sigma_0, \eta_0) D_y^2\phi/2 + \partial_\xi f(t, x, \xi_0 + \sigma_0, \eta_0) D_x\phi \\ + F_0(t, x, y, D_y)\phi = 0 \end{aligned} \tag{8.1}$$

modulo $\mathcal{O}(\lambda^{-\kappa})$ for some $\kappa > 0$ near $\gamma' = \{(t, x_0(t), y_0(t)) : t \in I'\}$ given by Proposition 6.1. Here $f \in S(\Lambda, g_k)$, $0 < \varrho < 1/2$, $x_0(t)$, $y_0(t)$, $\xi_0(t)$, $\eta_0(t)$ and σ_j are given by (6.3), (6.5) and (6.6), and $F_0(t, x, y, D_y)$ is a uniformly bounded first order differential operator. In fact, by Proposition 4.1 we have that $\partial_y F_1 \cong \partial_y c \partial_\eta F_1 \in S(\Lambda^{1/k}, g_k)$ modulo $S(1, g_k)$ which gives that $\partial_y \partial_\eta F_1(t, x, y, \lambda \partial_{x,y} \omega_\lambda)$ is uniformly bounded by Remark 5.1.

We shall choose the initial value of the amplitude $\phi = 1$ for $t = t_0$ such that $\operatorname{Im} w_0(t_0) = 0$, and because of (6.33) we only have to solve the equation modulo $\mathcal{O}(\lambda^\mu(|x - x_0(t)| + |y - y_0(t)|)^M)$ for some μ and any M. We first solve (8.1), but because of the lower order terms in (8.1) we will expand $\phi = \phi_0 + \lambda^{-\kappa}\phi_1 + \lambda^{-2\kappa}\phi_2 + \ldots$ in an asymptotic series with $\phi_j \in C^\infty$, which we will use in (5.1).

By making Taylor expansions in $\Delta x = x - x_0(t)$ and $\Delta y = y - y_0(t)$ of ϕ_0 and the coefficients of (8.1) we obtain a system of ODE's in the Taylor coefficients of ϕ_0. Observe that the Lagrange error terms of the Taylor expansions in the transport equation give terms that are $\mathcal{O}(\lambda^{1/k}(|x - x_0(t)| + |y - y_0(t)|)^{M+1})$ since $\varrho \le 1/k$. By taking ϱ small enough and using Lemma 6.1 with Remark 6.1 as in Sect. 6, we may assume that this system has uniformly integrable coefficients. Thus we get a uniformly bounded solution ϕ_0 to (8.1) modulo $\mathcal{O}(\lambda^{1/k}(|x - x_0(t)| + |y - y_0(t)|)^M + \lambda^{-\kappa})$ for any M such that $\phi_0(t_0) \equiv 1$. By induction we can successively make the lower order terms in (8.1) to be $\mathcal{O}(\lambda^{1/k}(|x - x_0(t)| + |y - y_0(t)|)^M + \lambda^{-\ell\kappa})$ by solving (5.3) for ϕ_ℓ with right hand side depending on ϕ_j, $j < \ell$, such that $\phi_\ell(t_0) \equiv 0$. Thus, we get a solution to (5.3) modulo $\mathcal{O}(\lambda^{1/k}(|x - x_0(t)| + |y - y_0(t)|)^M + \lambda^{-N})$ for any M and N.

In the case $\eta_0 = 0$, we use the phase function (7.2). By expanding (5.3) and using (7.4)–(7.7), the transport equation for ϕ becomes:

$$\begin{aligned} D_t\phi + \left(\partial_\eta F_1(t, x, \xi_0 + \sigma_0, 0) + \lambda^\varrho \partial_\eta^2 F_1(t, x, \xi_0 + \sigma_0, 0)(\eta_0 + \sigma_2)\right) D_y\phi \\ + \partial_\xi F_1(t, x, \xi_0 + \sigma_0, 0) D_x\phi_0 + \partial_\eta^2 F_1(t, x, \xi_0 + \sigma_0, 0) D_y^2\phi/2 \\ + F_0(t, x, y, D_y)\phi = 0 \quad (8.2) \end{aligned}$$

near $\gamma' = \{(t, x_0(t), y_0(t)) : t \in I'\}$ modulo $\mathcal{O}(\lambda^{-\kappa})$ for some $\kappa > 0$ if $\varrho \ll 1$. Since $\partial_y \partial_\eta F_1 \cong \partial_\eta(\partial_y c \partial_\eta F_1)$ is bounded we find that $F_0(t, x, y, D_y)$ is a uniformly first order bounded differential operator by Remark 5.3. On Σ_2 we have $\partial_\eta F_1 \in S^0$, $\partial_\eta^2 F_1 \in S^{-1}$ when $k > 2$, and $\partial_\eta^2 F_1 = \partial_\eta^2 f$ modulo S^{-1} when $k = 2$. We shall solve (8.2) with initial value $\phi \equiv 1$ when $t = t_0$.

As before we expand $\phi = \phi_0 + \lambda^{-\kappa}\phi_1 + \lambda^{-2\kappa}\phi_2 + \ldots$ in an asymptotic series with $\phi_j \in C^\infty$, which we will use in (5.1). Observe that the Lagrange term of the Taylor's expansions in the transport equation is $\mathcal{O}(\lambda^\varrho(|x - x_0(t)| + |y - y_0(t)|)^K)$ for any K. By taking the Taylor expansions in Δx and Δy of ϕ_0 and the coefficients of (8.2), we obtain a system of ODE's in the Taylor coefficients of ϕ_0. As in Sect. 7, we find from Lemma 6.1 that this system has uniformly integrable coefficients when $\varrho \ll 1$. So by choosing $\phi_0(t_0) \equiv 1$ we obtain a uniformly bounded solution to (8.2) modulo $\mathcal{O}(\lambda^\varrho(|x - x_0(t)| + |y - y_0(t)|)^M + \lambda^{-\kappa})$ for any M.

We can successively make the lower order terms in (5.3) to be $\mathcal{O}(\lambda^{\varrho}(|x - x_0(t)| + |y - y_0(t)|)^M + \lambda^{-\ell\kappa})$ by solving the Eq. (8.2) for ϕ_ℓ with right hand side depending on ϕ_j for $j < \ell$ such that $\phi_\ell(t_0) \equiv 0$. Thus we find that (5.3) holds modulo $\mathcal{O}(\lambda^{\varrho}(|x - x_0(t)| + |y - y_0(t)|)^M + \lambda^{-N})$ for any M and N and we have $\phi(t, x, y) = 1$ when $t = t_0$. Let $g_{1-\varrho}(dw) = \lambda^{2-2\varrho}|dw|^2$.

Proposition 8.1 *Assume that Propositions 4.1, 6.1 and 7.1 hold. Then for $\varrho \ll 1$ and any M and N we can solve the transport equations so that the expansion (5.3) is $\mathcal{O}(\lambda(|x - x_0(t)| + |y - y_0(t)|)^M + \lambda^{-N})$ near $\gamma' = \{ (t, x_0(t), y_0(t)) : t \in I' \}$. We have $\phi \in S(1, g_{1-\varrho})$ uniformly with support in a neighborhood of γ' where $x - x_0(t) = \mathcal{O}(\lambda^{\varrho-1/k})$, $y - y_0(t) = \mathcal{O}(\lambda^{-\varrho/4})$ and $\operatorname{Im} w_0(t) = \mathcal{O}(\lambda^{\varrho-1})$. We also have $\phi(t_0, x_0(t_0), y_0(t_0)) = 1$, $\lambda \gg 1$, for some $t_0 \in I'$ such that $\operatorname{Im} w_0(t_0) = 0$.*

In fact, we obtain this by cutting off the solution ϕ near γ'. The cutoff in (x, y) can be done for $\varrho \ll 1/k$ by the cutoff function

$$\psi((x - x_0(t))\lambda^{1/k-\varrho}, (y - y_0(t))\lambda^{\varrho/4}) \in S(1, g_{1-\varrho})$$

where $\psi(x, y) \in C_0^\infty$ such that $\psi = 1$ in a neighborhood of the origin. In fact, differentiation in x and y gives factors that are $\mathcal{O}(\lambda^{1/k-\varrho} + \lambda^{\varrho/4}) = \mathcal{O}(\lambda^{1-\varrho})$. Differentiation in t gives factors $x_0'\lambda^{1/k-\varrho}$ and $y_0'\lambda^{\varrho/4}$. Here $x_0' \in C^\infty$ uniformly by (6.11) and (7.10), and $y_0' = \mathcal{O}(\lambda^{1/k})$ by (6.17) and (7.17). Repeated differentiation of y_0' gives at most factors $\mathcal{O}(\lambda^{1/k})$ by (6.16), (6.17), (6.29), (7.15) and (7.19).

The cutoff in t can be done where $\operatorname{Im} w_0(t) \cong \lambda^{\varrho-1}$ by the cutoff function $\chi(\operatorname{Im} w_0(t)\lambda^{1-\varrho}) \in S(1, \lambda^{2-2\varrho}dt^2)$ with $\chi \in C_0^\infty(\mathbf{R})$ such that $\chi = 1$ near 0. By (6.33) and (7.22) the cutoff errors will be $\mathcal{O}(\lambda^{-N})$ for any N. We obtain that $\phi(t, x, y) \in S(1, g_{1-\varrho})$ uniformly, $\phi(t_0, x_0(t_0), y_0(t_0)) = 1$ and $\operatorname{Im} w_0(t_0) = 0$ for some $t_0 \in I'$.

9 Proof of Lemma 6.1

Observe that $k < \infty$ and that if Lemma 6.1 holds for some δ and C, then it trivially holds for smaller δ and κ and larger C. Assume that (6.21) (or (6.22) when $k = 2$) holds at t, by switching t and $-t$ we may assume $t > 0$. Assume that $\operatorname{Im} w_0(t) \geq 0$ satisfies $\operatorname{Im} w_0'(t) = -\operatorname{Im} f(t, x_0(t), \xi_0(t), \eta_0)$ and put

$$f_0(t) = |\operatorname{Im} f(t, x_0(t), \xi_0(t), \eta_0)| \tag{9.1}$$

$$f_1(t) = |\partial_{x,\xi}^\alpha \partial_\eta f(t, x_0(t), \xi_0(t), \eta_0)| \tag{9.2}$$

$$f_2(t) = |\partial_{x,\xi}^\alpha \partial_\eta^2 f(t, x_0(t), \xi_0(t), \eta_0)| \tag{9.3}$$

for a fixed α.

We shall first consider the case when $\operatorname{Im} w_0(t)$ has a zero of *finite order* at $t = 0$. Then since 0 is a minimum, $\operatorname{Im} w_0'(t)$ has a sign change of finite order from $-$ to $+$ at $t = 0$. Since $t \operatorname{Im} w_0'(t) \ge 0$ we have that $\operatorname{Im} w_0(t) = \int_0^t f_0(s)\,ds$ for $t > 0$ so (6.21), (4.15) and the Hölder inequality give

$$\lambda^{-1/k-\delta} \lesssim \int_0^t f_1(s)\,ds \lesssim \int_0^t f_0^{1/k+\varepsilon}(s)\,ds \lesssim \operatorname{Im} w_0(t)^{1/k+\varepsilon} \tag{9.4}$$

for $0 < t \ll 1$. Thus $\operatorname{Im} w_0(t) \gtrsim \lambda^{-(1+k\delta)/(1+k\varepsilon)}$ and since $\delta < \varepsilon$ we obtain $\lambda \operatorname{Im} w_0(t) \gtrsim \lambda^\kappa$ for some $\kappa > 0$.

In the case when $k = 2$ and (6.22) holds, we similarly find from (4.16) that

$$\lambda^{-\delta} \lesssim \int_0^t f_2(s)\,ds \lesssim \int_0^t f_0^\varepsilon(s)\,ds \lesssim \operatorname{Im} w_0(t)^\varepsilon \qquad 0 < t \ll 1 \tag{9.5}$$

which implies that $\lambda \operatorname{Im} w_0(t) \gtrsim \lambda^{1-\delta/\varepsilon} \gtrsim \lambda^\kappa$ for some $\kappa > 0$ since $\delta < \varepsilon$.

Next we consider the general case when $\operatorname{Im} w_0(t)$ vanishes of *infinite order* at $t = 0$, then $f_0(t)$ also vanishes of infinite order. For $\varepsilon \ge 0$ let I_ε be the maximal interval containing 0 such that $\operatorname{Im} w_0 \le \varepsilon$ on I_ε. By assumtion (4.15) (and (4.16) when $k = 2$) holds in a neighborhood I of I_0. By continuity, we have $I_\varepsilon \downarrow I_0$ when $\varepsilon \downarrow 0$. Since $\operatorname{Im} w_0 = \varepsilon$ on ∂I_ε where $\varepsilon \gtrsim \lambda^{\kappa-1} = o(1)$ for $\lambda \gg 1$ it suffices to prove the result in I for large enough λ. Observe that if $f_1 \ll \lambda^{-1/k-\delta}$ and $f_2 \ll \lambda^{1-2/k-\delta}$ in $[0, t]$ then neither (6.21) nor (6.22) can hold. If for some $s \in [0, t]$ we have that $f_1(s) \gtrsim \lambda^{-1/k-\delta}$ (or $f_2(s) \gtrsim \lambda^{-\delta}$ when $k = 2$) then by (4.15) we find that

$$\lambda^{-1/k-\delta} \lesssim f_1(s) \lesssim |f_0(s)|^{1/k+\varepsilon} \tag{9.6}$$

(or $\lambda^{-\delta} \lesssim f_2(s) \lesssim |f_0(s)|^\varepsilon$ by (4.16)). Since $\delta < \varepsilon$ we find that in both cases $f_0(s) \ge c\lambda^{-1+\varrho}$ for some $\varrho > 0$ and $c > 0$. Now we define t_0 as the smallest $t > 0$ such that $|\operatorname{Im} w_0'(t_0)| = f_0(t) = c\lambda^{-1+\varrho} = \kappa$. Since $f_0(t)$ vanishes of infinite order at $t = 0$, we find that $f_0(t_0) \le C_N |t_0|^N$ for any $N \ge 1$, which gives $|t_0| \gtrsim \kappa^{1/N}$. Thus, we can use Lemma 9.1 below with $\kappa = c\lambda^{-1+\varrho}$ for $\lambda \gg 1$ to obtain that

$$\max_{0 \le s \le t_0} \operatorname{Im} w_0(s) \gtrsim \kappa^{1+1/N} \cong \lambda^{(-1+\varrho)(1+1/N)} \tag{9.7}$$

where $(-1+\varrho)(1+1/N) = -1 + \varrho - (1-\varrho)/N > -1$ if we choose $N > 1/\varrho - 1$, which gives the result. □

Lemma 9.1 *Assume that* $0 \le F(t) \in C^\infty$ *has local minimum at* $t = 0$, *and let* I_{t_0} *be the closed interval joining* 0 *and* $t_0 \in \mathbf{R}$. *If*

$$\max_{I_{t_0}} |F'(t)| = |F'(t_0)| = \kappa \le 1$$

with $|t_0| \ge c\kappa^\varrho$ *for some* $\varrho > 0$ *and* $c > 0$*, then we have* $\max_{I_{t_0}} F(t) \ge C_{\varrho,c}\kappa^{1+\varrho}$*. The constant* $C_{\varrho,c} > 0$ *only depends on* ϱ*,* c *and the bounds on* F *in* C^∞*.*

Proof Let $f = F'$ then $F(t) = F(0) + \int_0^t f(s)\,ds \ge \int_0^t f(s)\,ds$ so assuming the minimum is $F(0) = 0$ only improves the estimate. By switching t to $-t$ we may assume $t_0 \le -c\kappa^\varrho < 0$. Let

$$g(t) = \kappa^{-1} f(t_0 + tc\kappa^\varrho) \tag{9.8}$$

then $|g(0)| = 1$, $|g(t)| \le 1$ for $0 \le t \le 1$ and

$$|g^{(N)}(t)| = c^N \kappa^{\varrho N-1} |f^{(N)}(t_0 + tc\kappa^\varrho)| \le C_N \qquad 0 \le t \le 1 \tag{9.9}$$

when $N \ge 1/\varrho$. By using the Taylor expansion at $t = 0$ for $N \ge 1/\varrho$ we find

$$g(t) = p(t) + r(t) \tag{9.10}$$

where p is the Taylor polynomial of order $N - 1$ of g at 0, and

$$r(t) = t^N \int_0^1 g^{(N)}(ts)(1-s)^{N-1}\,ds/(N-1)! \tag{9.11}$$

is uniformly bounded in C^∞ for $0 \le t \le 1$ and $r(0) = 0$. Since g also is bounded on the interval, we find that $p(t)$ is uniformly bounded in $0 \le t \le 1$. Since all norms on the finite dimensional space of polynomials of fixed degree are equivalent, we find that $p^{(k)}(0) = g^{(k)}(0)$ are uniformly bounded for $0 \le k < N$ which implies that $g(t)$ is uniformly bounded in C^∞ for $0 \le t \le 1$. Since $|g(0)| = 1$ there exists a uniformly bounded $\delta^{-1} \ge 1$ such that $|g(t)| \ge 1/2$ when $0 \le t \le \delta$, thus g has the same sign in that interval. Since $g(t) = \kappa^{-1} f(t_0 + tc\kappa^\varrho)$ we find

$$\delta/2 \le \left| \int_0^\delta g(s)\,ds \right| = \left| \kappa^{-\varrho} \int_{t_0}^{t_0 + c\delta\kappa^\varrho} \kappa^{-1} f(t)\,dt/c \right| \tag{9.12}$$

Since $t_0 + c\delta\kappa^\varrho \le 0$ we find that the variation of $F(t)$ on $[t_0, 0]$ is greater than $c\delta\kappa^{1+\varrho}/2$ and since $F \ge 0$ we find that the maximum of F on I_{t_0} is greater than $c\delta\kappa^{1+\varrho}/2$. □

10 The Proof of Theorem 2.1

We shall use the following modification of Lemma 26.4.15 in [13]. Recall that $\|u\|_{(k)}$ is the L^2 Sobolev norm of order k of $u \in C_0^\infty$ and let $\mathcal{D}'_\Gamma = \{ u \in \mathcal{D}' : \mathrm{WF}(u) \subset \Gamma \}$ for $\Gamma \subseteq T^*\mathbf{R}^n$.

Lemma 10.1 *Let*

$$u_\lambda(x) = \exp(i\lambda\omega_\lambda(x)) \sum_{j=0}^{M} \varphi_{j,\lambda}(x)\lambda^{-j\kappa} \qquad \lambda \geq 1 \tag{10.1}$$

with $\kappa > 0$, $\omega_\lambda \in C^\infty(\mathbf{R}^n)$ *satisfying* $\operatorname{Im}\omega_\lambda \geq 0$, $|\partial \operatorname{Re}\omega_\lambda| \geq c > 0$, *and* $\varphi_{j,\lambda} \in S(1, \lambda^{2-2\varrho}|dx|^2) = S(1, g_{1-\varrho})$, $\forall\, j\ \lambda$, *for some* $\varrho > 0$. *We assume that* $\omega_\lambda \to \omega_\infty$ *when* $\lambda \to \infty$, *and that* $\varphi_{j,\lambda}$ *has support in a compact set* Ω, $\forall\, j\ \lambda$. *Then we have*

$$\|u_\lambda\|_{(-N)} \leq C\lambda^{-N} \qquad \lambda \geq 1 \qquad \forall N \geq 0 \tag{10.2}$$

If $\lim_{\lambda\to\infty} \varphi_{0,\lambda}(x_0) \neq 0$ *and* $\operatorname{Im}\omega_\infty(x_0) = 0$ *for some* x_0 *then there exists* $c > 0$ *so that*

$$\|u_\lambda\|_{(-N)} \geq c\lambda^{-N-n/2} \qquad \lambda \geq 1 \qquad \forall N \geq 0 \tag{10.3}$$

Let $\Sigma = \lim_{\kappa\to\infty} \overline{\bigcup_{j,\lambda\geq\kappa} \operatorname{supp}\varphi_{j,\lambda}} \subset \Omega$ *and let* Γ *be the cone generated by*

$$\{\, (x, \partial\omega_\infty(x)),\ x \in \Sigma \,\} \tag{10.4}$$

Then for any m *we find* $\lambda^m u_\lambda \to 0$ *in* $\mathcal{D}'_\Gamma$ *so* $\lambda^m A u_\lambda \to 0$ *in* C^∞ *if* A *is a pseudodifferential operator such that* $\operatorname{WF}(A) \cap \Gamma = \emptyset$. *The estimates are uniform if* $\varphi_{j,\lambda}$ *is uniformly bounded in* $S(1, g_{1-\varrho})$ *with fixed compact support* $\forall\, j\ \lambda$ *and* $\omega_\lambda \in C^\infty$ *uniformly with fixed lower bound on* $|\partial \operatorname{Re}\omega_\lambda|$.

Observe that by Propositions 6.1 and 7.1 the phase functions ω_λ in (6.3) or (7.2) satisfy the conditions in Lemma 10.1 near $\{\, (t, x_0(t), y_0(t)) : t \in I' \,\}$ since $\xi_0(t) \neq 0$ and $\operatorname{Im}\omega_\lambda(t, x) \geq 0$. Also, the functions ϕ_j in the expansion (5.1) satisfy the conditions in Lemma 10.1 uniformly in λ by Proposition 8.1. Then $\Sigma = \{\, (t, x_0(t), y_0(t)) : t \in I' \,\}$ and the cone Γ is generated by

$$\{\, (t, x_0(t), y_0(t), 0, \xi_0(t), 0) : t \in I' \,\} \tag{10.5}$$

In fact, in both the expansions (6.3) and (7.2) we have that $\partial\omega_\infty(t, x, y) = (0, \xi_0(t) + \sigma_0(t, x), 0)$, and by Proposition 8.1 the supports of ϕ_j in (5.1) shrink to $\{\, (t, x_0(t), y_0(t)) : t \in I' \,\}$ as $\lambda \to \infty$ for any j.

Proof (*Proof of Lemma* 10.1) We shall modify the proof of [13, Lemma 26.4.15] to this case. We have that

$$\hat{u}_\lambda(\xi) = \sum_{j=0}^{M} \lambda^{-j\kappa} \int e^{i\lambda\omega_\lambda(x) - i\langle x, \xi\rangle} \varphi_{j,\lambda}(x)\, dx \tag{10.6}$$

Let U be a neighborhood of the projection on the second component of the set in (10.4). When $\xi/\lambda \notin U$ for $\lambda \gg 1$ we find that

$$\overline{\bigcup_j \operatorname{supp}\varphi_{j,\lambda}} \ni x \mapsto (\lambda\omega_\lambda(x) - \langle x, \xi\rangle)/(\lambda + |\xi|)$$

is in a compact set of functions with nonnegative imaginary part with a fixed lower bound on the gradient of the real part. Thus, by integrating by parts we find for any positive integer k that

$$|\hat{u}_\lambda(\xi)| \le C_k(\lambda + |\xi|)^{-k} \qquad \xi/\lambda \notin U \qquad \lambda \gg 1 \tag{10.7}$$

which gives any negative power of λ for k large enough. If V is bounded and $0 \notin \overline{V}$ then since u_λ is uniformly bounded in L^2 we find

$$\int_{\lambda V} |\hat{u}_\lambda(\xi)|^2 (1 + |\xi|^2)^{-N}\, d\xi \le C_V \lambda^{-2N} \tag{10.8}$$

which together with (10.7) gives (10.2). If $\chi \in C_0^\infty$ then we may apply (10.7) to χu_λ, thus we find for any positive integer k that

$$|\widehat{\chi u}_\lambda(\xi)| \le C(\lambda + |\xi|)^{-k} \qquad \xi \in W \qquad \lambda \gg 1 \tag{10.9}$$

if W is any closed cone with $(\operatorname{supp}\chi \times W) \bigcap \Gamma = \emptyset$. Thus we find that $\lambda^m u_\lambda \to 0$ in $\mathcal{D}'_\Gamma$ for every m. To prove (10.3) we may assume that $x_0 = 0$ and take $\psi \in C_0^\infty$. If $\operatorname{Im}\omega_\infty(0) = 0$ and $\lim_{\lambda\to\infty} \varphi_{0,\lambda}(0) \ne 0$ then since $\varphi_{j,\lambda}(x/\lambda) = \varphi_{j,\lambda}(0) + \mathcal{O}(\lambda^{-\varrho})$ in $\operatorname{supp}\psi$ $\forall\, j$ we find that

$$\begin{aligned} &\lambda^n e^{-i\lambda \operatorname{Re}\omega_\lambda(0)} \langle u_\lambda, \psi(\lambda\cdot)\rangle \\ &= \int e^{i\lambda(\omega_\lambda(x/\lambda) - \operatorname{Re}\omega_\lambda(0))} \psi(x) \sum_j \varphi_{j\lambda}(x/\lambda)\lambda^{-j\kappa}\, dx \\ &\to \int e^{i\langle \operatorname{Re}\partial_x\omega_\infty(0), x\rangle} \psi(x)\varphi_{0,\infty}(0)\, dx \qquad \lambda \to \infty \end{aligned} \tag{10.10}$$

which is not equal to zero for some suitable $\psi \in C_0^\infty$. Since

$$\|\psi(\lambda\,\cdot)\|_{(N)} \le C_N \lambda^{N-n/2} \tag{10.11}$$

we obtain from (10.10) that $0 < c \le \lambda^{N+n/2}\|u_\lambda\|_{(-N)}$ which gives (10.3) and the lemma. □

Proof (Proof of Theorem 2.1) By conjugating with elliptic Fourier integral operators and multiplying with pseudodifferential operators, we obtain that $P^* \in \Psi^2_{\mathrm{cl}}$ is of the form given by Proposition 4.1 microlocally near $\Gamma = \{\, (t, x_0, y_0, 0, \xi_0, 0) : t \in I \,\}$. Thus we may assume

$$P^* = D_t + F(t, x, y, D_x, D_y) + R \tag{10.12}$$

where $R \in \Psi^2_{cl}$ satisfies $\mathrm{WF}_{g_k}(R) \bigcap \Gamma \times \{\eta_0\} = \emptyset$ when $\kappa < \infty$, vanishes of infinite order at Σ_2 if $\kappa = \infty$, and the form of the symbol of F depends on whether $k < \infty$ or $k = \infty$.

Then we can construct approximate solutions u_λ to $P^* u_\lambda = 0$ of the form (5.1) for $\lambda \to \infty$ by using the expansion (5.3). The phase function ω_λ is given by (6.3) in the case when $k < \infty$ and $\eta_0 \neq 0$ or by (7.2) in the case when $\eta_0 = 0$.

First we solve the eikonal equation (6.1) modulo $\mathcal{O}\big(\lambda(|x - x_0(t)| + |y - y_0(t)|)^M\big)$ for any M by using Propositions 6.1 when $k < \infty$ and $\eta_0 \neq 0$ or Proposition 7.1 when $\eta_0 = 0$. By using Proposition 8.1 we can solve the transport equations so that the expansion (5.3) is $\mathcal{O}\big(\lambda(|x - x_0(t)| + |y - y_0(t)|)^M + \lambda^{-N}\big)$ for any M and N and $\phi_0(t_0, x_0(t_0), y_0(t_0)) = 1$ for some $t \in I'$. Because of the phase functions (6.33) or (7.22) this gives approximate solutions u_λ of the form (10.1) in Lemma 10.1. In fact, for any N we may choose M in Proposition 8.1 so that $|(D_t + F)u_\lambda| \lesssim \lambda^{-N}$. Now differentiation of $(D_t + F)u_\lambda$ can at most give a factor λ. In fact, differentiating the exponential gives a factor λ and differentiating the amplitude gives either a factor $\lambda^{1-\varrho}$, or a loss of a factor $x - x_0(t)$ or $y - y_0(t)$ in the expansion, which gives at most a factor $\lambda^{1/2-\varrho}$. Because of the bounds on the support of u_λ we obtain that

$$\|(D_t + F)u_\lambda\|_{(\nu)} = \mathcal{O}(\lambda^{-N-n}) \tag{10.13}$$

for any chosen ν. Since Propositions 6.1, 7.1 and 8.1 gives t_0 such that $\phi_0(t_0, x_0(t_0), y_0(t_0)) = 1$ and $\operatorname{Im} \omega_\lambda(t_0, x_0(t_0), y_0(t)) = 0$ when $\lambda \gg 1$, we find by (10.2) and (10.3) that

$$\lambda^{-N-n/2} \lesssim \|u_\lambda\|_{(-N)} \lesssim \lambda^{-N} \qquad \forall N \geq 0 \quad \lambda \gg 1 \tag{10.14}$$

Since u_λ has support in a fixed compact set, we find from Remark 5.2 and Lemma 10.1 that $\|Ru_\lambda\|_{(\nu)}$ and $\|Au_\lambda\|_{(0)}$ are $\mathcal{O}(\lambda^{-N-n})$ if $\mathrm{WF}(A)$ does not intersect Γ. Thus we find from (10.13) and (10.14) that (2.29) does not hold when $\lambda \to \infty$, so P is not solvable at Γ by Remark 2.7. □

References

1. Cardoso, F., Treves, F.: A necessary condition of local solvability for pseudo-differential equations with double characteristics. Ann. Inst. Fourier (Grenoble) **24**, 225–292 (1974)
2. Colombini, F., Pernazza, L., Treves, F.: Solvability and nonsolvability of second-order evolution equations. Hyperbolic Problems and Related Topics. Graduate Series in Analysis, pp. 111–120. International Press, Somerville (2003)
3. Dencker, N.: The resolution of the Nirenberg-Treves conjecture. Ann. Math. **163**, 405–444 (2006)
4. Dencker, N.: Solvability and limit bicharacteristics. J. Pseudo-Differ. Oper. Appl. **7**, 295–320 (2016)
5. Dencker, N.: Operators of subprincipal type. Anal. PDE **10**, 323–350 (2017)
6. Dencker, N.: Solvability and complex limit bicharacteristics. arXiv:1612.08680

7. Egorov, Ju.V.: Solvability conditions for equations with double characteristics. Dokl. Akad. Nauk SSSR **234**, 280–282 (1977)
8. Gilioli, A., Treves, F.: An example in the solvability theory of linear PDE's. Am. J. Math. **96**, 367–385 (1974)
9. Goldman, R.: A necessary condition for the local solvability of a pseudodifferential equation having multiple characteristics. J. Differ. Equ. **19**, 176–200 (1975)
10. Hörmander, L.: The Cauchy problem for differential equations with double characteristics. J. Anal. Math. **32**, 118–196 (1977)
11. Hörmander, L.: The Weyl calculus of pseudo-differential operators. Commun. Partial Differ. Equ. **32**, 359–443 (1979)
12. Hörmander, L.: Pseudodifferential operators of principal type. Singularities in Boundary Value Problems (Proceedings of the NATO Advanced Study Institute, Maratea, 1980). NATO Advanced Study Institutes Series C: Mathematical and Physical Sciences, vol. 65, pp. 69–96. Reidel, Dordrecht (1981)
13. Hörmander, L.: The Analysis of Linear Partial Differential Operators, vol. I–IV. Springer, Berlin (1983–1985)
14. Melrose, R.: The Cauchy problem for effectively hyperbolic operators. Hokkaido Math. J. **12**, 371–391 (1983)
15. Mendoza, G.A.: A necessary condition for solvability for a class of operators with involutive double characteristics. Microlocal Analysis (Boulder, Colorado, 1983). Contemporary Mathematics, vol. 27, pp. 193–197. American Mathematical Society, Providence (1984)
16. Mendoza, G.A., Uhlmann, G.A.: A necessary condition for local solvability for a class of operators with double characteristics. J. Funct. Anal. **52**, 252–256 (1983)
17. Mendoza, G.A., Uhlmann, G.A.: A sufficient condition for local solvability for a class of operators with double characteristics. Am. J. Math. **106**, 187–217 (1984)
18. Nishitani, T.: Effectively hyperbolic Cauchy problem. Phase Space Analysis of Partial Differential Equations. Pubblicazioni del Centro di Ricerca Matematica Ennio de Giorgi, vol. II, pp. 363–449. Scuola Normale Superiore di Pisa (2004)
19. Popivanov, P.: The local solvability of a certain class of pseudodifferential equations with double characteristics. C. R. Acad. Bulg. Sci. **27**, 607–609 (1974)
20. Treves, F.: Concatenations of second-order evolution equations applied to local solvability and hypoellipticity. Commun. Pure Appl. Math. **26**, 201–250 (1973)
21. Treves, F.: Introduction to Pseudodifferential and Fourier Integral Operators. The University Series in Mathematics, vol. 2. Plenum Press, New York (1980)
22. Wenston, P.R.: A necessary condition for the local solvability of the operator $P_m^2(x,D) + P_{2m-1}(x,D)$. J. Differ. Equ. **25**, 90–95 (1977)
23. Wenston, P.R.: A local solvability result for operators with characteristics having odd order multiplicity. J. Differ. Equ. **28**, 369–380 (1978)
24. Wittsten, J.: On some microlocal properties of the range of a pseudodifferential operator of principal type. Anal. PDE **5**, 423–474 (2012)
25. Yamasaki, A.: On a necessary condition for the local solvability of pseudodifferential operators with double characteristics. Commun. Partial Differ. Equ. **5**, 209–224 (1980)
26. Yamasaki, A.: On the local solvability of $D_1^2 + A(x_2, D_2)$. Math. Japon. **28**, 479–485 (1983)

Fractional-Order Operators: Boundary Problems, Heat Equations

Gerd Grubb

Abstract The first half of this work gives a survey of the fractional Laplacian (and related operators), its restricted Dirichlet realization on a bounded domain, and its nonhomogeneous local boundary conditions, as treated by pseudodifferential methods. The second half takes up the associated heat equation with homogeneous Dirichlet condition. Here we recall recently shown sharp results on interior regularity and on L_p-estimates up to the boundary, as well as recent Hölder estimates. This is supplied with new higher regularity estimates in L_2-spaces using a technique of Lions and Magenes, and higher L_p-regularity estimates (with arbitrarily high Hölder estimates in the time-parameter) based on a general result of Amann. Moreover, it is shown that an improvement to spatial C^∞-regularity at the boundary is not in general possible.

Keywords Fractional Laplacian · Stable process · Pseudodifferential operator · Dirichlet and Neumann conditions · Green's formula · Heat equation · Space-time regularity

1991 Mathematics Subject Classification 35K05 · 35K25 · 47G30 · 60G52

1 Introduction

This work is partly a survey of known results for the fractional Laplacian and its generalizations, with emphasis on pseudodifferential methods and local boundary conditions. Partly it brings new results for the associated heat equation.

Expanded version of a lecture given at the ISAAC Conference August 2017, Växjö University, Sweden.

G. Grubb (✉)
Department of Mathematical Sciences, Copenhagen University, Universitetsparken 5, 2100 Copenhagen, Denmark
e-mail: grubb@math.ku.dk

L. G. Rodino and J. Toft (eds.), *Mathematical Analysis and Applications—Plenary Lectures*, Springer Proceedings in Mathematics & Statistics 262,
https://doi.org/10.1007/978-3-030-00874-1_2

There is an extensive theory for boundary value problems and evolution problems for elliptic **differential** operators, developed through many years and including nonlinear problems, and problems with data of low smoothness. Boundary and evolution problems for **pseudodifferential** operators, such as fractional powers of the Laplacian, have been studied far less, and pose severe difficulties since the operators are **nonlocal**.

The presentation here deals with linear questions, since this is the basic knowledge one needs in any case. Our main purpose is to explain the application of pseudodifferential methods (with the fractional Laplacian as a prominent example). The boundary value theory has been established only in recent years. Plan of the paper:

(1) Fractional-order operators.
(2) Homogeneous Dirichlet problems on a subset of $\mathbb{R}^n$.
(3) Nonhomogeneous boundary value problems.
(4) Heat equations.

Remark 1.1 There are other strategies currently in use, such as methods for singular integral operators in probability theory and potential theory (cf. e.g. [10, 15, 17, 18, 24, 25, 45, 46, 48, 53, 55–59]), methods for embedding the problem in a degenerate elliptic differential operator situation ([13] and many subsequent studies, e.g. [14]), and exact calculations in polar coordinates for the ball (cf. e.g. [2, 21, 22, 64]). Each of the methods allow different types of generalizations of the fractional Laplacian, and solve a variety of problems, primarily in low-regularity spaces. It is perhaps surprising that the methods from the calculus of pseudodifferential operators have only entered the modern studies in the field in the last few years.

The new regularity results in Sect. 5 on heat problems, reaching beyond the recent works [25, 38, 58], are: Theorem 5.2 and its corollary giving a limitation on high spatial regularity, Theorems 5.6–5.8 on estimates in L_2-related spaces for x-dependent operators, and Theorems 5.14–5.19 on high estimates in Hölder spaces with respect to time, valued in L_2- or L_p-related spaces in x (including a Hölder-related space in x as a limiting case).

2 Fractional-Order Operators

The fractional Laplacian $P = (-\Delta)^a$ on $\mathbb{R}^n$, $0 < a < 1$, has linear and nonlinear applications in mathematical physics and differential geometry, and in probability and finance. (See e. g. Frank-Geisinger [26], Boulenger-Himmelsbach-Lenzmann [11], Gonzales-Mazzeo-Sire [28], Monard-Nickl-Paternain [52], Kulczycki [48], Chen-Song [17], Jakubowski [45], Bogdan-Burdzy-Chen [10], Applebaum [6], Cont-Tankov [18], and their references.)

The interest in probability and finance comes from the fact that $-P$ generates a semigroup e^{-tP} which is a stable Lévy process. Here P is viewed as a **singular integral operator**:

$$(-\Delta)^a u(x) = c_{n,a} PV \int_{\mathbb{R}^n} \frac{u(x) - u(x+y)}{|y|^{n+2a}}\, dy; \tag{2.1}$$

more general stable Lévy processes arise from operators

$$Pu(x) = PV \int_{\mathbb{R}^n} (u(x) - u(x+y))K(y)\, dy, \quad K(y) = \frac{K(y/|y|)}{|y|^{n+2a}}, \tag{2.2}$$

where the homogeneous kernel function $K(y)$ is locally integrable, **positive**, and **even**: $K(-y) = K(y)$, on $\mathbb{R}^n \setminus \{0\}$. Equation (2.2) can also be generalized to non-homogeneous kernel functions satisfying suitable estimates in terms of $|y|^{-n-2a}$. Usually, only real functions are considered in probability studies.

$(-\Delta)^a$ can instead be viewed as a **pseudodifferential operator** (ψdo) of order $2a$:

$$(-\Delta)^a u = \operatorname{Op}(|\xi|^{2a})u = \mathcal{F}^{-1}(|\xi|^{2a}\mathcal{F}u(\xi)), \tag{2.3}$$

using the Fourier transform $\mathcal{F}$, defined by $\hat{u}(\xi) = \mathcal{F}u(\xi) = \int_{\mathbb{R}^n} e^{-ix\cdot\xi}u(x)\, dx$ (extended from the Schwartz space $\mathcal{S}(\mathbb{R}^n)$ of rapidly decreasing C^∞-functions, to the temperate distributions $\mathcal{S}'(\mathbb{R}^n)$). Ψdo's are in general defined by

$$Pu = \operatorname{Op}(p(x,\xi))u = \mathcal{F}^{-1}_{\xi\to x}(p(x,\xi)\mathcal{F}u(\xi)); \tag{2.4}$$

note that this theory operates in a context of complex functions (and distributions). Moreover, (2.4) allows x-dependence in the symbol $p(x, \xi)$.

In (2.2), if $K \in C^\infty(\mathbb{R}^n \setminus \{0\})$, the operator it defines is the same as the operator defined by $p(\xi) = \mathcal{F}K(y)$ in (2.4); here $p(\xi)$ is homogeneous of degree $2a$, positive and even.

As a generalization of (2.3), we consider x-dependent classical ψdo's of order $2a \in \mathbb{R}_+$, with certain properties. That P is classical of order $2a$ means that there is an asymptotic expansion of the symbol $p(x, \xi)$ in a series of terms $p_j(x, \xi)$, $j \in \mathbb{N}_0$, that are homogeneous in ξ of order $2a - j$ for $|\xi| \geq 1$; the expansion holds in the sense that

$$|\partial_\xi^\alpha \partial_x^\beta [p(x,\xi) - \sum_{j<M} p_j(x,\xi)]| \leq C_{\alpha,\beta,M} \langle\xi\rangle^{2a-|\alpha|-M}, \text{ all } \alpha, \beta \in \mathbb{N}_0^n, M \in \mathbb{N}_0; \tag{2.5}$$

here $\langle\xi\rangle$ stands for $(|\xi|^2 + 1)^{\frac{1}{2}}$. To the operators defined from these symbols by (2.4) one adds the smoothing operators mapping $\mathcal{E}'(\mathbb{R}^n)$ to $C^\infty(\mathbb{R}^n)$ (also called negligible operators). We assume moreover that p is **even**, meaning that

$$p_j(x, -\xi) = (-1)^j p_j(x, \xi), \text{ all } j, \tag{2.6}$$

and **strongly elliptic**, meaning that for a positive constant c,

$$\operatorname{Re} p_0(x,\xi) \geq c|\xi|^{2a} \text{ for } |\xi| \geq 1. \tag{2.7}$$

Then $P = \mathrm{Op}(p(x,\xi))$ can be shown to have some of the same features as $(-\Delta)^a$. To sum up, we are assuming, with $a \in \mathbb{R}_+$:

Hypothesis 2.1 P is a classical pseudodifferential operator of order $2a$, even and strongly elliptic (cf. (2.4), (2.6), (2.7)).

In part of Sect. 5, we moreover assume $a < 1$. For some results we consider the subset of operators defined as in (2.2)ff. with a kernel function $K(y)$ that is smooth outside 0:

Hypothesis 2.2 P is as in (2.2), with $K(y)$ positive, homogeneous of degree $-n-2a$, even, and C^∞ on $\mathbb{R}^{n-1} \setminus \{0\}$.

This fits into the ψdo formulation, when we write $p(\xi) = \mathcal{F}K(y)$ as $(1-\chi(\xi))p(\xi) + \chi(\xi)p(\xi)$, with $\chi \in C_0^\infty(\mathbb{R}^n)$ and $\chi(\xi) = 1$ near 0. Here the first term is a symbol as under Hypothesis 2.1, and the operator defined from the second term maps large spaces of distributions (e.g. $\mathcal{E}'(\mathbb{R}^n)$) into C^∞-functions (hence is a negligible operator).

To give an example of an x-dependent operator, we can mention that $(-\Delta)^a$ will take the x-dependent form if it undergoes a smooth change of coordinates. As a more general example, P can be an operator defined as $P = A(x,D)^a$, where $A(x,D)$ is a second-order strongly elliptic differential operator. Here P is constructed via the resolvent (Seeley [61]). But of course, the symbol $p(x,\xi)$ can be taken much more general, not tied to differential operator considerations.

A difficult aspect of such operators is that they are **nonlocal**. This is a well-known feature in the pseudodifferential theory, where one can profit from pseudo-locality (namely, Pu is C^∞ on the set where u is C^∞). In a different approach, Caffarelli and Silvestre [13] showed that $(-\Delta)^a$ on $\mathbb{R}^n$ is the Dirichlet-to-Neumann operator for a degenerate elliptic **differential** boundary value problem on $\mathbb{R}^n \times \mathbb{R}_+$; **local** in dimension $n+1$. This observation was then used to obtain results by transforming problems for $(-\Delta)^a$ into problems for **local** operators in one more variable, e.g. in [14]. (However, in some cases where one needs to consider $(-\Delta)^a$ over a subset $\Omega \subset \mathbb{R}^n$, the transformation might lead to equally difficult problems in the new variables.)

3 Homogeneous Dirichlet Problems on a Subset of $\mathbb{R}^n$

How do we get P to act over Ω? There are several ways to answer this. Let us first introduce an appropriate scale of L_p-based Sobolev spaces.

The standard Sobolev–Slobodetskiĭ spaces $W^{s,p}(\mathbb{R}^n)$, $1 < p < \infty$ and $s \geq 0$, have a different character according to whether s is integer or not. Namely, for s integer, they consist of L_p-functions with derivatives in L_p up to order s, hence coincide with the Bessel-potential spaces $H_p^s(\mathbb{R}^n)$, defined for $s \in \mathbb{R}$ by

$$H_p^s(\mathbb{R}^n) = \{u \in \mathcal{S}'(\mathbb{R}^n) \mid \mathcal{F}^{-1}(\langle\xi\rangle^s \hat{u}) \in L_p(\mathbb{R}^n)\}. \tag{3.1}$$

For noninteger s, the $W^{s,p}$-spaces coincide with the Besov spaces, defined e.g. as follows: For $0 < s < 2$,

$$f \in B^s_p(\mathbb{R}^n) \iff \|f\|^p_{L_p} + \int_{\mathbb{R}^{2n}} \frac{|f(x) + f(y) - 2f((x+y)/2)|^p}{|x+y|^{n+ps}}\, dxdy < \infty; \tag{3.2}$$

and $B^{s+t}_p(\mathbb{R}^n) = (1-\Delta)^{-t/2} B^s_p(\mathbb{R}^n)$ for all $t \in \mathbb{R}$.

The Bessel-potential spaces H^s_p are important because they are most directly related to L_p; the Besov spaces B^s_p have other convenient properties, and are needed for boundary value problems in an H^s_p-context, because they are the correct range spaces for trace maps (both from H^s_p and B^s_p-spaces); see e.g. the overview in the introduction to [29]. For $p = 2$, the two scales are identical, and p is usually omitted. For $p \neq 2$ they are related by strict inclusions:

$$H^s_p \subset B^s_p \text{ when } p > 2, \quad H^s_p \supset B^s_p \text{ when } p < 2. \tag{3.3}$$

When working with operators of noninteger order, the use of the $W^{s,p}$-notation can lead to confusion since the definition depends on the integrality of s; moreover, this scale does not always interpolate well. In the following, we focus on the Bessel-potential scale H^s_p, but much of what we show is directly generalized to the Besov scale B^s_p, and to other scales (Besov–Triebel–Lizorkin spaces). There is a more general Besov scale $B^s_{p,q}$ (cf. e.g. Triebel [62]), where B^s_p equals the special case $B^s_{p,p}$.

There is an identification of $H^s_p(\mathbb{R}^n)$ with the dual space of $H^{-s}_{p'}(\mathbb{R}^n)$, $1/p + 1/p' = 1$, in a duality consistent with the L_2-duality, and there is a similar result for the Besov scale.

Let Ω be a open subset of $\mathbb{R}^n$ (we shall use it with C^∞-boundary, but much of the following holds under limited smoothness assumptions). One defines the two associated scales relative to Ω (the *restricted* resp. *supported* version):

$$\begin{aligned} \overline{H}^s_p(\Omega) &= r^+ H^s_p(\mathbb{R}^n), \\ \dot{H}^s_p(\overline{\Omega}) &= \{u \in H^s_p(\mathbb{R}^n) \mid \operatorname{supp} u \subset \overline{\Omega}\}; \end{aligned} \tag{3.4}$$

here $\operatorname{supp} u$ denotes the support of u (the complement of the largest open set where u is zero). Restriction from $\mathbb{R}^n$ to Ω is denoted r^+, extension by zero from Ω to $\mathbb{R}^n$ is denoted e^+ (it is sometimes tacitly understood). Restriction from $\overline{\Omega}$ to $\partial\Omega$ is denoted γ_0.

When $s > 1/p - 1$, one can identify $\dot{H}^s_p(\overline{\Omega})$ with a subspace of $\overline{H}^s_p(\Omega)$, closed if $s - 1/p \notin \mathbb{N}_0$ (equal if $1/p - 1 < s < 1/p$), and with a stronger norm if $s - 1/p \in \mathbb{N}_0$.

The space $\dot{H}^s_p$ is in some texts indicated with a ring, zero or twiddle, as e.g. $\overset{\circ}{H}{}^s_p$, $H^s_{p,0}$ or $\widetilde{H}^s_p$. In most current texts, $\overline{H}^s_p(\Omega)$ is denoted $H^s_p(\Omega)$ without the overline (that was introduced along with the notation $\dot{H}_p$ in [43, 44]), but we prefer to use it, since it is makes the role of the space more clear in formulas where both types occur.

Now let us present some operators associated with $(-\Delta)^a$ on Ω:

(a) The **restricted** Dirichlet fractional Laplacian P_{Dir}. It acts like $P = (-\Delta)^a$, defined on functions u that are 0 on $\mathbb{R}^n \setminus \Omega$, and followed by restriction r^+ to Ω:

$$P_{\mathrm{Dir}}u \text{ equals } r^+Pu \text{ when } \operatorname{supp} u \subset \overline{\Omega}. \tag{3.5}$$

In $L_2(\Omega)$, it is the operator defined variationally from the sesquilinear form

$$Q_0(u, v) = \tfrac{1}{2}c_{n,a} \int_{\mathbb{R}^{2n}} \frac{(u(x) - u(y))(\bar{v}(x) - \bar{v}(y))}{|x - y|^{n+2a}}\, dxdy, \ D(Q_0) = \dot{H}^a(\overline{\Omega}). \tag{3.6}$$

(b) The **spectral** Dirichlet fractional Laplacian $(-\Delta_{\mathrm{Dir}})^a$, defined e.g. via eigenfunction expansions of $-\Delta_{\mathrm{Dir}}$. It does not act like r^+Pe^+. It is not often used in probability applications. (Its regularity properties in L_p-Sobolev spaces are discussed in [36], which gives many references to the literature on it.)

(c) The **regional** fractional Laplacian, defined from the sesquilinear form

$$Q_1(u, v) = \tfrac{1}{2}c_{n,a} \int_{\Omega\times\Omega} \frac{(u(x) - u(y))(\bar{v}(x) - \bar{v}(y))}{|x - y|^{n+2a}}\, dxdy, \ D(Q_1) = \overline{H}^a(\Omega). \tag{3.7}$$

It acts like $r^+Pe^+ + w$ with a certain correction function w.

There are still other operators over Ω that can be defined from P, e.g. representing suitable Neumann problems. A local Neumann condition will be discussed below in Sect. 4. We refer to [36] Sect. 6, and its references, for an overview over the various choices. We shall now focus on **(a)**, where the operator acts like r^+P.

The homogeneous Dirichlet problem, for a smooth bounded open set Ω, is

$$r^+Pu = f \text{ in } \Omega, \quad \operatorname{supp} u \subset \overline{\Omega}. \tag{3.8}$$

As P we take $(-\Delta)^a$, or a more general ψdo as in Hypothesis 2.1 or 2.2.

P_{Dir} in $L_2(\Omega)$ is the realization of r^+P with domain

$$D(P_{\mathrm{Dir}}) = \{u \in \dot{H}^a(\overline{\Omega}) \mid r^+Pu \in L_2(\Omega)\}. \tag{3.9}$$

When P satisfies Hypothesis 2.2 (in particular, when $P = (-\Delta)^a$), then P_{Dir} is positive selfadjoint; for other P it is sectorial, with discrete spectrum in a sector. What can be said about the regularity of functions in the domain?

- Vishik and Eskin showed in the 1960s (see e.g. Eskin [23]):

$$D(P_{\mathrm{Dir}}) = \dot{H}^{2a}(\overline{\Omega}) \text{ if } a < \tfrac{1}{2}, \ D(P_{\mathrm{Dir}}) \subset \dot{H}^{a+\frac{1}{2}-\varepsilon}(\overline{\Omega}) \text{ if } a \geq \tfrac{1}{2}. \tag{3.10}$$

- Ros-Oton and Serra [55] showed in 2014 for $(-\Delta)^a$:

$$f \in L_\infty(\Omega) \implies u \in d^aC^\alpha(\overline{\Omega}) \text{ for small } \alpha; \tag{3.11}$$

here $d(x)$ equals $\operatorname{dist}(x, \partial\Omega)$ near $\partial\Omega$, and C^α is the Hölder space. They improved this later to $\alpha < a$ ($\alpha = a$ in some cases), and to more general P as in (2.2), and lifted the regularity conclusions to $f \in C^\gamma$, $u/d^a \in C^{a+\gamma}$ for small γ. For (3.11), Ω was assumed to be $C^{1,1}$.

- We showed in 2015 [34], for $1 < p < \infty$ and Ω smooth:

$$\begin{aligned} f \in \overline{H}_p^s(\Omega) &\iff u \in H_p^{a(s+2a)}(\overline{\Omega}), \text{ any } s \geq 0, \\ f \in C^\infty(\overline{\Omega}) &\iff u/d^a \in C^\infty(\overline{\Omega}); \end{aligned} \tag{3.12}$$

here $H_p^{a(s+2a)}(\overline{\Omega})$ is a space introduced by Hörmander [43] for $p = 2$. E.g. when $s = 0$,

$$H_p^{a(2a)}(\overline{\Omega}) \begin{cases} = \dot{H}_p^{2a}(\overline{\Omega}) \text{ if } a < 1/p, \\ \subset \dot{H}_p^{2a-\varepsilon}(\overline{\Omega}) \text{ if } a = 1/p, \\ \subset \dot{H}_p^{2a}(\overline{\Omega}) + d^a \overline{H}_p^a(\Omega) \text{ if } a > 1/p; \end{cases} \tag{3.13}$$

the spaces will be further explained below. Assertion (3.12) has corollaries in Hölder spaces by Sobolev embedding.

The contribution from Hörmander, accounted for in detail in [34], is in short the following: He defined the **μ-transmission property** in his book [44], Sect. 18.2:

Definition 3.1 A classical ψdo P of order m has the μ-transmission property at $\partial\Omega$, when

$$\partial_x^\beta \partial_\xi^\alpha p_j(x, -\nu) = e^{\pi i(m-2\mu-j-|\alpha|)} \partial_x^\beta \partial_\xi^\alpha p_j(x, \nu), \tag{3.14}$$

for all indices; here $x \in \partial\Omega$, and ν denotes the interior normal at x.

The property was formulated already in a photocopied lecture note from IAS Princeton 1965–1966 on ψdo boundary problems [43], handed out to a few people through the times, including Boutet de Monvel in 1968, the present author in 1980.

For P of order $2a$ and *even* (cf. (2.6)), it holds with $\mu = a$, for *any* smooth subset Ω (*all* normal directions are covered when (2.6) holds).

The case $\mu = 0$ is the transmission condition entering in the calculus of Boutet de Monvel, described e.g. in [12, 31, 32, 60].

Recalling that e^+ denotes extension by zero, let

$$\mathcal{E}_a(\overline{\Omega}) = e^+ d^a C^\infty(\overline{\Omega}). \tag{3.15}$$

Then by [44], Th. 18.2.18,

Theorem 3.2 *The a-transmission property at $\partial\Omega$ is necessary and sufficient in order that r^+P maps $\mathcal{E}_a(\overline{\Omega})$ into $C^\infty(\overline{\Omega})$.*

Note the importance of d^a.

The notation in [44] is slightly different from that in the notes [43], which we have adapted here. The notes moreover treated solvability questions, with $f \in C^\infty(\overline{\Omega})$ or in H^s-spaces. The space $H^{a(s)}(\overline{\Omega})$ was introduced. Originally it was defined as "the functions supported in $\overline{\Omega}$ that are mapped into $\overline{H}^{s-m}(\Omega)$ for any P that is elliptic of order m and has the a-transmission property", and the whole effort was to sort this out. We shall now explain the structure and its implications, for general H_p^s-spaces with $1 < p < \infty$.

Introduce first **order-reducing operators of plus/minus type.** For $\Omega = \mathbb{R}^n_+$, define for $t \in \mathbb{R}$:

$$\Xi^t_\pm = \operatorname{Op}((\langle \xi' \rangle \pm i\xi_n)^t) \text{ on } \mathbb{R}^n. \tag{3.16}$$

The symbols extend analytically in ξ_n to $\operatorname{Im} \xi_n \lessgtr 0$. Hence, by the Paley-Wiener theorem, $\Xi^t_\pm$ preserve support in $\overline{\mathbb{R}}^n_\pm$. Then for all $s \in \mathbb{R}$,

$$\begin{aligned} \Xi^t_+ &: \dot{H}^s_p(\overline{\mathbb{R}}^n_+) \xrightarrow{\sim} \dot{H}^{s-t}_p(\overline{\mathbb{R}}^n_+), \text{ with inverse } \Xi^{-t}_+, \\ r^+\Xi^t_- e^+ &: \overline{H}^s_p(\mathbb{R}^n_+) \xrightarrow{\sim} \overline{H}^{s-t}_p(\mathbb{R}^n_+), \text{ with inverse } r^+\Xi^{-t}_- e^+. \end{aligned} \tag{3.17}$$

Here the action of e^+ on spaces with $s < 0$ is understood such that the operators in the families Ξ^t_+ and $r^+\Xi^t_- e^+$ are adjoints for each $t \in \mathbb{R}$:

$$\Xi^t_+ : \dot{H}^s_p(\overline{\mathbb{R}}^n_+) \xrightarrow{\sim} \dot{H}^{s-t}_p(\overline{\mathbb{R}}^n_+) \text{ has the adjoint } r^+\Xi^t_- e^+ : \overline{H}^{-s+t}_{p'}(\mathbb{R}^n_+) \xrightarrow{\sim} \overline{H}^{-s}_{p'}(\mathbb{R}^n_+), \tag{3.18}$$

with respect to an extension of the duality $\int_{\mathbb{R}^n_+} u\bar{v}\, dx$ (more explanation in [34], Rem. 1.1).

Now define the *a***-transmission space** over $\mathbb{R}^n_+$:

$$H^{a(s)}_p(\overline{\mathbb{R}}^n_+) = \Xi^{-a}_+ e^+ \overline{H}^{s-a}_p(\mathbb{R}^n_+), \text{ for } s - a > -1/p'. \tag{3.19}$$

Here $e^+\overline{H}^{s-a}_p(\mathbb{R}^n_+)$ has a jump at $x_n = 0$ when $s - a > 1/p$; this is mapped by Ξ^{-a}_+ to a singularity of the type x_n^a.

In fact, we can show:

$$H^{a(s)}_p(\overline{\mathbb{R}}^n_+) \begin{cases} = \dot{H}^s_p(\overline{\mathbb{R}}^n_+) \text{ if } -1/p' < s - a < 1/p, \\ \subset \dot{H}^s_p(\overline{\mathbb{R}}^n_+) + e^+ x_n^a \overline{H}^{s-a}_p(\mathbb{R}^n_+) \text{ if } s - a - 1/p \in \mathbb{R}_+ \setminus \mathbb{N}, \end{cases} \tag{3.20}$$

with $\dot{H}^s_p(\overline{\mathbb{R}}^n_+)$ replaced by $\dot{H}^{s-\varepsilon}_p(\overline{\mathbb{R}}^n_+)$ if $s - a - 1/p \in \mathbb{N}$.

For example, for $1/p < s - a < 1 + 1/p$, $u \in H^{a(s)}_p(\overline{\mathbb{R}}^n_+)$ has the form

$$u = w + e^+ x_n^a K_0 \varphi, \tag{3.21}$$

where w and φ run through $\dot{H}^s_p(\overline{\mathbb{R}}^n_+)$ and $B^{s-a-1/p}_p(\mathbb{R}^{n-1})$, respectively, and K_0 is the Poisson operator $K_0 \colon \varphi \mapsto \mathcal{F}^{-1}_{\xi' \to x'}[\hat{\varphi}(\xi') r^+ e^{-\langle \xi' \rangle x_n}]$ solving the standard Dirichlet problem

$$(-\Delta + 1)v = 0 \text{ in } \mathbb{R}^n_+, \quad \gamma_0 u = \varphi \text{ on } \mathbb{R}^{n-1}.$$

The analysis hinges on the following formula for the inverse Fourier transform of $(\langle\xi'\rangle + i\xi_n)^{-a-1}$, where $e^+r^+x_n^a$ appears:

$$\mathcal{F}^{-1}_{\xi_n\to x_n}(\langle\xi'\rangle + i\xi_n)^{-a-1} = \Gamma(a+1)^{-1}e^+r^+x_n^a e^{-\langle\xi'\rangle x_n}.$$

The generalization to $\Omega \subset \mathbb{R}^n$ depends on finding suitable replacements of $\Xi^t_\pm$. They are a kind of generalized ψdo's (the symbols satisfy some but not all of the usual symbol estimates). It was important in [34] that we could rely on a truly pseudodifferential version $\Lambda^{(t)}_\pm$ found in [29].

The choice $P = (1-\Delta)^a$ on $\mathbb{R}^n$ with symbol $(1+|\xi|^2)^a$ serves as a model case with easy explicit calculations. Here one can factorize the symbol and operator directly:

$$(1+|\xi|^2)^a = (\langle\xi'\rangle - i\xi_n)^a(\langle\xi'\rangle + i\xi_n)^a, \quad (1-\Delta)^a = \Xi^a_-\,\Xi^a_+. \tag{3.22}$$

Let us show how to solve the model Dirichlet problem

$$r^+(1-\Delta)^a u = f \text{ on } \mathbb{R}^n_+, \quad \operatorname{supp} u \subset \overline{\mathbb{R}}^n_+. \tag{3.23}$$

Say, f is given in $\overline{H}^t_p(\mathbb{R}^n_+)$ for some $t \geq 0$, and u is a priori assumed to lie in $\dot{H}^a_p(\overline{\mathbb{R}}^n_+)$.

In view of the factorization (3.22),

$$r^+(1-\Delta)^a u = r^+\Xi^a_-\,\Xi^a_+ u = r^+\Xi^a_-(e^+r^+ + e^-r^-)\Xi^a_+ u = r^+\Xi^a_- e^+r^+\Xi^a_+ u,$$

since $r^-\Xi^a_+ u = 0$. (r^- denotes restriction from $\mathbb{R}^n$ to $\mathbb{R}^n_-$, e^- is extension by zero on $\mathbb{R}^n\setminus\mathbb{R}^n_-$.) By (3.17), the problem (3.23) is reduced by composition with $r^+\Xi^{-a}_- e^+$ to the left to the problem

$$r^+\Xi^a_+ u = g, \quad \operatorname{supp} u \subset \overline{\mathbb{R}}^n_+, \tag{3.24}$$

where $g = r^+\Xi^{-a}_- e^+ f \in \overline{H}^{t+a}_p(\mathbb{R}^n_+)$. Now there is an important observation, shown in Prop. 1.7 in [34]:

Lemma 3.3 *Let $s > a - 1/p'$. The mapping $\Xi^{-a}_+ e^+$ is a bijection from $\overline{H}^{s-a}_p(\mathbb{R}^n_+)$ to $H^{a(s)}_p(\overline{\mathbb{R}}^n_+)$ with inverse $r^+\Xi^a_+$.*

Then clearly, (3.24) is simply solved uniquely by

$$u = \Xi^{-a}_+ e^+ g. \tag{3.25}$$

Inserting the definition of g, we can conclude:

Proposition 3.4 *The problem (3.23) with f is given in $\overline{H}^t_p(\mathbb{R}^n_+)$ for some $t \geq 0$, and u sought in $\dot{H}^a_p(\overline{\mathbb{R}}^n_+)$, has the unique solution*

$$u = \Xi_+^{-a} e^+ r^+ \Xi_-^{-a} e^+ f, \tag{3.26}$$

lying in $\Xi_+^{-a}(e^+\overline{H}_p^{t+a}(\overline{\mathbb{R}}_+^n)) = H_p^{a(t+2a)}(\overline{\mathbb{R}}_+^n)$, *the a-transmission space.*

It is of course more difficult to treat variable-coefficient operators on curved domains. For such cases, the following result was shown in [34]:

Theorem 3.5 *Let P be a classical strongly elliptic ψ do on $\mathbb{R}^n$ of order $2a > 0$ with even symbol (i.e., P satisfies Hypothesis 2.1), and let Ω be a smooth bounded subset of $\mathbb{R}^n$. Let $s > a - 1/p'$. The homogeneous Dirichlet problem* (3.8)*, considered for $u \in \dot{H}_p^{a-1/p'+\varepsilon}(\overline{\Omega})$, satisfies:*

$$f \in \overline{H}_p^{s-2a}(\Omega) \implies u \in H_p^{a(s)}(\overline{\Omega}), \text{ the a-transmission space.} \tag{3.27}$$

Moreover, the mapping from u to f is a Fredholm mapping:

$$r^+ P\colon H_p^{a(s)}(\overline{\Omega}) \to \overline{H}_p^{s-2a}(\Omega) \text{ is Fredholm.} \tag{3.28}$$

A corollary for $s \to \infty$ is:

$$r^+ P\colon \mathcal{E}_a(\overline{\Omega}) \to C^\infty(\overline{\Omega}) \text{ is Fredholm.} \tag{3.29}$$

The big step forward by this theorem is that it describes the domain spaces in an exact way, and shows that they depend only on a, s, p, not on the operator P; and this works for all $s > a - 1/p'$.

The argumentation involves a reduction to problems belonging to the calculus of Boutet de Monvel, which is described e.g. in [12, 31, 32, 60]. We use techniques established more recently than [43, 44], in particular from [29]. The basic idea is to reduce the operator, on boundary patches, to the form

$$P \sim \Lambda_-^{(a)} Q \Lambda_+^{(a)}, \tag{3.30}$$

where $\Lambda_\pm^{(a)}$ are order-reducing pseudodifferential operators, preserving support in $\overline{\Omega}$ resp. $\mathbb{R}^n \setminus \Omega$, and Q is of order 0 and satisfies the 0-transmission condition, hence belongs to the Boutet de Monvel calculus. We shall not dwell on the proof here, but go on to some further developments of the theory.

Remark 3.6 The assumption that the ψdo P is even, was made for simplicity, and could everywhere be replaced by the assumption that P has the a-transmission property with respect to the chosen domain Ω.

Remark 3.7 In [33] (written after [34]), the results are extended to many other scales of spaces, such as Besov spaces $B^s_{p,q}$ and Triebel–Lizorkin spaces $F^s_{p,q}$. Of particular interest is the scale $B^s_{\infty,\infty}$, also denoted C^s_*, the Hölder–Zygmund scale. Here C^s_* identifies with the Hölder space C^s when $s \in \mathbb{R}_+ \setminus \mathbb{N}$, and for positive integer k satisfies $C^{k-\varepsilon} \supset C_*^k \supset C^{k-1,1} \supset C^k$ for small $\varepsilon > 0$; moreover, $C_*^0 \supset L_\infty$. Then Theorem 3.5 holds with H_p^s-spaces replaced by C_*^s-spaces.

Remark 3.8 The above applications of pseudodifferential theory require that the domain Ω has C^∞-boundary; in comparison, the results of e.g. Ros-Oton and coauthors in low-order Hölder spaces allow low regularity of Ω, using rather different methods. There exists a pseudodifferential theory with just Hölder-continuous x-dependence (see e.g. Abels [3, 4] and references), which may be useful to reduce the present smoothness assumptions, but non-smooth coordinate changes for ψdo's have not yet (to our knowledge) been established in a sufficiently useful way. At any rate, the results obtainable by ψdo methods can serve as a guideline for what one can aim for on domains with lower smoothness.

4 Nonhomogeneous Boundary Value Problems

When solutions u of the homogeneous Dirichlet problem lie in d^a times a Sobolev or Hölder space over $\overline{\Omega}$, there is a boundary value $\gamma_0(u/d^a)$, denoted

$$\gamma_0^a u = \gamma_0(u/d^a); \tag{4.1}$$

it is viewed as a Neumann boundary value. (We omit normalizing constants for now; they are described precisely in Remark 4.2 below.)

Ros-Oton and Serra [56, 57] showed the following integration-by-parts formula:

Theorem 4.1 *When u and u' are solutions of the homogeneous Dirichlet problem (3.8) for $(-\Delta)^a$ on Ω with $f, f' \in L_\infty(\Omega)$, Ω being $C^{1,1}$, $a > 0$, then*

$$\int_\Omega ((-\Delta)^a u\, \partial_j \bar{u}' + \partial_j u\, (-\Delta)^a \bar{u}')\, dx = c \int_{\partial\Omega} \nu_j(x)\, \gamma_0^a u\, \gamma_0^a \bar{u}'\, d\sigma; \tag{4.2}$$

here $\nu = (\nu_1, \dots, \nu_n)$ is the normal vector at $\partial\Omega$.

It is equivalent to a certain Pohozaev-type formula, which has important applications to uniqueness questions in nonlinear problems for $(-\Delta)^a$. It was generalized to other related x-independent singular integral operators in [59] (with Valdinoci), and we extended it to x-dependent ψdo's in [37]. (See the survey [54] for an introduction to fractional Pohozaev identities and their applications.)

Note that the collected order of the operators in the integral over Ω is $2a+1$; the formula generalizes a well-known formula for $a = 1$ where $\gamma_0^a u$ is replaced by the Neumann trace $\gamma_0(\partial_\nu u)$, and the Dirichlet trace $\gamma_0 u$ is 0.

What should a nonzero Dirichlet trace be in the context of fractional Laplacians? Look at the smoothest space:

$$\mathcal{E}_a(\overline{\mathbb{R}}^n_+) = \{u = e^+ x_n^a v \mid v \in C^\infty(\overline{\mathbb{R}}^n_+)\}. \tag{4.3}$$

By a Taylor expansion of v,

$$u(x) = x_n^a v(x',0) + x_n^{a+1}\partial_n v(x',0) + \tfrac{1}{2}x_n^{a+2}\partial_n^2 v(x',0) + \dots \text{ for } x_n > 0. \quad (4.4)$$

If $u \in \mathcal{E}_{a-1}(\overline{\mathbb{R}}^n_+)$, i.e., $u = e^+ x_n^{a-1} w$ with $w \in C^\infty(\overline{\mathbb{R}}^n_+)$, we have analogously:

$$u(x) = x_n^{a-1} w(x',0) + x_n^a \partial_n w(x',0) + \tfrac{1}{2}x_n^{a+1}\partial_n^2 w(x',0) + \dots \text{ for } x_n > 0. \quad (4.5)$$

Here $x_n^{a-1} w(x',0)$ is the only structural difference between (4.4) and (4.5). This term defines the Dirichlet trace: When $u \in \mathcal{E}_{a-1}(\overline{\mathbb{R}}^n_+)$, the Dirichlet trace is

$$\gamma_0^{a-1} u = \gamma_0(u/x_n^{a-1}), \text{ equal to } \gamma_0 w. \quad (4.6)$$

(Again we omit a normalizing constant.)

It is now natural to define a Neumann trace on $\mathcal{E}_{a-1}(\overline{\mathbb{R}}^n_+)$ from the second term in (4.5), by

$$\gamma_1^{a-1} u = \gamma_1(u/x_n^{a-1}), \text{ equal to } \gamma_1 w = \gamma_0(\partial_n w). \quad (4.7)$$

Note that it equals $\gamma_0^a u$ if $u \in \mathcal{E}_a(\overline{\mathbb{R}}^n_+)$.

Remark 4.2 Also higher order traces are defined on $\mathcal{E}_{a-1}(\overline{\mathbb{R}}^n_+)$, namely the functions $\partial_n^k w(x',0)$ in (4.5). With the correct normalizing constants they are:

$$\gamma_k^{a-1} u = \Gamma(a+k)\gamma_0(\partial_n^k(u/x_n^{a-1})), \quad k \in \mathbb{N}_0. \quad (4.8)$$

There are analogous definitions with $a-1$ replaced by $a-M$, $a \in \mathbb{R}_+$ and $M \in \mathbb{N}_0$; see details in [34], in particular Th. 5.1 showing mapping properties, and Th. 6.1 showing Fredholm solvability. For $(-\Delta)^a$ in the case where Ω is the unit ball in $\mathbb{R}^n$, related definitions are given by Abatangelo, Jarohs and Saldana in [2], with explicit solution formulas.

The above definitions can be carried over to Ω (where x_n is replaced by $d(x)$), and they extend to $H_p^{(a-1)(s)}(\overline{\Omega})$ spaces for sufficiently large s, cf. [34].

Now consider a P satisfying Hypothesis 2.1. We can define the *nonhomogeneous Dirichlet problem* for functions $u \in H_p^{(a-1)(s)}(\overline{\Omega})$ (hence supported in $\overline{\Omega}$), by

$$r^+ P u = f \text{ in } \Omega, \quad \gamma_0^{a-1} u = \varphi \text{ on } \partial\Omega. \quad (4.9)$$

For this we have the solvability result [33, 34]:

Theorem 4.3 *For $s > a - 1/p'$,*

$$\{r^+ P, \gamma_0^{a-1}\} \colon H_p^{(a-1)(s)}(\overline{\Omega}) \to \overline{H}_p^{s-2a}(\Omega) \times B_p^{s-a+1/p'}(\partial\Omega) \quad (4.10)$$

is a Fredholm mapping.

Here $B_p^{s-a+1/p'}(\partial\Omega)$ is the Besov space that usually appears as the range space for the standard Dirichlet trace operator γ_0 applied to $\overline{H}_p^{s-a+1}(\Omega)$. As in (3.20) (with a replaced by $a-1$), $H_p^{(a-1)(s)}(\overline{\Omega}) \subset \dot{H}_p^s(\Omega) + e^+ d^{a-1}\overline{H}_p^{s-a+1}(\Omega)$, when $s - a + 1/p' \in \mathbb{R}_+ \setminus \mathbb{N}$.

When $a < 1$, the factor $d(x)^{a-1}$ is unbounded, and the solutions of the form $u = d^{a-1}v$, for a nice v with nonzero boundary value, blow up at $\partial\Omega$ (a detailed analysis is given in [33] Rem. 2.10). Such solutions are called "large solutions" in Abatangelo [1]. Nevertheless, $u \in L_p(\Omega)$ if $1 < p < 1/(1-a)$.

Nonhomogeneous Dirichlet problems (also with consecutive sets of boundary data) are considered in [1, 33, 34, 37] and the recent [2].

We can moreover consider a boundary value problem where Neumann data are prescribed:

$$r^+ P u = f \text{ in } \Omega, \quad \gamma_1^{a-1} u = \psi \text{ on } \partial\Omega, \tag{4.11}$$

for $u \in H^{(a-1)(s)}(\overline{\Omega})$, $s > a + 1/p$. (The boundary condition here is *local*; there have also been defined other, nonlocal Neumann problems, see the overview in [36] Sect. 6.) To discuss the solvability of (4.11) we can construct a Dirichlet-to-Neumann operator [39]:

Theorem 4.4 *Let K_D be a parametrix of the mapping*

$$z \mapsto \gamma_0^{a-1} z, \text{ when } r^+ P z = 0 \text{ in } \Omega, \tag{4.12}$$

(an inverse when (4.10) *is a bijection). Then the mapping*

$$S_D = \gamma_1^{a-1} K_D, \tag{4.13}$$

the **Dirichlet-to-Neumann operator**, *is a classical pseudodifferential operator of order* 1 *on $\partial\Omega$, with principal symbol $s_{DN,0}$ derived from the principal symbol of P.*

In particular, $s_{DN,0}(x', \xi')$ is proportional to $|\xi'|$ for $|\xi'| \geq 1$, when $P = (-\Delta)^a$, considered in local coordinates at the boundary.

And then we have:

Theorem 4.5 *When S_{DN} is elliptic (i.e., $s_{DN,0}(x', \xi')$ is invertible for $|\xi'| \geq 1$), the Neumann problem* (4.11) *satisfies:*

$$\{r^+ P, \gamma_1^{a-1}\} \colon H_p^{(a-1)(s)}(\overline{\Omega}) \to \overline{H}_p^{s-2a}(\Omega) \times B_p^{s-a-1/p}(\partial\Omega) \tag{4.14}$$

is a Fredholm mapping, for $s > a + 1/p$.

Note that the ellipticity holds in the case where $P = (-\Delta)^a$.

It is remarkable that both the Dirichlet and the Neumann boundary operators γ_0^{a-1} and γ_1^{a-1} are **local**, in spite of the nonlocalness of the operator P.

The integration by parts formula (4.1), and the consequential Pohozaev formulas, hold on $H_p^{a(s)}$-spaces, where the Dirichlet trace $\gamma_0^{a-1}u$ vanishes and (consequently) the Neumann trace $\gamma_1^{a-1}u$ identifies with $\gamma_0^a u$.

The papers [1, 37, 56, 57, 59] did not establish formulas where both $\gamma_0^{a-1}u$ and $\gamma_1^{a-1}u$ can be nonvanishing. However, in comparison with the standard Laplacian Δ, it is natural to ask whether there are formulas generalizing the well-known full Green's formula with nonzero Dirichlet and Neumann data, to these operators. This question was answered in [39], where we showed:

Theorem 4.6 *When $u, v \in H^{(a-1)(s)}(\overline{\Omega})$, then for $s > a + \frac{1}{2}$,*

$$\int_\Omega (Pu\,\bar{v} - u\,\overline{P^*v})\,dx \\ = c_a \int_{\partial\Omega} (s_0\gamma_1^{a-1}u\,\gamma_0^{a-1}\bar{v} - s_0\gamma_0^{a-1}u\,\gamma_1^{a-1}\bar{v} + B\gamma_0^{a-1}u\,\gamma_0^{a-1}\bar{v})\,dx'; \tag{4.15}$$

here $s_0(x') = p_0(x', \nu(x'))$ for $x' \in \partial\Omega$, and B is a first-order ψdo on $\partial\Omega$.

Note that the only term in the right-hand side that may not be local, is the term with B, nonlocal in general. A closer study (work in progress) shows that B vanishes if $P = (-\Delta)^a$; we also find criteria under which B is local.

We shall not begin here to describe the method of proof; it consists of delicate localized ψdo considerations using the order-reduction operators, and elements of the Boutet de Monvel calculus.

5 Heat Equations

5.1 Anisotropic Spaces, Hölder Estimates, Counterexamples to High Spatial Regularity

For a given lower semibounded operator A in x-space it is of interest to study evolution problems with a time-parameter t,

$$Au(x,t) + \partial_t u(x,t) = f(x,t) \text{ for } t > 0, \quad u(x,0) = u_0(x). \tag{5.1}$$

Through many years, semigroup methods, as originally presented in Hille and Phillips [42], have been developed along with other methods from functional analysis to give interesting results for operators A acting like the Laplacian and other elliptic differential operators. Also nonlinear questions have been treated, e.g. where f or A are allowed to depend on u.

For A representing the fractional Laplacian and its generalizations, on $\mathbb{R}^n$ or a domain, the studies have begun more recently. A natural approach is here: To find the appropriate general strategies from the works on differential operators, and show the appropriate properties of A assuring that the methods can be applied to it.

Consider the evolution problem (heat equation) associated with an operator P as studied in Sects. 3 and 4, with a homogeneous Dirichlet condition:

$$\begin{aligned} Pu + \partial_t u &= f \text{ on } \Omega \times I, \quad I =]0, T[\,, \\ u &= 0 \text{ on } (\mathbb{R}^n \setminus \Omega) \times I, \\ u|_{t=0} &= u_0. \end{aligned} \tag{5.2}$$

Since P_{Dir} is a positive selfadjoint (or sectorial) operator in $L_2(\Omega)$, there is solvability in a framework of L_2-Sobolev spaces.

We are interested in the *regularity* of solutions.

This question has been treated recently by Leonori, Peral, Primo and Soria [50] in $L_r(I; L_q(\Omega))$-spaces, by Fernandez-Real and Ros-Oton [25] in anisotropic Hölder spaces, and by Biccari, Warma and Zuazua [9] for $(-\Delta)^a$ in local L_p-Sobolev spaces over Ω. Earlier results are shown e.g. in Felsinger and Kassmann [24] and Chang-Lara and Davila [15] (Hölder properties), and Jin and Xiong [46] (Schauder estimates). The references in the mentioned works give further information, also on related *heat kernel* estimates. Very recently (November 2017), the Hölder estimates were improved by Ros-Oton and Vivas [58].

We have a few contributions to this subject, that we shall describe in the following.

Let us first introduce **anisotropic spaces** of Sobolev or Hölder type. Let $d \in \mathbb{R}_+$. There are the Bessel-potential types, for $s \in \mathbb{R}$:

$$\begin{aligned} H_p^{(s,s/d)}(\mathbb{R}^n \times \mathbb{R}) &= \{u \in \mathcal{S}' \mid \mathcal{F}^{-1}((\langle\xi\rangle^{2d} + \tau^2)^{s/2d}\hat{u}(\xi,\tau)) \in L_p(\mathbb{R}^{n+1})\}, \\ \overline{H}_p^{(s,s/d)}(\Omega \times I) &= r_{\Omega\times I} H_p^{(s,s/d)}(\mathbb{R}^n \times \mathbb{R}), \end{aligned} \tag{5.3}$$

and there are related definitions of Besov-type with H_p replaced by B_p. There are the Hölder spaces:

$$\overline{C}^{(s,r)}(\Omega \times I) = L_\infty(I; \overline{C}^s(\Omega)) \cap L_\infty(\Omega; \overline{C}^r(I)), \text{ for } s, r \in \mathbb{R}_+; \tag{5.4}$$

including in particular the case $r = s/d$. (For s equal to an integer k, we denote by $\overline{C}^k(\Omega)$ the space of bounded continuous functions on $\overline{\Omega}$ with bounded derivatives up to order k, this includes the case $\Omega = \mathbb{R}^n$.) The spaces occur in many works; important properties are recalled e.g. in [30, 38] with further references.

The Hölder-type spaces are the primary objects in the investigations of Fernandez-Real and Ros-Oton in [25], Ros-Oton and Vivas in [58]. These authors have for the Dirichlet heat problem the following results, showing the role of d^a in Hölder estimates:

Theorem 5.1 *Let P be an operator of the form* (2.2)ff., $0 < a < 1$, *and consider solutions of the problem* (5.2).

1° [25], Cor. 1.6. *When Ω is a bounded open $C^{1,1}$ subset of $\mathbb{R}^n$, then the unique weak solution u with $f \in L_\infty(\Omega \times I)$ and $u_0 \in L_2(\Omega)$ satisfies*

$$\begin{aligned} \|u\|_{\overline{C}^{(a,1-\varepsilon)}(\Omega\times I')} + \|u/d^a\|_{\overline{C}^{(a-\varepsilon,1/2-\varepsilon/(2a))}(\Omega\times I')} \\ \leq C(\|f\|_{L_\infty(\Omega\times I)} + \|u_0\|_{L_2(\Omega)}), \end{aligned} \tag{5.5}$$

for any small $\varepsilon > 0$, $I' =]t_0, T[$ *with* $t_0 > 0$. *Moreover, if* $f \in \overline{C}^{(\gamma,\gamma/(2a))}(\Omega \times I)$ *with* $\gamma \in]0, a]$ *such that* $\gamma + 2a \notin \mathbb{N}$, *then* u *has the interior regularity:*

$$\|u\|_{\overline{C}^{(2a+\gamma,1+\gamma/(2a))}(\Omega' \times I')} \leq C' \|f\|_{C^{(\gamma,\gamma/(2a))}(\Omega \times I)}, \tag{5.6}$$

for any Ω' *with* $\overline{\Omega'} \subset \Omega$.

2° [58], Cor. 1.2. *Let* $\gamma \in]0, a[$, $\gamma + a \notin \mathbb{N}$. *When* Ω *is a bounded open* $C^{2,\gamma}$ *subset of* $\mathbb{R}^n$, *then* $f \in \overline{C}^{(\gamma,\gamma/(2a))}(\Omega \times I)$, $u_0 \in L_2(\Omega)$ *imply:*

$$\begin{aligned} \|u\|_{\overline{C}^{(\gamma,1+\gamma/(2a))}(\Omega \times I')} + \|u/d^a\|_{\overline{C}^{(a+\gamma,1/2+\gamma/(2a))}(\Omega \times I')} \\ \leq C(\|f\|_{\overline{C}^{(\gamma,\gamma/(2a))}(\Omega \times I)} + \|u_0\|_{L_2(\Omega)}), \end{aligned} \tag{5.7}$$

for any $I' =]t_0, T[$ *with* $t_0 > 0$.

In other words, if we take $\gamma = a - \varepsilon$ for a small $\varepsilon > 0$, (5.7) reads, with $\varepsilon' = \varepsilon/(2a)$,

$$\begin{aligned} \|u\|_{\overline{C}^{(a-\varepsilon,3/2-\varepsilon')}(\Omega \times I')} + \|u/d^a\|_{\overline{C}^{(2a-\varepsilon,1-\varepsilon')}(\Omega \times I')} \\ \leq C(\|f\|_{\overline{C}^{(a-\varepsilon,1/2-\varepsilon')}(\Omega \times I)} + \|u_0\|_{L_2(\Omega)}). \end{aligned} \tag{5.8}$$

An interesting question is whether the regularity of u can be lifted further, when f in 2° is replaced by a more regular function. As we shall see below, this is certainly possible with respect to the t-variable; this is also shown to some extent in [25]. However, there are limitations with respect to the boundary behavior in the x-variable. It is shown in [35] that when P is as in Hypothesis 2.1, any eigenfunction φ of P_{Dir} associated with a nonzero eigenvalue λ satisfies (more on the spaces in Theorem 5.2):

$$\varphi \in C_*^{a(3a)}(\overline{\Omega}) \subset d^a \overline{C}_*^{2a}(\overline{\Omega}) \begin{cases} = d^a \overline{C}^{2a}(\Omega) \text{ if } a \neq \frac{1}{2}, \\ \subset d^a \overline{C}^{2a-\varepsilon}(\Omega) \text{ if } a = \frac{1}{2}, \end{cases} \tag{5.9}$$

but, in the basic case $P = (-\Delta)^a$, φ is *not in* $d^a \overline{C}^\infty(\Omega)$ and not either in $\overline{C}^\infty(\Omega)$. The function $u(x, t) = e^{-\lambda t}\varphi(x)$ is clearly a solution of the heat equation with $f = 0$ (hence $f \in \overline{C}^\infty(I; \overline{C}^\infty(\Omega))$), but u and $u/d^a \notin L_\infty(I'; \overline{C}^\infty(\Omega))$. This shows a surprising contrast to the usual regularity rules for heat equations, and it differs radically from the stationary case, where we have (3.29).

The argument can be extended from $(-\Delta)^a$ to more general operators, and it can be sharpened to rule out also finite higher order regularities.

Theorem 5.2 *Let* P *satisfy Hypothesis* 2.2 *with* $0 < a < 1$, *or let* P *equal* $(-\Delta)^a$ *with* $a > 0$, *or the fractional Helmholtz operator* $(-\Delta + m^2)^a$ *with* $m > 0$ *and* $0 < a < 1$. *Let* Ω *be* C^∞. *Then any eigenfunction* φ *of* P_{Dir} *associated with a nonzero eigenvalue* λ *satisfies* (5.9), *but is not in* $\overline{C}^{a+\delta}(\Omega)$ *nor in* $C_*^{a(3a+\delta)}(\overline{\Omega})$ *for any* $\delta > 0$.

Proof Recall that C_*^s stands for the Hölder–Zygmund space, which identifies with C^s when $s \in \mathbb{R}_+ \setminus \mathbb{N}$, cf. Remark 3.7. As shown in [33], p. 1655 and Th. 3.2, $C_*^{a(2a+s)}(\overline{\Omega})$

is the solution space for the homogeneous Dirichlet problem with right-hand side in $\overline{C}^s_*(\Omega)$; here $C^{a(2a+s)}_*(\overline{\Omega}) \subset d^a\overline{C}^{a+s}_*(\Omega)$, but there is not equality. (One reason for the lack of equality is that the functions in $C^{a(2a+s)}_*(\overline{\Omega})$ are in C^{2a+s}_* over the interior, another is that $C^{a(2a+s)}_*(\overline{\Omega})$ only reaches a subspace of $d^a\overline{C}^{a+s}_*(\Omega)$ near the boundary, cf. [34] Th. 5.4.)

Assume that φ satisfies $P_{\rm Dir}\varphi = \lambda\varphi$ ($\lambda \neq 0$) and is in $\overline{C}^{a+\delta}(\Omega)$ for a positive $\delta < 1-a$; then in fact $\varphi \in \dot{C}^{a+\delta}(\overline{\Omega})$ since $\varphi \in C^{a(3a+\delta)}_*(\overline{\Omega}) \subset d^a\overline{C}^{2a+\delta}_*(\Omega)$ implies $\gamma_0\varphi = 0$. Now $\varphi \in \dot{C}^{a+\delta}(\overline{\Omega})$ implies $\gamma_0(\varphi/d^a) = 0$ since $\delta > 0$. It is shown in [59] for operators of the form (2.2)ff. with $0 < a < 1$, in [57] for $(-\Delta)^a$ with $a > 0$, and in [37], Ex. 4.10 for $(-\Delta + m^2)^a$, how it follows from Pohozaev identities that

$$P_{\rm Dir}v = \lambda v,\ \gamma_0(v/d^a) = 0 \implies v \equiv 0.$$

Thus $\varphi = 0$ and cannot be an eigenfunction. □

This allows us to conclude:

Corollary 5.3 *Consider the problem* (5.2). *For the operators P considered in Theorem* 5.2, *C^∞-regularity of f does not imply C^∞-regularity of u or u/d^a. In fact, there exist choices of $f(x,t) \in \overline{C}^\infty(\Omega \times I)$ with solutions $u(x,t)$ satisfying*

$$u \notin L_\infty(I'; \overline{C}^\infty(\Omega)),\quad u \notin L_\infty(I'; d^a\overline{C}^\infty(\Omega)). \tag{5.10}$$

More precisely, there exist solutions with $f(x,t) \in \overline{C}^\infty(\Omega \times I)$ such that

$$u \notin L_\infty(I'; \overline{C}^{a+\delta}(\Omega)),\quad u \notin L_\infty(I'; C^{a(3a+\delta)}_*(\overline{\Omega})),\ \textit{for any}\ \delta > 0. \tag{5.11}$$

Proof This follows by taking $u(x,t) = e^{-\lambda t}\varphi(x)$ with an eigenfunction φ as in Theorem 5.2. It solves the heat problem (5.2) with $f = 0$ (thus $f \in \overline{C}^\infty(\Omega \times I)$) and $u_0 = \varphi$, and it clearly satisfies (5.10) as well as (5.11). □

Note that the x-regularity obtained in (5.8) is close to the upper bound.

Corollary 5.3 gives counterexamples; one can show more systematically that $\gamma_0^a u$ being nonzero prevents the solution from being in C^∞ or d^aC^∞ at the boundary [40].

5.2 *Solvability in Sobolev Spaces*

Now let us consider estimates in Sobolev spaces.

In [41] (jointly with Solonnikov) and in [30] the author studied evolution problems for ψdo's P with the 0-transmission property at $\partial\Omega$, along with trace, Poisson and singular Green operators in the Boutet de Monvel calculus (cf. [12, 29, 31]), setting up a full calculus leading to existence, uniqueness and regularity theorems in anisotropic Bessel-potential and Besov spaces as mentioned in (5.3)ff.

These works take P of integer order, and do not cover the present case. We expect that a satisfactory generalization of the full boundary value theory in those works, to heat problems for our present operators, would be quite difficult to achieve. However their point of view on the ψdo P alone, considered on $\mathbb{R}^n$ without boundary conditions, can be extended, as follows:

For a classical strongly elliptic ψdo P of order $d \in \mathbb{R}_+$ on $\mathbb{R}^n$ (with global symbol estimates), we can construct an anisotropic symbol calculus on $\mathbb{R}^n \times \mathbb{R}$ that includes operators $P + \partial_t$ and their parametrices. It is not quite standard, since the typical strictly homogeneous symbol $|\xi|^d + i\tau$ is not C^∞ at points $(0, \tau)$ with $\tau \neq 0$. But this is a phenomenon handled in [31] by introducing classes of symbols with finite "regularity number" ν (essentially the Hölder regularity of the strictly homogeneous principal symbol at points $(0, \tau)$), and keeping track of how the value of ν behaves in compositions and parametrix constructions.

The calculus gives, on $\mathbb{R}^{n+1}$ and locally in $\Omega \times I$ [38]:

Theorem 5.4 *Let P be a classical strongly elliptic ψdo of order $d \in \mathbb{R}_+$. Then $P + \partial_t$ maps $H_p^{(s,s/d)}(\mathbb{R}^n \times \mathbb{R})$ continuously into $H_p^{(s-d,s/d-1)}(\mathbb{R}^n \times \mathbb{R})$ for any $s \in \mathbb{R}$. Moreover:*

1° If $u \in H_p^{(r,r/d)}(\mathbb{R}^n \times \mathbb{R})$ for some large negative r (this holds in particular if $u \in \mathcal{E}'(\mathbb{R}^{n+1})$ or e.g. $L_p(\mathbb{R}; \mathcal{E}'(\mathbb{R}^n))$), then

$$(P + \partial_t)u \in H_p^{(s,s/d)}(\mathbb{R}^n \times \mathbb{R}) \implies u \in H_p^{(s+d,s/d+1)}(\mathbb{R}^n \times \mathbb{R}). \tag{5.12}$$

2° Let $\Sigma = \Omega \times I$, and let $u \in H_p^{(s,s/d)}(\mathbb{R}^n \times \mathbb{R})$. Then

$$(P + \partial_t)u|_\Sigma \in H_{p,\mathrm{loc}}^{(s,s/d)}(\Sigma) \implies u \in H_{p,\mathrm{loc}}^{(s+d,s/d+1)}(\Sigma). \tag{5.13}$$

This theorem works for *any* strongly elliptic classical ψdo of positive order, not just fractional Laplacians, but for example also $-\Delta + (-\Delta)^{1/2}$ or $(-\Delta)^{1/2} + b(x) \cdot \nabla + c(x)$ with real C^∞-coefficients. The result extends by standard localization methods to the case where $\mathbb{R}^n$ is replaced by a closed manifold.

Note that in 2°, the regularity of u is only lifted by 1 in t, and the hypothesis on u concerns all $x \in \mathbb{R}^n$; the necessity of this is pointed out in related situations in [15] and [25].

By use of embedding theorems, we can moreover derive from the above a local regularity result in anisotropic Hölder spaces:

Theorem 5.5 *Let P and Σ be as in Theorem 5.4. Let $s \in \mathbb{R}_+$, and let $u \in C^{(s,s/d)}(\mathbb{R}^n \times \mathbb{R}) \cap \mathcal{E}'(\mathbb{R}^n \times \mathbb{R})$. Then*

$$(P + \partial_t)u|_\Sigma \in C_{\mathrm{loc}}^{(s,s/d)}(\Sigma) \implies u|_\Sigma \in C_{\mathrm{loc}}^{(s+d-\varepsilon,(s-\varepsilon)/d+1)}(\Sigma), \tag{5.14}$$

for small $\varepsilon > 0$.

Observe the similarity with (5.6) (where $d = 2a$). On one hand we have a loss of ε; on the other hand we have general x-dependent operators P just required to be strongly elliptic (albeit with smooth symbols), and no upper limitations on s. The ε might possibly be removed by working with our operators on the Hölder–Zygmund scale (cf. e.g. [33]).

Still other spaces could be examined. There is the work of Yamazaki [63] on ψdo's acting in anisotropic Besov–Triebel–Lizorkin spaces (defined in his work); these include the H_p and B_p spaces as special cases. However, the operators in [63] seem to be more regular (their quasi-homogeneous symbols being smooth outside of 0) than the operators $\partial_t + P$ that we study here. There is yet another type of spaces, the so-called modulation spaces on $\mathbb{R}^n$, where there are very recent results on heat equations (with nonlinear generalizations) by Chen, Wang, Wang and Wong [16].

For the case where boundary conditions at $\partial\Omega$ are imposed, there is not (yet) a systematic boundary-ψdo theory as in the stationary case. But using suitable functional analysis results we can make some progress, showing how the heat equation solutions behave in terms of Sobolev-type spaces involving the factor d^a. We henceforth restrict the attention to the case $0 < a < 1$, and to problems with initial value 0,

$$\begin{aligned} Pu + \partial_t u &= f \text{ on } \Omega \times I, \quad I =]0, T[\,, \\ u &= 0 \text{ on } (\mathbb{R}^n \setminus \Omega) \times I, \\ u|_{t=0} &= 0. \end{aligned} \tag{5.15}$$

There is a straightforward result in the L_2-framework:

Theorem 5.6 *Let P satisfy Hypothesis* 2.1 *with $a < 1$, and let Ω be a smooth bounded subset of $\mathbb{R}^n$. For f given in $L_2(\Omega \times I)$, there is a unique solution u of* (5.15) *satisfying*

$$u \in L_2(I; H^{a(2a)}(\overline{\Omega})) \cap \overline{H}^1(I; L_2(\Omega)); \tag{5.16}$$

here $H^{a(2a)}(\overline{\Omega}) = D(P_{\mathrm{Dir},2})$ equals $\dot{H}^{2a}(\overline{\Omega})$ if $a < \frac{1}{2}$, and is as described in (3.13)*ff. (with $p = 2$) when $a \geq \frac{1}{2}$.*

Moreover, $u \in \overline{C}^0(I; L_2(\Omega))$.

Proof Define the sesquilinear form

$$Q_0(u, v) = (r^+ Pu, v)_{L_2(\Omega)},$$

first for $u, v \in C_0^\infty(\Omega)$ and then extended by closure in $\dot{H}^a$-norm to a bounded sesquilinear form with domain $\dot{H}^a(\overline{\Omega})$. In view of the strong ellipticity, it is coercive:

$$\operatorname{Re} Q_0(u, u) \geq c_0 \|u\|^2_{\dot{H}^a} - \xi \|u\|^2_{L_2} \text{ with } c_0 > 0, \xi \in \mathbb{R}, \tag{5.17}$$

and hence defines via the Lax–Milgram lemma a realization of $r^+ P$ with domain (3.9); for precision we shall denote the operator $P_{\mathrm{Dir},2}$. (The Lax–Milgram construction is described e.g. in [32], Sect. 12.4.) The adjoint is defined similarly from $Q_0^*(u, v) = \overline{Q_0(v, u)}$, and it follows from (5.17) that the spectrum and numerical range is contained in a set $\{z \in \mathbb{C} \mid |\operatorname{Im} z| \leq C(\operatorname{Re} z + \xi), \operatorname{Re} z > \xi_0\}$, and

$$\|(P_{\mathrm{Dir},2}-\lambda)^{-1}\|_{\mathcal{L}(L_2)} \le c\langle\lambda\rangle^{-1} \text{ for } \operatorname{Re}\lambda \le -\xi_0. \tag{5.18}$$

We can then apply Lions and Magenes [51] Th. 4.3.2, which shows that there is a unique solution u of (5.15) in $L_2(I; D(P_{\mathrm{Dir},2}))$. Here, moreover, $\partial_t u = f - r^+Pu \in L_2(\Omega \times I)$, so $u \in \overline{H}^1(I; L_2(\Omega))$.

The last statement follows since $\overline{H}^1(I; L_2(\Omega)) \subset \overline{C}^0(I; L_2(\Omega))$. □

The domain of $P_{\mathrm{Dir},2}$ is a Sobolev space $\dot{H}^{2a}(\overline{\Omega})$ when $a < \frac{1}{2}$, but not a standard Sobolev space when $a \ge \frac{1}{2}$. However, it is then contained in one. Namely, if we take $r \ge 0$ such that

$$r \begin{cases} = 2a \text{ if } 0 < a < \frac{1}{2}, \\ < a + \frac{1}{2} \text{ if } \frac{1}{2} \le a < 1, \end{cases} \tag{5.19}$$

then

$$H^{a(2a)}(\overline{\Omega}) \subset \dot{H}^r(\overline{\Omega}) \subset \overline{H}^r(\Omega), \text{ hence } \|u\|_{\overline{H}^r} \le c\|u\|_{H^{a(2a)}}, \tag{5.20}$$

for all $0 < a < 1$. This follows for $\Omega = \mathbb{R}^n_+$ since $\overline{H}^a(\mathbb{R}^n_+) = \dot{H}^a(\overline{\mathbb{R}}^n_+)$ if $a < \frac{1}{2}$, and $\overline{H}^a(\mathbb{R}^n_+) \subset \dot{H}^{\frac{1}{2}-\varepsilon}(\overline{\mathbb{R}}^n_+)$ if $a \ge \frac{1}{2}$, so that by (3.19),

$$H^{a(2a)}(\overline{\mathbb{R}}^n_+) = \Xi_+^{-a} e^+ \overline{H}^a(\mathbb{R}^n_+) \subset \Xi_+^{-a} \dot{H}^{r-a}(\overline{\mathbb{R}}^n_+) = \dot{H}^r(\overline{\mathbb{R}}^n_+); \tag{5.21}$$

there is a similar proof for general Ω. Observe that $r \le 2a$ in all cases.

This can be used to jack up the regularity result of Theorem 5.6 by one derivative in t and an improved x-regularity, when f is H^r in x and H^1 in t. Higher t-regularity can also be obtained. To do this, we shall apply the more refined Th. 4.5.2 in [51], introduced there for the purpose of showing higher regularities. For the convenience of the reader, we list a slightly reformulated version:

Theorem 5.7 (From [51] Th. 4.5.2.) *Let X and $\mathcal{H}$ be Hilbert spaces, with $X \subset \mathcal{H}$, continuous injection. Let A be an unbounded linear operator in X such that $A - \lambda$ is a bijection from the domain $D_X(A) = \{u \in X \mid Au \in X\}$ onto X, for all $\lambda \in \mathbb{C}$ with* $\operatorname{Re}\lambda \le -\xi_0$.

Assume moreover that for all such λ, and for $u \in D_X(A)$,

$$\|(A-\lambda)u\|_X + \langle\lambda\rangle^\beta \|(A-\lambda)u\|_{\mathcal{H}} \ge c(\|u\|_{D_X(A)} + \langle\lambda\rangle^{\beta+1}\|u\|_{\mathcal{H}}), \tag{5.22}$$

where $\beta > 0$ and $c > 0$ are given.

Let $\beta \notin \frac{1}{2} + \mathbb{N}$. The problem $Au + \partial_t u = f$ for $t \in I$, $u(0) = 0$, with f given in $L_2(I; X) \cap H^\beta(I; \mathcal{H})$ with $f^{(j)}(0) = 0$ for $j < \beta - \frac{1}{2}$, has a unique solution

$$u \in L_2(I; D_X(A)) \cap H^{\beta+1}(I; \mathcal{H}). \tag{5.23}$$

This allows us to show:

Theorem 5.8 *Assumptions as in Theorem* 5.6.

1° *If* $f \in L_2(I; \overline{H}^r(\Omega)) \cap \overline{H}^1(I; L_2(\Omega))$ *for some* r *satisfying* (5.19), *with* $f|_{t=0} = 0$, *then the solution of* (5.15) *satisfies*

$$u \in L_2(I; H^{a(2a+r)}(\overline{\Omega})) \cap \overline{H}^2(I; L_2(\Omega)). \tag{5.24}$$

2° *For any integer* $k \geq 2$, *if* $f \in L_2(I; \overline{H}^r(\Omega)) \cap \overline{H}^k(I; L_2(\Omega))$ *with* $\partial_t^j f|_{t=0} = 0$ *for* $j < k$, *then*

$$u \in L_2(I; H^{a(2a+r)}(\overline{\Omega})) \cap \overline{H}^{k+1}(I; L_2(\Omega)). \tag{5.25}$$

It follows in particular that

$$f \in \bigcap_k \overline{H}^k(I; \overline{H}^r(\Omega)),\ \partial_t^j f|_{t=0} = 0 \text{ for } j \in \mathbb{N}_0 \implies u \in \bigcap_k \overline{H}^k(I; H^{a(2a+r)}(\overline{\Omega})). \tag{5.26}$$

Proof With A acting like r^+P, denote

$$D_r(A) = \{v \in H^{a(2a)}(\overline{\Omega}) \mid Av \in \overline{H}^r(\Omega)\}. \tag{5.27}$$

Here $D_r(A) = H^{a(2a+r)}(\overline{\Omega})$ in view of Theorem 3.5 (since $r \geq 0$). Moreover, it equals $D_{\overline{H}^r}(A) = \{v \in \overline{H}^r(\Omega) \mid Av \in \overline{H}^r(\Omega)\}$, since $H^{a(2a)}(\overline{\Omega}) \subset \dot{H}^r(\overline{\Omega}) \subset \overline{H}^r(\Omega)$. When $\operatorname{Re}\lambda \leq -\xi_0$, the bijectiveness of $A - \lambda$ from $H^{a(2a)}(\overline{\Omega})$ to $L_2(\Omega)$ implies bijectiveness from $D_r(A)$ to $\overline{H}^r(\Omega)$. All this shows that $D_r(A) = D_{\overline{H}^r}(A)$ is as in the start of Theorem 5.7 with $X = \overline{H}^r(\Omega)$, $\mathcal{H} = L_2(\Omega)$. Moreover, there is an equivalence of norms

$$\|(A + \xi_0)v\|_{\overline{H}^r} \simeq \|v\|_{D_r(A)}, \text{ for } v \in D_r(A). \tag{5.28}$$

Note also that besides the inequality (5.18), that may be written

$$\langle\lambda\rangle \|(A - \lambda)^{-1} g\|_{L_2} \leq c\|g\|_{L_2}, \text{ for } g \in L_2(\Omega), \quad \operatorname{Re}\lambda \leq -\xi_0, \tag{5.29}$$

we have that by (5.20), (5.29) for $r = 0$ and (5.2),

$$\begin{aligned}\|(A - \lambda)^{-1} g\|_{\overline{H}^r} &\leq c\|(A - \lambda)^{-1} g\|_{H^{a(2a)}} \simeq \|(A + \xi_0)(A - \lambda)^{-1} g\|_{L_2} \\ &\leq \|g\|_{L_2} + |\lambda + \xi_0| \|(A - \lambda)^{-1} g\|_{L_2} \leq c'\|g\|_{L_2}, \text{ for } g \in L_2(\Omega),\end{aligned}$$

so altogether,

$$\langle\lambda\rangle \|(A - \lambda)^{-1} g\|_{L_2} + \|(A - \lambda)^{-1} g\|_{\overline{H}^r} \leq c_1\|g\|_{L_2}, \text{ for } g \in L_2(\Omega). \tag{5.30}$$

For $v \in D_r(A)$, with $(A - \lambda)v$ denoted g, (5.28) implies

$$\begin{aligned}\|v\|_{D_r(A)} &\simeq \|(A+\xi_0)v\|_{\overline{H}^r} \le c_2(\|(A-\lambda)v\|_{\overline{H}^r} + |\lambda+\xi_0|\|v\|_{\overline{H}^r}) \\ &\le c_3(\|(A-\lambda)v\|_{\overline{H}^r} + \langle\lambda\rangle\|g\|_{L_2}),\end{aligned} \tag{5.31}$$

where we used (5.30) in the last step. Moreover, by (5.30),

$$\langle\lambda\rangle^2\|v\|_{L_2} \le c_1\langle\lambda\rangle\|g\|_{L_2},$$

so we altogether find the inequality for $v \in D_r(A)$:

$$\|v\|_{D_r(A)} + \langle\lambda\rangle^2\|v\|_{L_2} \le c_4(\|(A-\lambda)v\|_{\overline{H}^r} + \langle\lambda\rangle\|(A-\lambda)v\|_{L_2}), \quad \operatorname{Re}\lambda \le -\xi_0. \tag{5.32}$$

We can now apply Theorem 5.7, with $\beta = 1$, $X = \overline{H}^r(\Omega)$ and $\mathcal{H} = L_2(\Omega)$. It follows that the solution of (5.15) with $f(x,t)$ given in $L_2(I; \overline{H}^r(\Omega)) \cap \overline{H}^1(I; L_2(\Omega))$, $f|_{t=0} = 0$, satisfies

$$u \in L_2(I; D_r(A)) \cap \overline{H}^2(I; L_2(\Omega)),$$

from which (5.19) follows since $D_r(A) = H^{a(2a+r)}(\overline{\Omega})$. This shows 1°.

Now let $k \ge 2$. By (5.29) and (5.31), since $v = (A-\lambda)^{-1}g$,

$$\begin{aligned}\|v\|_{D_r(A)} + \langle\lambda\rangle^{k+1}\|v\|_{L_2} &\le c_3(\|(A-\lambda)v\|_{\overline{H}^r} + \langle\lambda\rangle\|g\|_{L_2}) + c\langle\lambda\rangle^k\|g\|_{L_2} \\ &\le c_5(\|(A-\lambda)v\|_{\overline{H}^r} + \langle\lambda\rangle^k\|(A-\lambda)v\|_{L_2}),\end{aligned} \tag{5.33}$$

which allows an application of Theorem 5.7 with $\beta = k$, $X = \overline{H}^r(\Omega)$, $\mathcal{H} = L_2(\Omega)$, giving the conclusion (5.25). Equation (5.26) follows when $k \to \infty$. This shows 2°. □

Note in particular that $D_r(A) = H^{a(4a)}(\overline{\Omega}))$ when $a < \frac{1}{2}$.

From the point of view of anisotropic Sobolev spaces, the solutions in (5.24) and (5.25) satisfy, since $H^{a(2a+r)}(\overline{\Omega}) \subset H^{a(2a)}(\overline{\Omega}) \subset \dot{H}^r(\overline{\Omega}) \subset \overline{H}^r(\Omega)$,

$$u \in L_2(I; \overline{H}^r(\Omega)) \cap \overline{H}^2(I; L_2(\Omega)) = \overline{H}^{(r,2)}(\Omega \times I), \text{ resp. } u \in \overline{H}^{(r,k+1)}(\Omega \times I),$$

but (5.24)–(5.26) give a more refined information.

Note that $r < 3/2$ in all these cases. We think that a lifting to higher values of r, of the conclusion of 1° concerning regularity in x at the boundary, would demand very different methods, or may not even be possible, because of the incompatibility of $H^{a(s)}(\overline{\Omega})$ with standard high-order Sobolev spaces when s is high. Cf. also Corollary 5.3, which excludes higher smoothness in a related situation.

Next, we turn to L_p-related Sobolev spaces with general p. The following result was shown for x-independent operators in [38]:

Theorem 5.9 *Let P satisfy Hypothesis* 2.2 *with $a < 1$. Then* (5.15) *has for any $f \in L_p(\Omega \times I)$ a unique solution $u(x,t) \in \overline{C}^0(I; L_p(\Omega))$ $(1 < p < \infty)$; it satisfies:*

$$u \in L_p(I; H_p^{a(2a)}(\overline{\Omega})) \cap \overline{H}_p^1(I; L_p(\Omega)). \tag{5.34}$$

The result is sharp, since it gives the exact domain for u mapped into $f \in L_p(\Omega \times I)$.

The proof relies on a theorem of Lamberton [49] which was used earlier in the work of Biccari, Warma and Zuazua [9]; they showed a local version of Theorem 5.9 for $P = (-\Delta)^a$, where $H_p^{a(2a)}(\overline{\Omega})$ in (5.34) is replaced by $B^{2a}_{p,\mathrm{loc}}(\Omega)$ if $p \geq 2$, $a \neq \frac{1}{2}$, by $H^1_{p,\mathrm{loc}}(\Omega)$ if $a = \frac{1}{2}$, and by $B^{2a}_{p,2,\mathrm{loc}}(\Omega)$ if $p < 2$, $a \neq \frac{1}{2}$.

Remark 5.10 Another paper [7], preparatory for [9], is devoted to a computational proof of local regularity (regularity in compact subsets of Ω) of the solutions of the stationary Dirichlet problem (3.8) for $P = (-\Delta)^a$ with $f \in L_p(\Omega)$. Here the authors were apparently unfamiliar with the pseudodifferential elliptic regularity theory (mentioned in [34]) that gave the answer many years ago, see the addendum [8]. The addendum also corrects some mistakes connected with the definition of $W^{s,p}$-spaces, cf. (3.3)ff. above. The proof of local regularity is repeated in [9].

Let us recall the proof of Theorem 5.9 from [38]: The Dirichlet realization in $L_p(\Omega)$, namely the operator $P_{\mathrm{Dir},p}$, acting like r^+P with domain

$$D(P_{\mathrm{Dir},p}) = \{u \in \dot{H}_p^a(\overline{\Omega}) \mid r^+Pu \in L_p(\Omega)\}, \tag{5.35}$$

coincides on $L_2(\Omega) \cap L_p(\Omega)$ with $P_{\mathrm{Dir},2}$ defined variationally from the sesquilinear form

$$Q_0(u,v) = \int_{\mathbb{R}^{2n}} (u(x) - u(y))(\bar{v}(x) - \bar{v}(y))K(x-y)\,dxdy \text{ on } \dot{H}^a(\overline{\Omega}),$$

where $K(y) = c\mathcal{F}^{-1}p(\xi)$, positive and homogeneous of degree $-2a - n$.

The form has the Markovian property: When u_0 is defined from a real function u by $u_0 = \min\{\max\{u, 0\}, 1\}$, then $Q_0(u_0, u_0) \leq Q_0(u, u)$. It is a so-called Dirichlet form, as explained in Fukushima, Oshima and Takeda [27], pp. 4–5 and Example 1.2.1, and Davies [19]. Then, by [27] Th. 1.4.1 and [19] Th. 1.4.1–1.4.2, $-P_{\mathrm{Dir},p}$ generates a strongly continuous contraction semigroup $T_p(t)$ not only in $L_2(\Omega)$ for $p = 2$ but also in $L_p(\Omega)$ for any $1 < p < \infty$, and $T_p(t)$ is bounded holomorphic. Then the statements in Theorem 5.9 follow from Lamberton [49] Th. 1. □

Remark 5.11 An advantage of the above results is that we have a precise characterization of the Dirichlet domain, namely $D(P_{\mathrm{Dir},p}) = H_p^{a(2a)}(\overline{\Omega})$. This space can be further described, as shown in [34], cf. Th. 5.4. When $p < 1/a$ then $H_p^{a(2a)}(\overline{\Omega}) = \dot{H}_p^{2a}(\overline{\Omega})$. Since $\dot{H}_p^{2a}(\overline{\Omega}) \subset \overline{H}_p^{2a}(\Omega)$, the solutions u are in the anisotropic space

$$\overline{H}_p^{(2a,1)}(\mathbb{R}^n \times I), \text{ supported for } x \in \overline{\Omega}.$$

When $p > 1/a$, $H_p^{a(2a)}(\overline{\mathbb{R}}^n_+)$ consists, as recalled earlier in (3.21)ff., of the functions $v = w + x_n^a K_0 \varphi$ with $w \in \dot{H}_p^{2a}(\overline{\mathbb{R}}^n_+)$, $\varphi \in B_p^{a-1/p}(\mathbb{R}^{n-1})$; here K_0 is the Poisson operator $K_0 \colon \varphi \mapsto e^{-\langle D' \rangle x_n}\varphi$ solving the Dirichlet problem for $1-\Delta$ with $(1-\Delta)u = 0$ and nontrivial boundary data φ. Also in the curved case, $H_p^{a(2a)}(\overline{\Omega})$ consists of $\dot{H}_p^{2a}(\overline{\Omega})$ plus a space of Poisson solutions in $\overline{H}_p^a(\Omega)$ multiplied by d^a, cf. [40].

We observe moreover that if we take $r \geq 0$ such that

$$r \begin{cases} = 2a \text{ if } 0 < a < 1/p, \\ < a + 1/p \text{ if } 1/p \leq a < 1, \end{cases} \tag{5.36}$$

(note that $r \leq 2a$ in all cases), then $H_p^{a(2a)}(\overline{\Omega}) \subset \overline{H}_p^r(\Omega)$, for all $0 < a < 1$ (in view of (3.19), cf. also (5.21)). The statement (5.34) then implies that

$$u \in \overline{H}_p^{(r,r/(2a))}(\Omega \times I). \tag{5.37}$$

When f has a higher regularity than $L_p(\Omega \times I)$, the *interior* regularity can be improved a little by use of Theorem 5.4 (for the boundary regularity, see Theorem 5.18 and 5.19):

Theorem 5.12 *Let u be as in Theorem* 5.9*, and let r satisfy* (5.36)*. Then u satisfies* (5.37)*, and moreover, for $0 < s \leq r$,*

$$f \in \overline{H}_p^{(s,s/(2a))}(\Omega \times I) \implies u \in H_{p,\mathrm{loc}}^{(s+2a,s/(2a)+1)}(\Omega \times I). \tag{5.38}$$

In particular, if $a < 1/p$ and $f \in \overline{H}_p^{(2a,1)}(\Omega \times I)$, then $u \in H_{p,\mathrm{loc}}^{(4a,2)}(\Omega \times I)$.

5.3 Higher Time-Regularity

The result of Theorem 5.9 can be considerably extended by use of the theory of Amann [5], in the question of time-regularity. Fix $p \in]1, \infty[$. The fact that $-P_{\mathrm{Dir},p}$ is the generator of a bounded holomorphic semigroup $T_p(t)$ in $L_p(\Omega)$ (for t in a sector around $\mathbb{R}_+$ depending on p, cf. [19] Th. 1.4.2), assures that there is an obtuse sector

$$V_\delta = \{\lambda \in \mathbb{C} \mid \arg \lambda \in]\pi/2 - \delta, 3\pi/2 + \delta[\,\}, \tag{5.39}$$

where the resolvent $(P_{\mathrm{Dir},p} - \lambda)^{-1}$ exists and satisfies an inequality

$$|\lambda| \|(P_{\mathrm{Dir},p} - \lambda)^{-1}\|_{\mathcal{L}(L_p)} \leq M, \tag{5.40}$$

cf. Hille and Phillips [42] Th. 17.5.1, or e.g. Kato [47] Th. IX.1.23. Since $P_{\mathrm{Dir},p}$ has a bounded inverse, $\|(P_{\mathrm{Dir},p} - \lambda)^{-1}\|_{\mathcal{L}(L_p)} \leq c$ also holds for λ in a neighborhood of 0, so we can replace $|\lambda|$ by $\langle\lambda\rangle$ in (5.40) (with a larger constant M'). Note that furthermore,

$$\|P_{\mathrm{Dir},p}(P_{\mathrm{Dir},p}-\lambda)^{-1}f\|_{L_p}=\|f+\lambda(P_{\mathrm{Dir},p}-\lambda)^{-1}f\|_{L_p}\le(1+M)\|f\|_{L_p},$$

so that $(P_{\mathrm{Dir},p}-\lambda)^{-1}$ is bounded uniformly in λ from $L_p(\Omega)$ to $D(P_{\mathrm{Dir},p})$ with the graph-norm.

Thus, if we set

$$E_0=L_p(\Omega),\quad E_1=D(P_{\mathrm{Dir},p})=H_p^{a(2a)}(\overline{\Omega}), \tag{5.41}$$

we have that $A=P_{\mathrm{Dir},p}$ satisfies (with $B_\varepsilon=\{|\lambda|<\varepsilon\}$ for a small $\varepsilon>0$)

$$\langle\lambda\rangle\|(A-\lambda)^{-1}\|_{\mathcal{L}(E_0)}+\|(A-\lambda)^{-1}\|_{\mathcal{L}(E_0,E_1)}\le c\text{ for }\lambda\in V_\delta\cup B_\varepsilon. \tag{5.42}$$

We can then apply Theorem 8.8 of Amann [5]. It is formulated with vector-valued Besov spaces $B^s_{q,r}(\mathbb{R};X)$ (valued in a Banach space X, a function space in the applications), where the case $q=r=\infty$ is particularly interesting for our purposes, since $B^s_{\infty,\infty}(\mathbb{R};X)$ equals the vector-valued Hölder–Zygmund space $C^s_*(\mathbb{R};X)$. This coincides with the vector-valued Hölder space $C^s(\mathbb{R};X)$ when $s\in\mathbb{R}_+\setminus\mathbb{N}$, see Remark 3.7 for further information.

Theorem 5.13 (From [5] Th. 8.8.) *Let $s\in\mathbb{R}$ and $q,r\in[1,\infty]$. Let E_0 and E_1 be Banach spaces, with E_1 continuously injected in E_0, and let A be a linear operator in E_0 with domain E_1 and range E_0, satisfying* (5.42).

For f given in $B^s_{q,r}(\mathbb{R};E_0)\cap L_{1,\mathrm{loc}}(\mathbb{R};E_0)$ and supported in $\overline{\mathbb{R}}_+$, the problem

$$Au+\partial_t u=f\text{ for }t\in\mathbb{R},\quad \operatorname{supp}u\subset\overline{\mathbb{R}}_+,$$

has a unique solution u in the space of distributions in $B^{s+1}_{q,r}(\mathbb{R};E_0)\cap B^s_{q,r}(\mathbb{R};E_1)$ supported in $\overline{\mathbb{R}}_+$; it is described by

$$u(t)=\int_0^t e^{-(t-\tau)A}f(\tau)\,d\tau,\quad t>0.$$

An application of this theorem with $q=r=\infty$ and E_0, E_1 defined by (5.41) gives:

Theorem 5.14 *Assumptions as in Theorem 5.9. The solution u satisfies: When $f\in\dot C^s_*(\overline{\mathbb{R}}_+;L_p(\Omega))\cap L_{1,\mathrm{loc}}(\mathbb{R},L_p(\Omega))$ for some $s\in\mathbb{R}$, then*

$$u\in\dot C^s_*(\overline{\mathbb{R}}_+;H_p^{a(2a)}(\overline{\Omega}))\cap\dot C^{s+1}_*(\overline{\mathbb{R}}_+;L_p(\Omega)). \tag{5.43}$$

In particular, for $s\in\mathbb{R}_+\setminus\mathbb{N}$,

$$f\in\dot C^s(\overline{\mathbb{R}}_+;L_p(\Omega))\iff u\in\dot C^s(\overline{\mathbb{R}}_+;H_p^{a(2a)}(\overline{\Omega}))\cap\dot C^{s+1}(\overline{\mathbb{R}}_+;L_p(\Omega)). \tag{5.44}$$

We here use conventions as in (3.4): The set of $u(x,t) \in C^s_*(\mathbb{R}; X)$ that vanish for $t < 0$ is denoted $\dot{C}^s_*(\overline{\mathbb{R}}_+; X)$. The conclusion $\Longrightarrow$ is a special case of Theorem 5.13 with $A = P_{\mathrm{Dir},p}$; the converse $\Longleftarrow$ follows immediately by application of $A + \partial_t$ to u.

Remark 5.15 Amann's theorem can of course also be applied with other choices of $B^s_{q,r}$, e.g. Sobolev–Slobodetskiĭ spaces $W^{s,q} = B^s_{q,q}$ when $s \in \mathbb{R}_+ \setminus \mathbb{N}$; this leads to analogous statements. (Note that integer cases $s \in \mathbb{N}_0$ are not covered, in particular not the result of Theorem 5.9 for $p \neq 2$, since the H^s_p-spaces are not in the scale $B^s_{q,r}$ when $p \neq 2$.)

Observe the consequences:

Corollary 5.16 *Assumptions as in Theorem 5.9. The solution u satisfies:*

$$f \in \dot{C}^\infty(\overline{\mathbb{R}}_+; L_p(\Omega)) \iff u \in \dot{C}^\infty(\overline{\mathbb{R}}_+; H_p^{a(2a)}(\overline{\Omega})). \tag{5.45}$$

Moreover,

$$\begin{aligned} f \in \dot{C}^s(\overline{\mathbb{R}}_+; L_\infty(\Omega)) &\Longrightarrow u \in \dot{C}^s(\overline{\mathbb{R}}_+; d^a\overline{C}^{a-\varepsilon}(\Omega)), \\ f \in \dot{C}^\infty(\overline{\mathbb{R}}_+; L_\infty(\Omega)) &\Longrightarrow u \in \dot{C}^\infty(\overline{\mathbb{R}}_+; d^a\overline{C}^{a-\varepsilon}(\Omega)), \end{aligned} \tag{5.46}$$

for $s \in \mathbb{R}_+ \setminus \mathbb{N}$ and $\varepsilon \in]0, a]$.

Proof When $f \in \dot{C}^\infty(\overline{\mathbb{R}}_+; L_p(\Omega))$, it is in particular in $\dot{C}^s(\overline{\mathbb{R}}_+; L_p(\Omega))$ for all $s \in \mathbb{R}_+ \setminus \mathbb{N}$, where (5.44) implies that $u \in \dot{C}^s(\overline{\mathbb{R}}_+; H_p^{a(2a)}(\overline{\Omega}))$. Taking intersections over s, we find (5.45).

Assertion (5.46) follows from (5.44) resp. (5.45), since $L_\infty(\Omega) \subset L_p(\Omega)$ for $1 < p < \infty$, and $\bigcap_p H_p^{a(2a)}(\overline{\Omega}) \subset d^a\overline{C}^{a-\varepsilon}(\Omega)$ (as already observed in [34] Sect. 7). □

The statements in (5.46) improve the results of [25, 58] concerning time-regularity.

Amann's theorem can also be used in situations with a higher x-regularity of f. The crucial point is to obtain an estimate (5.42) on the desired pair of Banach-spaces E_0, E_1 with $E_1 \subset E_0$.

Lemma 5.17 *Let $1 < p < \infty$, and let A be an operator acting like $P_{\mathrm{Dir},p}$ in Theorem 5.9. When $2a < 1/p$, A is bijective from $H_p^{a(4a)}(\overline{\Omega})$ to $\dot{H}_p^{2a}(\overline{\Omega})$, and we have a resolvent inequality for $\lambda \in V_\delta \cup B_\varepsilon$ as in (5.42):*

$$\langle\lambda\rangle \|(A-\lambda)^{-1} f\|_{\dot{H}_p^{2a}} + \|(A-\lambda)^{-1} f\|_{H_p^{a(4a)}} \leq c\|f\|_{\dot{H}_p^{2a}}, \text{ for } f \in \dot{H}_p^{2a}(\overline{\Omega}). \tag{5.47}$$

Proof Define $D_{s,p}(A)$ by

$$D_{s,p}(A) = H_p^{a(2a+s)}(\overline{\Omega}); \tag{5.48}$$

it will be used for $s = 0$ and $2a$. Since $2a < 1/p$, $\overline{H}_p^{2a}(\Omega)$ identifies with $\dot{H}_p^{2a}(\overline{\Omega})$, so A has the asserted bijectiveness property in view of Theorem 3.5 plus [33] Th. 3.5 on the invariance of kernel and cokernel.

By (5.42), we have the inequality

$$\langle\lambda\rangle\|(A-\lambda)^{-1}f\|_{L_p}+\|(A-\lambda)^{-1}f\|_{D_{0,p}}\le c\|f\|_{L_p}\text{ for }f\in L_p(\Omega),\qquad(5.49)$$

and we want to lift it to the case where $L_p(\Omega)$ is replaced by $E_0=\dot H_p^{2a}(\overline{\Omega})$ and $D_{0,p}(A)$ is replaced by $E_1=D_{2a,p}(A)$. Let $f\in\dot H_p^{2a}(\overline{\Omega})=D_{0,p}(A)$, and denote $Af=g$; it lies in $L_p(\Omega)$, and $\|g\|_{L_p}\simeq\|f\|_{\dot H_p^{2a}}$. Then

$$\begin{aligned}\|\lambda(A-\lambda)^{-1}f\|_{\dot H_p^{2a}}&=\|-f+A(A-\lambda)^{-1}f\|_{\dot H_p^{2a}}=\|-f+(A-\lambda)^{-1}Af\|_{\dot H_p^{2a}}\\&\le\|f\|_{\dot H_p^{2a}}+\|(A-\lambda)^{-1}g\|_{\dot H_p^{2a}}\le\|f\|_{\dot H_p^{2a}}+c\|g\|_{L_p}\le c'\|f\|_{\dot H_p^{2a}},\end{aligned}\qquad(5.50)$$

where we applied (5.49) to g. Next,

$$\begin{aligned}\|(A-\lambda)^{-1}f\|_{D_{2a,p}(A)}&\simeq\|A(A-\lambda)^{-1}f\|_{\overline{H}_p^{2a}}\simeq\|A(A-\lambda)^{-1}f\|_{\dot H_p^{2a}}\\&=\|f+\lambda(A-\lambda)^{-1}f\|_{\dot H_p^{2a}}\le(1+c')\|f\|_{\dot H_p^{2a}},\end{aligned}$$

using (5.50). Together with (5.50), this proves (5.47). □

This leads to a supplement to Theorem 5.14:

Theorem 5.18 *Assumptions as in Theorem* 5.9. *If* $2a<1/p$, *the solution satisfies, for* $s\in\mathbb{R}_+\setminus\mathbb{N}$:

$$\begin{aligned}f\in\dot C^s(\overline{\mathbb{R}}_+;\dot H_p^{2a}(\overline{\Omega}))&\iff u\in\dot C^s(\overline{\mathbb{R}}_+;H_p^{a(4a)}(\overline{\Omega}))\cap\dot C^{s+1}(\overline{\mathbb{R}}_+;\dot H_p^{2a}(\overline{\Omega})),\\f\in\dot C^\infty(\overline{\mathbb{R}}_+;\dot H_p^{2a}(\overline{\Omega}))&\iff u\in\dot C^\infty(\overline{\mathbb{R}}_+;H_p^{a(4a)}(\overline{\Omega})).\end{aligned}\qquad(5.51)$$

Proof We apply Theorem 5.13 with $E_0=\dot H_p^{2a}(\overline{\Omega})$, $E_1=H_p^{a(4a)}(\overline{\Omega})$, where the required resolvent inequality is shown in Lemma 5.17. □

Let us also apply Amann's theorem to the L_2-operators studied in Theorem 5.6, listing just the resulting Hölder estimates. Here when $I=]0,T[$, we denote by $C_+^s(I;X)$ the space of functions in $\overline{C}^s(]-\infty,T[;X)$ with support in $[0,T]$.

Theorem 5.19 *Assumptions as in Theorem* 5.6. *With* ξ_0 *as in* (5.18), *there is a set* $V_\delta\cup B_\varepsilon$ *(cf.* (5.39)*) such that* $A=P_{\mathrm{Dir},2}+\xi_0$ *satisfies an inequality* (5.42), *both for the choice* $\{E_0,E_1\}=\{L_2(\Omega),H^{a(2a)}(\overline{\Omega})\}$ *when* $a<1$, *and for the choice* $\{E_0,E_1\}=\{\dot H^{2a}(\overline{\Omega}),H^{a(4a)}(\overline{\Omega})\}$ *when* $a<\frac14$.

The solution of (5.15) *satisfies, for* $s\in\mathbb{R}_+\setminus\mathbb{N}$, $I=]0,T[$:

$$f\in C_+^s(I;L_2(\Omega))\iff u\in C_+^s(I;H^{a(2a)}(\overline{\Omega}))\cap C_+^{s+1}(I;L_2(\Omega)),\qquad(5.52)$$

$$f\in C_+^\infty(I;L_2(\Omega))\iff u\in C_+^\infty(I;H^{a(2a)}(\overline{\Omega})).\qquad(5.53)$$

Moreover, if $a < \frac{1}{4}$,

$$
\begin{aligned}
f \in C_+^s(I; \dot H^{2a}(\overline{\Omega})) &\iff u \in C_+^s(I; H^{a(4a)}(\overline{\Omega})) \cap C_+^{s+1}(I; \dot H^{2a}(\overline{\Omega})),\\
f \in C_+^\infty(I; \dot H^{2a}(\overline{\Omega})) &\iff u \in C_+^\infty(I; H^{a(4a)}(\overline{\Omega})).
\end{aligned}
\tag{5.54}
$$

Proof The estimate of $(A-\lambda)^{-1}$ with $E_0 = L_2(\Omega)$, $E_1 = H^{a(2a)}(\overline{\Omega})$, is assured by the information in the proof of Theorem 5.6 that the numerical range and spectrum of $P_{\mathrm{Dir},2}$ is contained in an angular set $\{z \in \mathbb{C} \mid |\operatorname{Im} z| \le C(\operatorname{Re} z + \xi), \operatorname{Re} z > \xi_0\}$. (This is a standard fact in the theory of operators defined from sesquilinear forms, see e.g. Cor. 12.21 in [32].) Then the resolvent estimate holds for $\lambda \to \infty$ on the rays in a closed sector disjoint from the sector $\{|\operatorname{Im} z| \le C \operatorname{Re}(z+\xi)\}$.

The estimate of $(A-\lambda)^{-1}$ with $E_0 = \dot H^{2a}(\overline{\Omega})$, $E_1 = H^{a(4a)}(\overline{\Omega})$ now follows exactly as in Lemma 5.17, when $a < \frac{1}{4}$.

For the solvability assertions, note that $u(x,t)$ satisfies $P_{\mathrm{Dir},2}u + \partial_t u = f$ if and only if $v(x,t) = e^{-\xi_0 t}u(x,t)$ satisfies $(P_{\mathrm{Dir},2} + \xi_0)v + \partial_t v = e^{-\xi_0 t} f$.

An application of Theorem 5.13 to $A = P_{\mathrm{Dir},2} + \xi_0$ leads to a solvability result like (5.44) for a solution v with right-hand side f_1. When the problem (5.15) is considered for a given function $f \in C_+^s(I; E_0)$, we extend f to a function $\tilde f \in \dot C^s(\overline{\mathbb{R}}_+; E_0)$, and apply the solvability result with $f_1 = e^{-\xi_0 t}\tilde f$; this gives a solution v, and we set $u = (e^{\xi_0 t}v)|_I$, solving (5.15). It has the regularity claimed in (5.52), and conversely, application of $P_{\mathrm{Dir},2} + \partial_t$ to such a function gives a right-hand side in $C_+^s(I; E_0)$. (5.53) follows immediately.

The statements in (5.54) follow by a similar application of Theorem 5.13 with $E_0 = \dot H^{2a}(\overline{\Omega})$, $E_1 = H^{a(4a)}(\overline{\Omega})$. □

It seems plausible that (5.51) and (5.54) can be extended to all $a < 1$ when $\dot H_p^{2a}(\overline{\Omega})$ is replaced by the possibly weaker space $\dot H_p^r(\overline{\Omega})$, cf. (5.36), and $H_p^{a(4a)}(\overline{\Omega})$ is replaced by $H_p^{a(2a+r)}(\overline{\Omega})$ (for $p = 2$, it is to some extent obtained in (5.27)). This might be based on an extension of Amann's strategy. The Fourier transform used in [51] is in [5] replaced by multiplier theorems in a suitable sense; it could be investigated whether something similar might be done based on resolvent inequalities in the style of (5.32), (5.33).

On the other hand, we expect that the limitations on r are essential in some sense, since the domain and range for the Dirichlet problem are not compatible in higher-order spaces, as noted also earlier.

Remark 5.20 The results in this subsection on higher t-regularity based on Theorem 5.13 from [5], are in the case $p \ne 2$ restricted to the x-independent case $P = \operatorname{Op}(p(\xi))$ with homogeneous symbol. However, since the proofs rely entirely on general resolvent inequalities, there is a leeway to extend the results to perturbations $P + P'$ when P' is so small that such resolvent inequalities still hold; this will allow some x-dependence and lower-order terms. The idea can be further developed.

It is very likely that other methods could be useful and bring out further perspectives for the heat problem. One possibility could be to try to establish an H^∞-calculus for the realizations of fractional Laplacians and their generalizations. This was done for elliptic differential operators with boundary conditions in a series of works, see e.g. Denk, Hieber and Prüss [20] for an explanation of the theory and many references. However, pseudodifferential problems pose additional difficulties. The only contribution treating an operator within the Boutet de Monvel calculus, that we know of, is Abels [4]. The present pseudodifferential operators do not even belong to the Boutet de Monvel calculus, but have a different boundary behavior.

It will be interesting to see to what extent the results can be further improved by this or other tools from the extensive literature on differential heat operator problems.

References

1. Abatangelo, N.: Large s-harmonic functions and boundary blow-up solutions for the fractional Laplacian. Discret. Contin. Dyn. Syst. **35**, 5555–5607 (2015)
2. Abatangelo, N., Jarohs, S., Saldana, A.: Integral representation of solutions to higher-order fractional Dirichlet problems on balls. Commun. Contemp. Math. to appear. arXiv:1707.03603
3. Abels, H.: Pseudodifferential boundary value problems with non-smooth coefficients. Commun. Partial Differ. Equ. **30**, 1463–1503 (2005)
4. Abels, H.: Reduced and generalized Stokes resolvent equations in asymptotically flat layers. II. H_∞-calculus. J. Math. Fluid Mech. **7**, 223–260 (2005)
5. Amann, H.: Operator-valued Fourier multipliers, vector-valued Besov spaces, and applications. Math. Nachr. **186**, 5–56 (1997)
6. Applebaum, D.: Lévy processes - from probability to finance and quantum groups. Not. Am. Math. Soc. **51**, 1336–1347 (2004)
7. Biccari, U., Warma, M., Zuazua, E.: Local elliptic regularity for the Dirichlet fractional Laplacian. Adv. Nonlinear Stud. **17**, 387–409 (2017)
8. Biccari, U., Warma, M., Zuazua, E.: Addendum: local elliptic regularity for the Dirichlet fractional Laplacian. Adv. Nonlinear Stud. **17**, 837 (2017)
9. Biccari, U., Warma, M., Zuazua, E.: Local regularity for fractional heat equations. SEMA-SIMAI Springer Series to appear. arXiv:1704.07562
10. Bogdan, K., Burdzy, K., Chen, Z.-Q.: Censored stable processes. Probab. Theory Relat. Fields **127**, 89–152 (2003)
11. Boulenger, T., Himmelsbach, D., Lenzmann, E.: Blowup for fractional NLS. J. Funct. Anal. **271**, 2569–2603 (2016)
12. Boutet de Monvel, L.: Boundary problems for pseudo-differential operators. Acta Math. **126**, 11–51 (1971)
13. Caffarelli, L., Silvestre, L.: An extension problem related to the fractional Laplacian. Commun. Partial Differ. Equ. **32**, 1245–1260 (2007)
14. Caffarelli, L.A., Salsa, S., Silvestre, L.: Regularity estimates for the solution and the free boundary of the obstacle problem for the fractional Laplacian. Invent. Math. **171**, 425–461 (2008)
15. Chang-Lara, H., Davila, G.: Regularity for solutions of non local parabolic equations. Calc. Var. Partial Differ. Equ. **49**, 139–172 (2014)
16. Chen, M., Wang, B., Wang, S., Wong, M.W.: On dissipative nonlinear evolutional pseudo-differential equations. arXiv:1708.09519
17. Chen, Z.-Q., Song, R.: Estimates on Green functions and Poisson kernels for symmetric stable processes. Math. Ann. **312**, 465–501 (1998)
18. Cont, R., Tankov, P.: Financial Modelling with Jump Processes. Financial Mathematics Series. Chapman & Hall/CRC, Boca Raton (2004)

19. Davies, E.B.: Heat Kernels and Spectral Theory. Cambridge Tracts in Mathematics, vol. 92. Cambridge University Press, Cambridge (1989)
20. Denk, R., Hieber, M., Prüss, J.: R-boundedness, Fourier multipliers and problems of elliptic and parabolic type. Mem. Am. Math. Soc. **166**(788) (2003), viii+114 pp
21. Dipierro, S., Grunau, H.: Boggio's formula for fractional polyharmonic Dirichlet problems. Ann. Mat. Pura Appl. **196**, 1327–1344 (2017)
22. Dyda, B., Kuznetsov, A., Kwasnicki, M.: Eigenvalues of the fractional Laplace operator in the unit ball. J. Lond. Math. Soc. **95**(2), 500–518 (2017)
23. Eskin, G.: Boundary Value Problems for Elliptic Pseudodifferential Equations, AMS Translations. American Mathematical Society, Providence (1981)
24. Felsinger, M., Kassmann, M.: Local regularity for parabolic nonlocal operators. Commun. Partial Differ. Equ. **38**, 1539–1573 (2013)
25. Fernandez-Real, X., Ros-Oton, X.: Regularity theory for general stable operators: parabolic equations. J. Funct. Anal. **272**, 4165–4221 (2017)
26. Frank, R., Geisinger, L.: Refined semiclassical asymptotics for fractional powers of the Laplace operator. J. Reine Angew. Math. **712**, 1–37 (2016)
27. Fukushima, M., Oshima, Y., Takeda, M.: Dirichlet Forms and Symmetric Markov Processes. De Gruyter Studies in Mathematics, vol. 19. Walter de Gruyter & Co., Berlin (1994)
28. Gonzalez, M., Mazzeo, R., Sire, Y.: Singular solutions of fractional order conformal Laplacians. J. Geom. Anal. **22**, 845–863 (2012)
29. Grubb, G.: Pseudo-differential boundary problems in Lp spaces. Commun. Partial Differ. Equ. **15**, 289–340 (1990)
30. Grubb, G.: Parameter-elliptic and parabolic pseudodifferential boundary problems in global Lp Sobolev spaces. Math. Z. **218**, 43–90 (1995)
31. Grubb, G.: Functional Calculus of Pseudodifferential Boundary Problems. Progress in Mathematics, vol. 65, 2nd edn. Birkhäuser, Boston (1996)
32. Grubb, G.: Distributions and Operators. Graduate Texts in Mathematics, vol. 252. Springer, New York (2009)
33. Grubb, G.: Local and nonlocal boundary conditions for μ-transmission and fractional elliptic pseudodifferential operators. Anal. P.D.E. **7**, 1649–1682 (2014)
34. Grubb, G.: Fractional Laplacians on domains, a development of Hörmander's theory of μ-transmission pseudodifferential operators. Adv. Math. **268**, 478–528 (2015)
35. Grubb, G.: Spectral results for mixed problems and fractional elliptic operators. J. Math. Anal. Appl. **421**, 1616–1634 (2015)
36. Grubb, G.: Regularity of spectral fractional Dirichlet and Neumann problems. Math. Nachr. **289**, 831–844 (2016)
37. Grubb, G.: Integration by parts and Pohozaev identities for space-dependent fractional-order operators. J. Differ. Equ. **261**, 1835–1879 (2016)
38. Grubb, G.: Regularity in L_p Sobolev spaces of solutions to fractional heat equations. J. Funct. Anal. **274**, 2634–2660 (2018)
39. Grubb, G.: Green's formula and a Dirichlet-to-Neumann operator for fractional-order pseudodifferential operators. arXiv:1611.03024, to appear in Commun. Partial Differ. Equ
40. Grubb, G.: Limited regularity of solutions to fractional heat equations. arXiv:1806.1002
41. Grubb, G., Solonnikov, V.A.: Solution of parabolic pseudo-differential initial-boundary value problems. J. Differ. Equ. **87**, 256–304 (1990)
42. Hille, E., Phillips, R.S.: Functional Analysis and Semi-groups. American Mathematical Society Colloquium Publications, vol. 31, Rev. edn. American Mathematical Society, Providence (1957)
43. Hörmander, L.: Seminar notes on pseudo-differential operators and boundary problems. Lectures at IAS Princeton 1965–1966. Available from Lund University, https://lup.lub.lu.se/search/
44. Hörmander, L.: The Analysis of Linear Partial Differential Operators, III. Springer, Berlin (1985)
45. Jakubowski, T.: The estimates for the Green function in Lipschitz domains for the symmetric stable processes. Probab. Math. Stat. **22**, 419–441 (2002)

46. Jin, T., Xiong, J.: Schauder estimates for solutions of linear parabolic integro-differential equations. Discret. Contin. Dyn. Syst. **35**, 5977–5998 (2015)
47. Kato, T.: Perturbation Theory for Linear Operators. Die Grundlehren der mathematischen Wissenschaften, vol. 132. Springer, New York (1966)
48. Kulczycki, T.: Properties of Green function of symmetric stable processes. Probab. Math. Stat. **17**, 339–364 (1997)
49. Lamberton, D.: Équations d'évolution linéaires associées à des semi-groupes de contractions dans les espaces Lp. J. Funct. Anal. **72**, 252–262 (1987)
50. Leonori, T., Peral, I., Primo, A., Soria, F.: Basic estimates for solutions of a class of nonlocal elliptic and parabolic equations. Discret. Contin. Dyn. Syst. **35**, 6031–6068 (2015)
51. Lions, J.-L., Magenes, E.: Problèmes aux limites non homogènes et applications, vol. 1, 2. Editions Dunod, Paris (1968)
52. Monard, F., Nickl, R., Paternain, G.P.: Efficient nonparametric Bayesian inference for X-ray transforms. Ann. Stat. to appear. arXiv:1708.06332
53. Ros-Oton, X.: Nonlocal elliptic equations in bounded domains: a survey. Publ. Mat. **60**, 3–26 (2016)
54. Ros-Oton, X.: Boundary regularity, Pohozaev identities and nonexistence results. arXiv:1705.05525, to appear as a chapter in "Recent developments in the nonlinear theory", pp. 335–358, De Gruyter, Berlin (2018)
55. Ros-Oton, X., Serra, J.: The Dirichlet problem for the fractional Laplacian: regularity up to the boundary. J. Math. Pures Appl. **101**, 275–302 (2014)
56. Ros-Oton, X., Serra, J.: The Pohozaev identity for the fractional Laplacian. Arch. Rational Mech. Anal. **213**, 587–628 (2014)
57. Ros-Oton, X., Serra, J.: Local integration by parts and Pohozaev identities for higher order fractional Laplacians. Discret. Contin. Dyn. Syst. **35**, 2131–2150 (2015)
58. Ros-Oton, X., Vivas, H.: Higher-order boundary regularity estimates for nonlocal parabolic equations. Calc. Var. Partial Differ. Equ. 57, no. 5, Art. **111**, 20 (2018)
59. Ros-Oton, X., Serra, J., Valdinoci, E.: Pohozaev identities for anisotropic integro-differential operators. Commun. Partial Differ. Equ. **42**, 1290–1321 (2017)
60. Schrohe, E.: A short introduction to Boutet de Monvel's calculus. Approaches to Singular Analysis. Operator Theory: Advances and Applications, vol. 125, pp. 85–116. Birkhäuser, Basel (2001)
61. Seeley, R.: The resolvent of an elliptic boundary problem. Am. J. Math. **91**, 889–920 (1969)
62. Triebel, H.: Interpolation Theory, Function Spaces, Differential Operators. North-Holland Publishing Company, Amsterdam (1978)
63. Yamazaki, M.: A quasihomogeneous version of paradifferential operators. I. Boundedness on spaces of Besov type. J. Fac. Sci. Univ. Tokyo Sect. IA Math. **33**, 131-74 (1986)
64. Zaba, M., Garbaczewski, P.: Ultrarelativistic bound states in the spherical well. J. Math. Phys. **57**, 26 (2016)

A Class of Planar Hypocomplex Vector Fields: Solvability and Boundary Value Problems

A. Meziani

Abstract This paper deals with the solvability of planar vector fields $L = A\partial_x + B\partial_y$, with A and B complex-valued function in a domain $\Omega \subset \mathbb{R}^2$. We assume that L has a first integral Z that is a homeomorphism in Ω. To such a vector field, we associate a Cauchy–Pompeiu type operator and investigate the Hölder solvability of $Lu = f$ and of a related Riemann–Hilbert problem when f is in L^p with $p > 2 + \sigma$, where σ is a positive number associated to L.

1 Introduction

This survey paper deals with partial differential equations related to planar vector fields with degeneracies. We give a presentation of classical results and recent developments. The presentation focuses mainly on the extent to which some important properties of the $\overline{\partial}$-operator extend to complex vector fields. This work is heavily inspired by results of the classical theory of generalized analytic functions (see [4, 30]) and by the more recent theory of involutive structures (see [6, 28]).

The basic question in PDE is whether the equation has local solution. In the mid 1950s, Hans Lewy [15] published a short paper which proves that there are functions $f \in C^\infty(\mathbb{R}^3)$ such that the equation

$$\frac{\partial u}{\partial x} + i\frac{\partial u}{\partial y} - (x + iy)\frac{\partial u}{\partial z} = f$$

has no distribution solution in any open set of $\mathbb{R}^3$. This example had a profound influence on the development of PDE. In the early 1960s, Louis Nirenberg and

The author would like to express his deep gratitude to the organizers of the 11th ISAAC Congress for the invitation to participate. This paper is an expanded version of the talk delivered by the author at the ISAAC's meeting.

A. Meziani (✉)
Department of Mathematics, Florida International University, Miami, FL 33199, USA
e-mail: meziani@fiu.edu

L. G. Rodino and J. Toft (eds.), *Mathematical Analysis and Applications—Plenary Lectures*, Springer Proceedings in Mathematics & Statistics 262,
https://doi.org/10.1007/978-3-030-00874-1_3

François Treves [23–25] established a necessary and sufficient condition for local solvability of C^∞ linear differential operators of principal type. This condition known as the *Condition* (P) can be stated as follows: The operator P with principal symbol $a(x,\xi)+ib(x,\xi)$, with a, b real valued, is locally solvable in the C^∞ category if and only if $b(x,\xi)$ does not change sign along the null bicharacteristics of $a(x,\xi)$.

It should be stressed that *Condition* (P) insures existence of C^∞ local solutions of the equation $Pu = f$ when f is C^∞ and does not address existence or properties of the solutions when f is not smooth. On the other hand, we know that if in addition P is elliptic, the equation is well understood when f (or the coefficients of P) are in L^p spaces and existence of Hölder continuous solutions are obtained. In this paper, we distinguish a class of linear operators of first order in $\mathbb{R}^2$ that are elliptic except along certain curves. We show that such operators share many properties with the Cauchy–Riemann operator. Most of the proofs of the results stated here can be found in the recent papers [8–10, 18–20].

To give a comprehensive presentation, we start in Sect. 2 by describing why there are C^∞ functions f for which the Mizohata equation

$$\frac{\partial u}{\partial y} - 2iy\frac{\partial u}{\partial x} = f$$

has no solutions. In Sect. 3, we set the stage for the introduction of our class of vector fields. We illustrate *Condition* (P) by examples and distinguish hypocomplex vector fields. Then within hypocomplex vector field consider those that satisfy a partial Łojasiewicz type condition. In Sect. 4, we associate to such a vector field L with first integral $Z(x,y)$ defined in an open set $\Omega \Subset \mathbb{R}^2$ a Cauchy–Pompeiu type operator

$$T_Z f(x,y) = \frac{1}{2\pi i}\int_\Omega \frac{f(\xi,\eta)}{Z(\xi,\eta)-Z(x,y)}d\xi d\eta\,.$$

This operator plays a crucial role in characterizing the properties of the solutions of the equation $Lu = f$. In particular, we can associate with L a positive number $\sigma > 0$ such that if $f \in L^p(\Omega)$ with $p > 2+\sigma$, then the solutions of $Lu = f$ are in C^α with

$$\alpha = \frac{p-\sigma-2}{p(\sigma+1)}\,.$$

The study of the semilinear equation $Lu = F(x,y,u)$ in Sect. 5 leads to a similarity principle for the vector field L. More precisely, if a, $b \in L^p(\Omega)$ with $p > 2+\sigma$, then any solution of the equation $Lu = au + b\overline{u}$ can be represented as $u = w\,\mathrm{e}^s$ where s is a Hölder continuous function w satisfies and $Lw = 0$. Section 6 deals with the generalized Riemann–Hilbert problem for L in an open set Ω:

$$\begin{cases} Lu = au + b\overline{u} + f & \text{in } \Omega \\ \mathrm{Re}(\overline{\Lambda}u) = \phi & \text{on } \partial\Omega \end{cases}.$$

We use the Fredholm alternative to a modification of the operator T_Z together with a Schwarz operator for L to prove that the boundary value problem has a solution provided that the index of Λ is nonnegative. The last section deals a Cauchy problem associated with the vector field L. It is essentially the Cauchy–Kovalevsky Theorem within the kernel of L. More precisely, if T is any vector field such that

$$T : \mathrm{Ker}(L) \longrightarrow \mathrm{Ker}(L)$$

consider the Cauchy problem

$$\begin{cases} \dfrac{\partial u}{\partial t}(t, x, y) = A(t, x, y)\, Tu + B(t, x, y)u + C(t, x, y) \\ u(0, x, y) = \psi(x, y) \end{cases}$$

If for every t fixed the functions $A,\ B,\ C$ and ψ satisfy

$$LA = LB = LC = L\psi = 0\,,$$

then the initial value problem has a unique solution u with $Lu = 0$. It should be noted that this result is new and more general results dealing with these type of Cauchy problems associated with vector fields will appear in a forthcoming paper.

2 Nonsolvability: The Mizohata Vector Field

The two-dimensional version of the famous counterexample by Hans Lewy [15] to the solvability of differential operators is given by the Mizohata operator

$$M = \frac{\partial}{\partial y} - 2iy\frac{\partial}{\partial x} \tag{2.1}$$

with first integral $Z = x + iy^2$. It is proved (see [6, 22, 28]) that there are C^∞ functions defined near $0 \in \mathbb{R}^2$ such that equation

$$Mu = f \tag{2.2}$$

has no distribution solution in any neighborhood of $0 \in \mathbb{R}^2$.

The idea behind the construction of such a function f is based on the following observation. Suppose that $\Omega \subset \mathbb{R}^2$ is a connected open set symmetric with respect to the x-axis (i.e. if $(x, y) \in \Omega$, then $(x, -y) \in \Omega$), and such that $\Omega \cap \{y = 0\}$ contains an interval I. The we have the following.

Lemma 2.1 *Let $\Omega \subset \mathbb{R}^2$ be as above and suppose that $v \in C^0(\Omega)$ satisfies $Mv = 0$, then there is a function H holomorphic in the interior of $Z(\Omega)$, continuous up to the interval $Z(I) \subset \partial Z(\Omega)$ such that $v = H \circ Z$.*

Proof Consider the map Φ defined on $\mathbb{R}^2$ by $\Phi(x, y) = (x, y^2)$ and its restrictions Φ^+ and Φ^- to $\mathbb{R}^2_+ = \{y > 0\}$ and $\mathbb{R}^2_- = \{y < 0\}$, respectively. Then the pushforward of M via $\Phi^\pm$ are

$$\Phi^\pm_* M = \mp 4\sqrt{s}\,\frac{\partial}{\partial \overline{\zeta}}\,,$$

where $\zeta = x + is,\ s > 0$. Hence, if $v \in C^0(\Omega)$ satisfies $Mv = 0$, the the functions $H^\pm = v \circ (\Phi^\pm)^{-1}$ are continuous on $Z(\Omega)$ and holomorphic functions in the interior of $Z(\Omega)$. In addition $H^+ = H^-$ on the interval $Z(I)$. Therefore $H^+ = H^-$ in $Z(\Omega)$. □

Let $p_0 = (x_0, y_0) \in \mathbb{R}^2$ with $y_0 > 0, 0 < r_0 < y_0$, and let $g \in C^\infty(\mathbb{R}^2)$ such that $g \geq 0$, supp(g) is contained in the disc $D = D(p_0, r_0)$, with center p_0 and radius r_0, and $g(p_0) > 0$.

Lemma 2.2 *Let g be a function as above and Ω be an open subset in $\mathbb{R}^2$ symmetric with respect to the x-axis, such that $\overline{D} \subset \Omega$. Then equation $Mu = g$ has no continuous solution in Ω.*

Proof Let $\hat{D}$ be the symmetric of the disc D with respect to the x-axis and let $\Omega_0 = \Omega \backslash (D \cup \hat{D})$. Then Ω_0 is symmetric with respect to x-axis and $\Omega_0 \cap \{y = 0\}$ is non empty open set of the x-axis. If $u \in C^0(\Omega)$ solves equation $Mu = g$, and supp$(g) \subset D$, then $Mu = 0$ in Ω_0 and by Lemma 2.1, $u = H \circ Z$ with $H \in C^0(Z(\Omega_0))$ and holomorphic in the interior of $Z(\Omega_0)$. Note also that since $Mu = 0$ in $\Omega_- = \Omega \cap \mathbb{R}^2_-$, then $u = \tilde{H} \circ Z$ with $\tilde{H}$ holomorphic in $Z(\Omega_-) = Z(\Omega)$. Hence H extends to $Z(D)$ as $\hat{H}$. We have

$$\int_{\partial D} u\, dZ(x, y) = \int\!\!\int_D Mu\, dxdy = \int\!\!\int_D g\, dxdy > 0$$

on the other hand

$$\int_{\partial D} u\, dZ(x, y) = \int_{\partial Z(D)} \hat{H}(\zeta)\, d\zeta = 0\,.$$

This is a contradiction and such a function u cannot exist. □

Proposition 2.1 *There exist a function $f \in C^\infty(\mathbb{R}^2)$ such that equation $Mu = f$ has no continuous solution in any neighborhood of $0 \in \mathbb{R}^2$.*

Proof Let $\{p_k\}$, $k \in \mathbb{Z}^+$ be a sequence of points in the upper half plane $\mathbb{R}^2_+$, such that $p_k \longrightarrow 0$ as $k \longrightarrow \infty$. Let $\{r_k\}$ be a sequence of positive real numbers with $r_k \longrightarrow 0$ such that the disks $D(p_k, r_k)$ are pairwise disjoint. Let $f \in C^\infty(\mathbb{R}^2)$ such that $f \geq 0$,

$$\operatorname{supp}(f) \subset \bigcup_{k\in\mathbb{Z}^+} D(p_k, r_k)$$

and for each $k \in \mathbb{Z}^+$ there exists a point $m_k \in D(p_k, r_k)$ with $f(m_k) > 0$. It follows immediately from Lemma 2.2 that equation $Mu = f$ cannot have a continuous solution in any open set $\Omega \ni 0$ □

Let $\Omega \subset \mathbb{R}^2$ be open symmetric with respect to the x-axis with a nonempty intersection with the x-axis. For $p > 2$, denote by $S^p(\Omega)$ the space of functions f in Ω such that

$$\int\int_{\Omega} \left|\frac{f(x,y)}{y}\right|^p y\, dxdy < \infty \tag{2.3}$$

For $f \in S^p(\Omega)$ we associate the function $\hat{f}$ defined on $\mathbb{R}$ by

$$\hat{f}(x) = \int\int_{\Omega^+} \frac{f(p,q)}{p+iq^2-x}\, dpdq - \int\int_{\Omega^-} \frac{f(p,q)}{p+iq^2-x}\, dpdq\,, \tag{2.4}$$

where $\Omega^\pm = \Omega \cap \mathbb{R}^2_\pm$. We have the following theorem (see [16] for a C^∞ version)

Theorem 2.1 *Let $f \in S^p(\Omega)$. Equation $Mu = f$ has a solution in $C^\alpha(\Omega)$ with $\alpha = \dfrac{p-2}{p}$ if and only if the function $\hat{f}$ is the trace on the x-axis of a holomorphic function defined in the interior of $Z(\Omega)$.*

Proof As in the proof of Lemma 2.1, let Φ^+ and Φ^- be the restrictions to Ω^+ and Ω^- of the map $\Phi(x, y) = (x, y^2)$. For a function g defined in Ω, let

$$\tilde{g}^+(\xi,\eta) = g\circ(\Phi^+)^{-1}(\xi,\eta) = g(\xi,\sqrt{\eta})\,,$$
$$\tilde{g}^-(\xi,\eta) = g\circ(\Phi^-)^{-1}(\xi,\eta) = g(\xi,-\sqrt{\eta})\,.$$

Note that if $f \in S^p(\Omega)$, then $\dfrac{\tilde{f}^\pm(s,t)}{\sqrt{t}} \in L^p(Z(\Omega))$. Now if u solves $Mu = f$ in Ω, then we have

$$\mp 4\sqrt{\eta}\frac{\partial \tilde{u}^\pm}{\partial\overline{\zeta}}(\xi,\eta) = \tilde{f}^\pm(\xi,\eta) \quad \text{in} \quad Z(\Omega)$$

where $\zeta = \xi + i\eta$. Hence, it follows from the solvability of the $\overline{\partial}$ that $\tilde{u}^\pm \in C^\alpha(Z(\Omega)$ are given by

$$\tilde{u}^\pm(\xi,\eta) = \pm\frac{1}{4i\pi}\int_{Z(\Omega)} \frac{\tilde{f}^\pm(s,t)}{\sqrt{t}(s+it-\xi-i\eta)}\, dsdt + H^\pm(\zeta)\,,$$

where $H^\pm$ are holomorphic functions defined in $Z(\Omega)$.

In order for u to be continuous on the $\Omega \cap \{y = 0\}$, we need

$$\tilde{u}^+(\xi,0) = \tilde{u}^-(\xi,0)\,.$$

This implies

$$\frac{1}{4i\pi}\int_{Z(\Omega)} \frac{\tilde{f}^+(s,t)+\tilde{f}^-(s,t)}{\sqrt{t}(s+it-\xi)}\,dsdt \;=H^-(\xi+i0)-H^+(\xi+i0)\,.$$

Going back to the coordinates $t=\tau^2$ in the integral, we get

$$2\pi i\,\hat{f}(\xi) = \iint_{\Omega^+}\frac{f(s,\tau)}{s+i\tau-\xi}\,dsd\tau - \iint_{\Omega^-}\frac{f(s,\tau)}{s+i\tau-\xi}\,dsd\tau = H^-(\xi+i0)-H^+(\xi+i0).$$

□

3 A Class of Hypocomplex Vector Fields

As we saw in the previous section, even for the simple Mizohata vector field M with a linear degeneracy along a line, the equation $Mu = f$ might not be locally solvable. In fact a necessary and sufficient condition for local solvability is given the Nirenberg–Treves *Condition* (P) (see [6, 23, 28]). The *Condition* (P) deals with the local solvability of more general linear differential operators of arbitrary orders. In the case of vector fields in two variables this condition has a very simple interpretation.

Let

$$L = A(x,y)\frac{\partial}{\partial x} + B(x,y)\frac{\partial}{\partial y} \tag{3.1}$$

with C^∞ coefficients A and B defined in an open set $\Omega \in \mathbb{R}^2$ with $|A|+|B|>0$ For every point $p\in\Omega$ We can find local coordinates (s,t) defined near p such that

$$L = m(s,t)\left(\frac{\partial}{\partial t} - ib(s,t)\frac{\partial}{\partial s}\right)$$

with b an $\mathbb{R}$-valued function. Condition (P) just means that for every s, the function $t \longrightarrow b(s,t)$ does not change sign. Condition (P) guarantees local solvability and local integrability in the C^∞ category:

Theorem 3.1 *Let $f\in C^\infty(\Omega)$ and $p_0\in\Omega$. Then equation $Lu=f$ has a C^∞ solution u defined in some open neighborhood $U\ni p_0$ if and only if L satisfies Condition (P) at p_0. Moreover, in this case L has a first integral defined near p_0. That is there exist $Z\in C^\infty(U)$ such that $dZ(p_0)\neq 0$ and $LZ=0$.*

Remark 3.2 Condition (P) can also be formulated in terms of the fibers of the first integral. A vector field in two variables satisfies *Condition* (P) if and only if its orbits are locally connected. If Z is a local first integral of a vector field L, this means that locally $Z^{-1}(Z(p))$ are connected sets for any point p.

Example 3.3 We illustrate the above result by the following simple vector fields

(1) The Mizohata vector field M with first integral $Z(x, y) = x + iy^2$ satisfies $Z^{-1}(Z(x, y)) = \{(x, y), (x, -y)\}$. Thus M does not satisfy *Condition* (P)at each point $(x, 0)$.
(2) The vector field
$$H = \frac{\partial}{\partial y} - 3iy^2\frac{\partial}{\partial x}$$
whose first integral $Z(x, y) = x + iy^3$ is a global homeomorphism satisfies *Condition* (P).
(3) The vector field
$$X = \frac{\partial}{\partial y} - ix\frac{\partial}{\partial x} \tag{3.2}$$
has $Z(x, y) = xe^{iy}$ as a first integral. If $x \neq 0$, then
$$Z^{-1}(Z(x, y)) = \{(x, y + 2k\pi),\ k \in \mathbb{Z}\}$$
and $Z^{-1}(Z(0, y)) = \{x = 0\}$. Thus X satisfies *Condition* (P).

Remark 3.4 It should be noted that *Condition* (P) guarantees existence of smooth local solutions when the right hand side is smooth. However, near one-dimensional fibers non-smooth solutions exist when the right hand side is smooth. This is the case for the vector field X given by (3.2) with a one-dimensional orbit $\{x = 0\}$. The equation $Xu = 0$ has unbounded solutions for example $u = \ln|x| + iy$.

Remark 3.5 For a vector field L satisfying *Condition* (P) at a point p, the space of solutions of the homogeneous equation $Lu = 0$ can be described as follows. Let Z be a first integral of L near p. If u is continuous solution of the equation $Lu = 0$ near p, then there exists an open set $V \ni p$ and a continuous function h defined on $Z(V)$ such that h is holomorphic in the interior of $Z(V)$ and $u = h \circ Z$ in V (see [3]).

If we want to obtain regularity of solutions, we need to discard vector fields with one dimensional orbits. That is, consider only those vector fields such that for every $p \in \Omega$ there exists an open set $U \subset \Omega$, with $p \in U$ and a first integral Z in U with the property that $Z : U \longrightarrow Z(U) \subset \mathbb{C}$ is a homeomorphism. Such vector fields are called *hypocomplex*. For a hypocomplex C^∞ vector field L, all solutions of the equation $Lu = f$ are C^∞ when f is C^∞ (see [28]).

Even within the class of hypocomplex vector fields, in general Hölder solvability does not hold as the following example shows.

Example 3.6 The vector field

$$L = \frac{\partial}{\partial y} - i\left(\frac{1}{y^2}\exp\left(\frac{-1}{|y|}\right)\right)\frac{\partial}{\partial x}$$

is C^∞ in $\mathbb{R}^2$ with first integral

$$Z(x, y) = x + i\frac{|y|}{y}\exp\left(\frac{-1}{|y|}\right).$$

Note that $Z : \mathbb{R}^2 \longrightarrow \mathbb{C}$ is a C^∞ homeomorphism and so L is hypocomplex. For $\epsilon > 0$, consider the open set

$$U = \{(x, y) \in \mathbb{R}^2 : |Z(x, y)| < \epsilon, \ 0 < \arg(Z(x, y)) < \frac{\pi}{4}\}.$$

Let $f \in L^\infty(\mathbb{R}^2)$ be the characteristic function of U, i.e. $f = 1$ on U and $f = 0$ on $\mathbb{R}^2 \backslash U$. It is proved in [6] that equation $Lu = f$ has no bounded solutions.

If we want to extend the solvability of $Lu = f$ to include Hölder regularity when f is in L^p space, it is then necessary to restrict further the subclass of hypocomplex vector fields. It appears that the right condition to impose on L is to satisfy a partial Łojasiewicz type condition. Given a vector field L in Ω, let Σ be the characteristic set of L:

$$\Sigma = \{p : L_p \wedge \overline{L}_p = 0\}$$

L satisfies the Łojasiewicz type condition (LC) if for every $p \in \Sigma$, L has a local first integral $Z = x + i\phi(x, t)$ (with $Z(p) = 0$) defined in an open set $p \in U \subset \mathbb{R}^2$ such that Z is a homeomorphism and satisfies

$$|\phi(x, t)|^\sigma \leq C|\phi_t(x, t)|$$

for some $C > 0$ and $\sigma \in (0, 1)$.

Remark 3.7 The (LC) condition is invariant under change of coordinates or choice of a local first integral. Also if Σ is 1-dimensional manifold, then the (LC) condition can be written as

$$\operatorname{dis}(Z(m), Z(\Sigma))^\sigma \leq C|\operatorname{Im}(Z_x\overline{Z}_y)|.$$

Example 3.8 Let $\phi(x, y)$ be a homogeneous polynomial of degree $n + 1$ with real coefficients such that $\dfrac{\partial\phi}{\partial y} > 0$ for $(x, y) \neq 0$. Consider the vector field

$$L = (1 + i\phi_x(x, y))\frac{\partial}{\partial y} - i\phi_y(x, y)\frac{\partial}{\partial x}$$

with first integral $Z(x, y) = x + i\phi(x, y)$. Then L satisfies (LC) with $\sigma = \dfrac{n}{n+1}$. Indeed, we can rewrite ϕ and ϕ_y in polar coordinates (r, θ) as

$$\phi(r, \theta) = r^{n+1} Q(\theta) \quad \phi_y(r, \theta) = r^n P(\theta)$$

with Q and P trigonometric polynomials and $P(\theta) > 0 \;\; \forall \theta \in [0, 2\pi]$. Thus

$$|\phi(r, \theta)|^{n/(n+1)} \leq C|\phi_y(r, \theta)|$$

Example 3.9 The hypocomplex vector field

$$L = \frac{\partial}{\partial y} - i\left(\frac{1}{y^2}\exp\left(\frac{-1}{|y|}\right)\right)\frac{\partial}{\partial x}$$

is C^∞ in $\mathbb{R}^2$ with first integral

$$Z(x, y) = x + i\frac{|y|}{y}\exp\left(\frac{-1}{|y|}\right).$$

does not satisfy (LC) since

$$\lim_{y\to 0} y^2 \exp\frac{1-\sigma}{|y|} = \infty \quad \text{if } \sigma < 1 .$$

From now on and for the sake of clarity of exposition, we will consider vector fields L given by (3.1) defined in an open neighborhood of $\overline{\Omega} \subset\subset \mathbb{R}^2$ and satisfying the following properties.

(i) L is hypocomplex in $\overline{\Omega}$
(ii) The characteristic set $\Sigma \subset \overline{\Omega}$ is a $C^{1+\epsilon}$ curve nontangent to L
(iii) For every $p \in \Sigma$, there exists an open set U, with $p \in U$, such that $\Sigma \cap U$ is given by a defining $C^{1+\epsilon}$ function $\rho(x, y)$ such that

$$\mathrm{Im}(A\overline{B})(x, y) = |\rho(x, y)|^\sigma g(x, y),$$

for some continuous function g in U satisfying $g(x, y) \neq 0$ for all $(x, y) \in \mathcal{U}$.

It is proved in [9] that vector fields satisfying the above conditions have simple local normal.

Proposition 3.1 ([9]) *Suppose that L satisfies* (i), (ii), *and* (iii). *Then, for every $p \in \Sigma$, there exist an open neighborhood U such that $U\backslash\Sigma$ consists of two connected components U^+ and U^-, and local coordinates (x^+, t^+) (respectively (x^-, t^-)) centered at p such that L is a multiple of the following vector field in U^+ (respectively U^-):*

$$L_\sigma = \partial/\partial t^\pm - i|t^\pm|^\sigma \partial/\partial x^\pm, \tag{3.3}$$

with first integral

$$Z_\sigma(x,t) = x^\pm + i\frac{t^\pm|t^\pm|^\sigma}{\sigma+1}. \tag{3.4}$$

Remark 3.10 It should be noted that under the above assumptions, the vector field L has a global first integral Z defined in $\overline{\Omega}$ such that $Z : \overline{\Omega} \longrightarrow Z(\overline{\Omega})$ is a global homeomorphism (see [9]).

4 A Cauchy–Pompeiu Operator

Let $\Omega \subset \mathbb{R}^2$ be relatively compact such that $\partial\Omega$ is piecewise of class C^1 and let

$$Z : \overline{\Omega} \to Z(\overline{\Omega}) \subset \mathbb{C} \tag{4.1}$$

be a homeomorphism of class $C^{1+\epsilon}$ (for some $\epsilon > 0$). We assume that the associated Hamiltonian vector field:

$$L = Z_x\frac{\partial}{\partial y} - Z_y\frac{\partial}{\partial x}. \tag{4.2}$$

satisfies conditions (i), (ii), (iii) given in the previous section. Let $\Sigma_1, \cdots \Sigma_N$ be the connected components of the 1-dimensional manifold Σ. It follows from Proposition (3.1) that for each $j = 1, \cdots, N$ there exists $\sigma_j > 0$ satisfying (3.4). Let

$$\sigma = \max \sigma_j. \tag{4.3}$$

For $f \in L^1(\Omega)$, define a generalized Cauchy–Pompeiu integral operator associated with L as follows

$$T_Z f(x,y) = \frac{1}{2\pi i}\int_\Omega \frac{f(\xi,\eta)}{Z(\xi,\eta) - Z(x,y)} d\xi d\eta. \tag{4.4}$$

It is proved in [9] that this operator satisfies many properties similar to those of the classical operator. More precisely, we have the following results. The detailed proofs can be found in the referenced paper.

Theorem 4.1 *Let σ be given by (4.3) and let $f \in L^p(\Omega)$, with $p > 2 + \sigma$. Then, there exists a constant $M(p,\sigma,\Omega) > 0$ such that the operator T_Z satisfies*

$$|T_Z f(x,y)| \le M(p,\sigma,\Omega)\,||f||_p, \quad \forall (x,y) \in \Omega.$$

Theorem 4.2 *Let σ be given by (4.3) and let $f \in L^1(\Omega)$. Then, $Tf \in L^q(\Omega)$, for any $1 \le q < 2 - \sigma/(\sigma+1)$.*

To establish that $T_Z f$ satisfies $LT_Z f = f$, we need the following proposition.

Proposition 4.1 *Let $w \in C(\overline{\Omega}) \cap C^1(\Omega)$. Then, for all $(x, y) \in \Omega$, we have*

$$w(x,y) = \frac{1}{2\pi i}\int_{\partial\Omega} \frac{w(\alpha,\beta)}{Z(\alpha,\beta) - Z(x,y)} dZ(\alpha,\beta) + \frac{1}{2\pi i}\int_{\Omega} \frac{Lw(\alpha,\beta)}{Z(\alpha,\beta) - Z(x,y)} d\alpha d\beta.$$

Proof First note that as a direct consequence of the Green's formula, if $w \in C(\overline{\Omega}) \cap C^1(\Omega)$. Then

$$\int_{\Omega} Lw\, dxdy = -\int_{\partial\Omega} w\, dZ(x,y). \tag{4.5}$$

Now, let $(x_0, y_0) \in \Omega$ fixed. Set $z_0 = Z(x_0, y_0)$ and let $\epsilon > 0$ such that $\overline{D_\epsilon} \subset Z(\Omega)$, where $D_\epsilon = D(z_0, \epsilon)$. Define $K_\epsilon = Z^{-1}(D_\epsilon)$ and $\Omega_\epsilon = \Omega \setminus K_\epsilon$. Let

$$f(x,y) = \frac{w(x,y)}{Z(x,y) - z_0}.$$

We have $f \in C(\overline{\Omega_\epsilon}) \cap C^1(\Omega_\epsilon)$. Hence, by (4.5),

$$\begin{aligned}\int_{\Omega_\epsilon} \frac{Lw(x,y)}{Z(x,y) - z_0} dxdy &= -\int_{\partial\Omega_\epsilon} \frac{w(x,y)}{Z(x,y) - z_0} dZ(x,y) \\ &= -\int_{\partial\Omega} \frac{w(x,y)}{Z(x,y) - z_0} dZ(x,y) + \int_{\partial K_\epsilon} \frac{w(x,y)}{Z(x,y) - z_0} dZ(x,y);\end{aligned} \tag{4.6}$$

Since

$$\int_{\partial K_\epsilon} f(x,y) dZ(x,y) = \int_{Z(\partial K_\epsilon)} (f \circ Z^{-1})(\zeta) d\zeta = \int_{\partial B_\epsilon} \frac{\tilde{w}(\zeta)}{\zeta - z_0} d\zeta,$$

where $\tilde{w} = w \circ Z^{-1}$, then by using polar coordinates $\zeta = z_0 + \epsilon e^{i\theta}$, $\theta \in [0, 2\pi]$, we obtain

$$\int_{\partial D_\epsilon(z_0)} \frac{\tilde{w}(\zeta)}{\zeta - z_0} d\zeta = \int_0^{2\pi} \frac{\tilde{w}(z_0 + \epsilon e^{i\theta})}{\epsilon e^{i\theta}} i\epsilon e^{i\theta} d\theta \to 2\pi i \tilde{w}(z_0), \text{ as } \epsilon \to 0.$$

Therefore,

$$\int_{\partial K_\epsilon} \frac{w(x,y)}{Z(x,y) - z_0} dZ(x,y) \to 2\pi i w(x_0, y_0), \text{ as } \epsilon \to 0. \tag{4.7}$$

On the other hand, as done in the proof of Theorem 4.1,

$$(x,y) \mapsto \frac{1}{Z(x,y) - z_0} \in L^q(\Omega) \subset L^1(\Omega).$$

Hence,

$$\int_{\Omega_\epsilon} \frac{Lw(x,y)}{Z(x,y)-z_0}\,dxdy \to \int_{\Omega} \frac{Lw(x,y)}{Z(x,y)-z_0}\,dxdy, \quad as \quad \epsilon \to 0. \tag{4.8}$$

The proposition follows from (4.6)–(4.8). □

Theorem 4.3 *If $f \in L^1(\Omega)$ then $T_Z f$ satisfies $L(T_Z f) = f$ in Ω.*

Proof Let $f \in L^1(\Omega)$. By Theorem 4.2, $T_Z f \in L^q(\Omega)$, $1 < q < 2 - \sigma/(\sigma+1)$. Hence, by applying Proposition 4.1 we have, for $\phi \in C_0^\infty(\Omega)$

$$\begin{aligned}
\langle L(T_Z f), \phi\rangle &= -\int_\Omega T_Z f(x,y) L\phi(x,y)dxdy \\
&= -\frac{1}{2\pi i}\int_\Omega \left(\int_\Omega \frac{f(\xi,\eta)}{Z(\xi,\eta)-Z(x,y)}\,d\xi d\eta\right) L\phi(x,y)dxdy \\
&= \int_\Omega f(\xi,\eta)\left(-\frac{1}{2\pi i}\int_\Omega \frac{L\phi(x,y)}{Z(x,y)-Z(\xi,\eta)}\,dxdy\right) d\xi d\eta \\
&= \int_\Omega f(\xi,\eta)\phi(\xi,\eta)d\xi d\eta = \langle f, \phi\rangle .
\end{aligned}$$

Therefore, $L(T_Z f) = f$ in Ω. □

Remark 4.4 The condition $f \in L^p$ with $p > 2+\sigma$ is optimal for the existence of bounded solutions. Indeed, for the following simple vector field $L = \partial_t - i|t|^\sigma \partial_x$, the function $v : \overline{\Omega} \to \mathbb{C}$ defined by $v(x,t) = \ln|\ln|Z(x,t)||$ is not bounded but solves $Lv = f$ with

$$f(x,t) = \frac{-i|t|^\sigma}{\overline{Z(x,t)}\,\ln|Z(x,t)|} \in L^p(\Omega), \quad \text{for any} \quad 1 \le p \le 2+\sigma.$$

That $f \in L^p$ with $p \le 2+\sigma$ follows from the fact that $\displaystyle\int_0^a \frac{dr}{r^s|\ln r|^p} < \infty$ if and only if $s \le 1$.

Let $\tau = \dfrac{\sigma}{\sigma+1}$, where σ is defined above. Note that for $p > 2+\sigma$ and $q = \dfrac{p}{p-1}$ we have $q < 2-\tau$. The following theorem establish the Hölder continuity of T_Z.

Theorem 4.5 *Let $f \in L^p(\Omega)$, with $p > 2+\sigma$. Let q and τ be as given above. If u satisfies $Lu = f$ in Ω, then $u \in C^\alpha(\Omega)$ with $\alpha = \dfrac{2-q-\tau}{q}$.*

5 A Similarity Principle

The similarity principle for the Bers–Vekua equation

$$\frac{\partial w}{\partial \overline{z}} = A(x, y)w + B(x, y)\overline{w} \tag{5.1}$$

in an open set $\Omega \subset \mathbb{R}^2$ states that if $A,\ B \in L^p(\Omega)$ with $p > 2$, then the solutions of (5.1) can be represented in Ω as

$$w(x, y) = H(z)\mathrm{e}^{s(x,y)}\,,$$

with $s \in C^\gamma(\Omega)$, $\gamma = (p-2)/p$, and H is a holomorphic function in Ω (see [7, 30]). Thus in many aspects, the solutions of (5.1) have the same behavior as holomorphic functions. This principle has deep consequences on the solutions of first and second order elliptic PDEs. The classical similarity principle for generalized analytic functions was investigated in [5, 17] for solutions of complex vector fields.

In this section we extend the similarity principle to the class of vector fields under consideration. In order to establish this principle, we need first to consider the solvability of the more general semilinear equation

$$Lu = F(x, y, u)\,. \tag{5.2}$$

We assume that L is defined in an open set $\tilde{\Omega} \subset \mathbb{R}^2$. Let Ω be a relatively compact open subset of $\tilde{\Omega}$ and let $\Psi \in L^p(\overline{\Omega}; \overline{\mathbb{R}_+})$, $p > 2+\sigma$ and σ is given by (4.3). We define the space $\mathcal{F}^\alpha_\Psi$ to be the set of functions $F : \overline{\Omega} \times \mathbb{C} \to \mathbb{C}$ satisfying

(a) $F(.,\zeta) \in L^p(\overline{\Omega})$, for every $\zeta \in \mathbb{C}$;
(b) $|F(x,t,\zeta_1) - F(x,t,\zeta_2)| \le \Psi(x,t)|\zeta_1-\zeta_2|^\alpha, 0 < \alpha \le 1$, for all $\zeta_1, \zeta_2 \in \mathbb{C}$.

Throughout this section, we will assume that q is the Hölder conjugate of p, $\tau = \dfrac{\sigma}{\sigma+1}$, and $\beta = \dfrac{2-q-\tau}{q}$. By using the Shauder Fixed Point Theorem, we get the following results.

Theorem 5.1 ([9]) *Let $F \in \mathcal{F}^\alpha_h$. Then:*

(1) If $0 < \alpha < 1$, Eq. (5.2) has a solution $u \in C^\beta(\overline{\Omega})$.
(2) If $\alpha = 1$, for every $(x,t) \in \Omega$, there exist an open subset $\mathcal{U} \subset \Omega$, with $(x,t) \in \mathcal{U}$, such that Eq. (5.2) has a solution $u \in C^\beta(\mathcal{U})$. If, moreover, the constant $M(p,\sigma,\Omega)$ appearing in Theorem 4.1 satisfies $M(p,\sigma,\Omega)||\Psi||_p < 1$, then (5.2) has a solution in $C^\beta(\overline{\Omega})$.

Remark 5.2 It follows at once from this theorem that if $a,\ b,\ f \in L^p_{\mathrm{loc}}(\Omega)$, with $p > 2+\sigma$, then for every $p \in \Omega$, there exists an open set $U \ni p$ such that equation

$$Lu = au + b\overline{u} + f$$

has solution $u \in C^\beta(U)$.

Now, consider F given by

$$F(x, y, u) = g(x, y)H(x, y, u) + f(x, y) \tag{5.3}$$

where $f, g \in L^p(\Omega)$, $p > 2 + \sigma$, and $H : \overline{\Omega} \times \mathbb{C} \to \mathbb{C}$ is continuous and bounded, with $||H||_\infty < K$ for some positive constant K.

Theorem 5.3 ([9]) *Let F be given by* (5.3). *Then, equation $Lu = F(x, y, u)$ has a solution $u \in C^\beta(\overline{\Omega})$.*

As a consequence of Theorems 4.5 and 5.3 we give here a strong version of the similarity principle for the operator L:

Theorem 5.4 ([9]) *Let $a, b \in L^p(\Omega)$, $p > 2 + \sigma$, with $\sigma > 0$ given by (4.3). Then for every $u \in L^\infty(\Omega)$ solution of equation*

$$Lu = au + b\overline{u} \tag{5.4}$$

there exists a holomorphic function h defined in $Z(\Omega)$ and a function $s \in C^\beta(\overline{\Omega})$ such that

$$u(x, y) = h(Z(x, y))e^{s(x,y)}, \quad \forall (x, y) \in \Omega. \tag{5.5}$$

Conversely, for every holomorphic function h in $Z(\Omega)$ there is $s \in C^\beta(\overline{\Omega})$ such that the function u given by (5.5) *solves* (5.4).

Proof The proof is an adaptation of that found in [18]-Theorem 4.1. Suppose that $u \in L^\infty(\Omega)$ and that u is not identically zero. Since L is smooth and elliptic in $\Omega \setminus \Sigma$ we know that L is locally equivalent to a multiple of the Cauchy–Riemann operator $\partial/\partial\overline{z}$ in $\Omega \setminus \Sigma$ (see, for instance, [6]). The classical similarity principle (see [7, 30]) applies and the function u has the representation (5.5) in the neighborhood of each point $(x, y) \notin \Sigma$. Hence, u has isolated zeros in $\Omega \setminus \Sigma$. Define the function ϕ in Ω by $\phi = \overline{u}/u$ at the points where u is not zero and by $\phi = 0$ at the points where $u = 0$ and on Σ. Note that $\phi \in L^\infty(\Omega)$. It follows that $a + b\phi \in L^p(\Omega)$. Consider the equation

$$Ls = -(a + b\phi). \tag{5.6}$$

By Theorem 4.5 this equation has a solution $s \in C^\beta(\overline{\Omega})$. Define $v = ue^s$. A simple calculation shows that $Lv = 0$. Then, v can be factored as $v = h \circ Z$, with h holomorphic on $Z(\Omega)$. This proves the first part of the theorem.

Next, let h be a holomorphic function in $Z(\Omega)$. Define the function φ in $Z(\Omega)$ by $\varphi = \overline{h}/h$ at the points where h is not zero and by $\varphi = 0$ at the points where $h = 0$. Then $\tilde{\varphi} = \varphi \circ Z \in L^\infty(\Omega)$. Consequently, $b\tilde{\varphi} \in L^p(\Omega)$, $p > 2 + \sigma$. Hence, by Theorem 5.3, equation

$$Ls = a + b\tilde{\varphi}e^{\overline{s}-s}$$

has a solution $s \in C^{\beta}(\overline{\Omega})$. It follows at once that u given by

$$u(x, y) = h(Z(x, y))e^{s(x,y)}, \quad (x, y) \in \Omega,$$

solves (5.4) in Ω. $\square$

6 A Boundary Value Problem

The Riemann–Hilbert problem for generalized analytic functions $\overline{\partial} w = aw + b\overline{w}$ in an open set Ω, with a boundary condition $\text{Re}\,(\Lambda w) = \phi$ on $\partial\Omega$ is extensively investigated [4, 12, 21, 26, 30]. It's study is motivated by its relation to other areas in mathematics and its use to model many problems in physical applications [1, 11, 27]. In this section, we consider the analogue problem for our class of hypocomplex vector fields.

First we set the problem in standard form, define the index of a function with respect to a vector field over a closed curve, and modify the Cauchy–Pompeiu operator T_Z into a more convenient operator T_Z^k.

Let L be a hypocomplex vector vector field satisfying conditions (i), (ii), and (iii) given in Sect. 3. Let σ be the associated number given by (4.3). We assume that L is defined in an open set $\tilde{\Omega}$ and that Ω is a simply connected, relatively compact subset of $\tilde{\Omega}$ with a boundary $\partial\Omega$ of class $C^{1+\epsilon}$ with $\epsilon > 0$. We know that L has a global first integral

$$Z_1 : \overline{\Omega} \longrightarrow Z_1(\overline{\Omega}) \subset \mathbb{C}$$

so that Z_1 is global homeomorphism (see Remark 3.10 and [9]). Note that for any holomorphic function h defined in $Z_1(\overline{\Omega})$, the function $h \circ Z_1$ is again a first integral near each point q where $h'(q) \neq 0$. In particular if

$$H : Z_1(\overline{\Omega}) \longrightarrow \overline{\mathbb{D}}$$

is a conformal map onto the unit disc $\mathbb{D}$, then

$$Z : \overline{\Omega} \longrightarrow \overline{\mathbb{D}} \tag{6.1}$$

defined by $Z = H \circ Z_1$ is a global first integral and a global homeomorphism of class $C^{1+\epsilon}$.

Remark 6.1 For $L = A\partial_x + B\partial_y$, the Jacobian determinant J_Z, of its first integral Z, satisfies

$$\text{sgn}\,(J_Z) = -\text{sgn}\,\big(\text{Im}(A\overline{B})\big),$$

where $\text{sgn}(x) = 1$ if $x > 0$ and $\text{sgn}(x) = -1$ if $x < 0$. Note that under our hypotheses, the quantity $\text{Im}(A\overline{B})$ does not change sign in Ω. Hence, any first integral of L preserves orientation if $\text{Im}(A\overline{B}) \leq 0$ and reverses it if $\text{Im}(A\overline{B}) \geq 0$.

Let Γ be a smooth closed curve in $\tilde{\Omega}$. Let $\varphi \in C(\Gamma; \partial\mathbb{D})$, we denote by $\text{Wind}_\Gamma\varphi$ the winding number of φ. In particular, $\text{Wind}_\Gamma Z = -\text{sgn}\left(\text{Im}(A\overline{B})\right)$. Define the index of $\Lambda \in C(\Gamma; \partial\mathbb{D})$ over Γ with respect to the vector field L as

$$\text{Ind}_{L,\Gamma}\Lambda = \text{Wind}_\Gamma Z \cdot \text{Wind}_\Gamma\Lambda = -\text{sgn}\left(\text{Im}(A\overline{B})\right)\text{Wind}_\Gamma\Lambda$$

Thus, $\text{Ind}_{L,\Gamma}\Lambda$ coincides with $\text{Wind}_\Gamma\Lambda$ when Z is orientation preserving and are opposite when Z is orientation reversing.

Let $\kappa \in \mathbb{Z}^+$ and Z as in (6.1). We define the operator T_Z^κ by

$$T_Z^\kappa f(x,y) = \frac{1}{2\pi i}\int_\Omega \left\{\frac{f(\xi,\eta)}{Z(\xi,\eta) - Z(x,y)} - Z^{2\kappa}(x,y)\frac{Z(x,y)\overline{f(\xi,\eta)}}{1 - Z(x,y)\overline{Z(\xi,\eta)}}\right\} d\xi d\eta. \tag{6.2}$$

The following theorem is proved in [8].

Theorem 6.2 ([8]) *Let $f \in L^p(\Omega)$, with $p > 2 + \sigma$, and let $\beta = \dfrac{2 - q - \tau}{q}$. Then $L(T_Z^\kappa f) = f$ and there exist positive constants $M = M(p,\sigma,\Omega)$ and $C = C(p,\sigma,\Omega)$ such that*

$$|T_Z^\kappa f(x,y)| \leq M\,||f||_p, \quad \forall (x,y) \in \overline{\Omega} \tag{6.3}$$

and

$$|T_Z^\kappa f(p_1) - T_Z^\kappa f(p_2)| \leq C\,||f||_p\,|Z(p_1) - Z(p_2)|^\beta, \quad \forall p_1, p_2 \in \overline{\Omega}. \tag{6.4}$$

We use the operator T_Z^κ together with the Fredholm alternative to establish solvability of the Riemann–Hilbert problem.

$$\begin{cases} Lu = au + b\overline{u} + f & \text{in } \Omega \\ \text{Re}(\overline{\Lambda}u) = \phi & \text{on } \partial\Omega \end{cases}, \tag{6.5}$$

with $a, b, f \in L^p(\Omega)$, $p > 2 + \sigma$, $\Lambda \in C^\alpha(\partial\Omega)$, $|\Lambda| = 1$, $\phi \in C^\alpha(\partial\Omega; \mathbb{R})$, $0 < \alpha < 1$.

First we start by the following property of T_Z^κ.

Proposition 6.1 ([8]) *Let $f \in L^p(\Omega)$, with $p > 2 + \sigma$. Then, $T_Z^\kappa f$ (given by (6.2)) satisfies*

$$\begin{cases} LT_Z^\kappa f = f \text{ in } \Omega \\ Re(\overline{Z^\kappa}\,T_Z^\kappa f) = 0 \text{ on } \partial\Omega. \end{cases} \tag{6.6}$$

Proof Let $(x,y) \in \overline{\Omega}$. We can write

$$T_Z^{\kappa} f(x, y) = T_Z f(x, y) - Z^{2\kappa}(x, y) B_Z f(x, y), \tag{6.7}$$

where

$$B_Z f(x, y) \doteq \frac{1}{2\pi i} \int_{\Omega} \frac{Z(x, y)\overline{f(\xi, \eta)}}{1 - Z(x, y)\overline{Z(\xi, \eta)}} d\xi d\eta.$$

We already know from Theorem 6.2 that $L(T_Z^{\kappa} f) = f$. We need only verify the boundary condition. It follows from (6.7) and from $|Z^{\kappa}| = 1$ on $\partial\Omega$ that for $(x, y) \in \partial\Omega$, we have

$$\overline{Z^{\kappa}} T_Z^{\kappa} f(x, y) = \overline{Z^{\kappa}} T_Z f(x, y) - Z^{\kappa} B_Z f(x, y),$$

and

$$B_Z f(x, y) = \overline{\frac{1}{2\pi i} \int_{\Omega} \frac{f(\xi, \eta)}{Z(\xi, \eta) - Z(x, y)} d\xi d\eta} = \overline{T_Z f(x, y)}.$$

Thus $\overline{Z^{\kappa}}\, T_Z^{\kappa} f(x, y) = 2i\mathrm{Im}(\overline{Z^{\kappa}} T_Z f(x, y))$ and the proposition follows. □

Next we use a Schwarz operator for the vector field L to the particular boundary value problem

$$\begin{cases} L(\Phi) = 0 \text{ in } \Omega \\ \mathrm{Re}(\overline{Z^{\kappa}}\Phi) = \varphi \text{ on } \partial\Omega \end{cases} \tag{6.8}$$

Define the Schwarz operator on $\overline{\Omega}$ by

$$(S_Z \varphi)(x, y) \doteq \frac{1}{2\pi i} \int_{\partial\Omega} \varphi(\xi, \eta) \frac{Z(\xi, \eta) + Z(x, y)}{Z(\xi, \eta) - Z(x, y)} \frac{dZ(\xi, \eta)}{Z(\xi, \eta)}. \tag{6.9}$$

Proposition 6.2 ([8]) *For $\varphi \in C(\partial\Omega; \mathbb{R})$ and κ a nonnegative integer, the general solution of problem* (6.8) *is of the form*

$$\Phi(x, y) = Z^{\kappa}(x, y)(S_Z \varphi)(x, y) + \sum_{j=0}^{2\kappa} b_j Z^j(x, y), \tag{6.10}$$

with $b_j \in \mathbb{C}$, $j = 0, 1, \ldots, 2\kappa$, and $b_{2\kappa - j} = -\overline{b}_j$. Moreover, if $\varphi \in C^{\alpha}(\partial\Omega; \mathbb{R})$, $0 < \alpha < 1$, then $\Phi \in C^{\alpha}(\partial\Omega; \mathbb{C})$.

Proof We know from the classical theory of the Riemann–Hilbert problem (see [4, 12, 21, 30]) that the general solution of the problem

$$\frac{\partial h}{\partial \overline{\zeta}} = 0 \text{ in } \mathbb{D} \setminus \{0\} \text{ and } \mathrm{Re}(h) = \psi \text{ on } \partial\mathbb{D}$$

(with $\psi \in C(\partial\mathbb{D}, \mathbb{R})$) having a pole at most of order κ at $\zeta = 0$ has the form

$$h(\zeta) = \frac{1}{2\pi i}\int_{\partial\mathbb{D}} \psi(\omega)\frac{\omega+\zeta}{\omega-\zeta}\frac{d\omega}{\omega} + ic_0 + \sum_{j=1}^{\kappa}(c_j\zeta^j - \overline{c_j}\zeta^{-j}),$$

with $c_j \in \mathbb{C}$, $c_0 \in \mathbb{R}$. It can be seen at once that for $\psi(\zeta) = \varphi \circ Z^{-1}(\zeta)$, the general solution of (6.8) is

$$\Phi(x, y) = Z(x, y)^{\kappa} h(Z(x, y))$$

and has the form (6.10). □

To continue, we need a compactness property of an operator associated to T_Z^{κ}. For $a, b \in L^p(\Omega)$ with $p > 2+\sigma$, define the operator G on the Banach space $C^0(\overline{\Omega})$ (equipped with the supremum norm) by

$$Gu = T_Z^{\kappa}(au + b\overline{u}) . \tag{6.11}$$

We have the following:

Proposition 6.3 *The operator* $G : C^0(\overline{\Omega}) \longrightarrow C^0(\overline{\Omega})$ *is compact and*

$$Ker(I - G) = \{0\} .$$

The proof of these properties can be found in [8].
As a consequence we get the solvability of the following problem.

$$\begin{cases} Lu = au + b\overline{u} + f & \text{in } \ \Omega \\ \mathrm{Re}(\overline{Z^{\kappa}}\, u) = \varphi & \text{on } \ \partial\Omega \end{cases} . \tag{6.12}$$

Indeed, problem (6.12) can be transformed into solving the integral equation

$$u = G(u) + T_Z^{\kappa} f + \Phi$$

where Φ satisfies $L\Phi = 0$ in Ω and $\mathrm{Re}(\overline{Z^{\kappa}}\,\Phi) = \varphi$ on $\partial\Omega$. Thanks to Proposition 6.3, the Fredholm alternative give the existence of a solution.

Proposition 6.4 ([8]) *Let* $a, b, f \in L^p(\Omega)$ *and* $\phi \in C(\partial\Omega, \mathbb{R})$. *The boundary value problem* (6.12) *has a solution* $u \in C^{\alpha_0}(\overline{\Omega})$, *with* $\alpha_0 = \min\{\alpha, (2 - q - \tau)/q\}$, *where* $q = p/(p-1)$.

Finally, for the general Riemann–Hilbert problem (6.5) we have the following theorem.

Theorem 6.3 ([8]) *Let* $a, b, f \in L^p(\Omega)$, $p > 2+\sigma$, $\Lambda \in C^{\alpha}(\partial\Omega)$, $|\Lambda| = 1$, $\phi \in C^{\alpha}(\partial\Omega; \mathbb{R})$, $0 < \alpha < 1$. *If* $Ind_{L,\partial\Omega}\Lambda \geq 0$, *then the Riemann–Hilbert problem (6.5) has a solution* $u \in C^{\alpha_0}(\overline{\Omega})$ *with* α_0 *as in Proposition 6.4.*

Proof Let $\kappa = \mathrm{Ind}_{L,\partial\Omega}\Lambda \geq 0$. We use the standard techniques of the classical Riemann–Hilbert problem (see [4]) to transform the boundary function Λ into the function Z^κ. For this we make use of the Schwarz operator S_Z defined in (6.9). Define $\vartheta \in C^\alpha(\partial\Omega)$ by

$$\vartheta(x, y) = \arg\left(Z^{-\kappa}(x, y)\Lambda(x, y)\right)$$

and $\gamma \in C^\alpha(\overline{\Omega})$ by

$$\gamma(x, y) = (S_Z(\vartheta))(x, y).$$

Then $L\gamma = 0$ in Ω and $\mathrm{Re}(\gamma) = \vartheta$ on $\partial\Omega$. Hence,

$$e^{-i\gamma} = e^{\mathrm{Im}(\gamma)+i\kappa\arg(Z)-i\arg(\Lambda)} = e^{\mathrm{Im}(\gamma)}Z^\kappa\overline{\Lambda}. \tag{6.13}$$

Let $\tilde{b} = be^{-2\mathrm{Im}(\gamma)}$ and $\tilde{f} = fe^{-i\gamma}$. Then $\tilde{b}, \tilde{f} \in L^p(\Omega)$, $p > 2+\sigma$ and from Proposition 6.4 the problem

$$\begin{cases} LU = aU + \tilde{b}\overline{U} + \tilde{f} & \text{in } \Omega \\ \mathrm{Re}(\overline{Z^\kappa}U) = e^{\mathrm{Im}(\gamma)}\phi & \text{on } \partial\Omega \end{cases} \tag{6.14}$$

has solution $U \in C^{\alpha_0}(\overline{\Omega})$, where $\alpha_0 = \min\{\alpha, (2-q-\tau)/q\}$. It is seen at once that the function $u = e^{i\gamma}U$ in $\overline{\Omega}$ satisfies (6.5). □

7 Cauchy–Kovalevski Theorem in a Hypocomplex Structure

The two dimensional Cauchy–Kovalevski Theorem guarantees the existence of a local solution of the initial value problem

$$\begin{cases} \dfrac{\partial u}{\partial t} = A(t, z, u)\dfrac{\partial u}{\partial z} + B(t, z, u) \\ \\ u(0, z) = \phi(z) \end{cases}$$

provided the data are continuous and holomorphic with respect to z and u. In general the Cauchy problem is not well posed in the elliptic setting. There is an interesting example in [13] that shows that $u_t + uu_x = 0$, $u(0, x) = f(x)$, with f a C^∞, has no solutions. More recently it was proved in [14] that the Cauchy–Kovalevski of a quasilinear operators is unstable with respect to initial data under C^∞-perturbations. This result was generalized in [2] to fully nonlinear operators. Initial value problems for generalized analytic functions are considered in [29].

In this section, we investigate the Cauchy problem when the data and solutions are in the kernel of a vector field. More precisely, we study the Cauchy problem when $\overline{\partial}_z$ is replaced by a vector field L and ∂_z is replaced by another vector field T that keeps invariant the kernel of L. General versions of the Cauchy problem in hypoanalytic structures will be considered separately in a forthcoming paper. The initial value problem we consider here is therefore

$$\begin{cases} \dfrac{\partial u}{\partial t} = p(t,x,y)\,Tu + q(t,x,y,u) \\ \\ u(0,x,y) = \phi(x,y) \end{cases}$$

where $Lp = Lq = L\phi = 0$ and the solutions u is sought in $\mathrm{Ker}(L)$. It should be noted right away that without further hypotheses, the initial value problem is not solvable as the following example shows.

Example 7.1 In $\mathbb{R}\times\mathbb{R}^2$ consider the problem

$$\begin{cases} \dfrac{\partial u}{\partial t} = it\dfrac{\partial u}{\partial x} + f(t,x) \\ u(0,x,y) = 0 \end{cases}$$

Since $\partial_t - it\partial_x$ is a Mizohata operator, there are C^∞ functions $f(t,x)$ such that the problem has no distribution solution (see Sect. 2).

Let Ω be an open subset in $\mathbb{R}^2$, $Z \in C^\infty(\overline{\Omega})$ such that $Z : \overline{\Omega} \longrightarrow Z(\overline{\Omega}) \subset \mathbb{C}$ is a homeomorphism, $dZ(p) \neq 0$ for every $p \in \overline{\Omega}$, and such that the vector field

$$L = Z_x\frac{\partial}{\partial y} - Z_y\frac{\partial}{\partial x}$$

is hypocomplex and satisfies conditions (i), (ii), and (iii) of Sect. 2. We start by a characterisation of the vector fields that keep invariant $\mathrm{Ker}(L)$. First since dZ is nowhere zero, then there is a C^∞ vector field T defined on $\overline{\Omega}$ such that $TZ = 1$ in $\overline{\Omega}$.

Lemma 7.1 *A C^∞ vector field $\tilde{T}$ in $\overline{\Omega}$ is such that*

$$\tilde{T} : Ker(L) \longrightarrow Ker(L)$$

if and only if there exist $P,\ Q \in C^\infty(\overline{\Omega})$ with $P \in Ker(L)$ such that

$$\tilde{T} = PT + QL$$

where $TZ = 1$ in Ω.

Proof Since T and L are independent, then there exist $P, Q \in C^\infty(\overline{\Omega})$ such that $\tilde{T} = PT + QL$. If $u \in C^0(\overline{\Omega}) \cap \mathrm{Ker}(L)$, then it follows from the hypotheses on L that $u = H \circ Z$ for some holomorphic function H in $Z(\Omega)$. Therefore $Tu = H'(Z)$ and $L\tilde{T}u = LP\,H'(Z) = 0$ holds for any H if and only if $P \in \mathrm{Ker}(L)$. □

Now we proceed as in [29, 31] and define a cone shaped domain over Ω on which the initial value problem will be solved. Let $\eta > 0$ and for $(x, y) \in \Omega$, let $D(x, y) = \text{dist}((x, y), \partial\Omega)$. Define an open set Ω^η by

$$\Omega^\eta = \{(t, (x, y)) \in \mathbb{R} \times \Omega : |t| < \eta D(x, y)\}$$

For $t \in \mathbb{R}$, let

$$D_t(x, y) = D(x, y) - \frac{|t|}{\eta}$$

and Ω_t^η be the t-slice in Ω^η:

$$\begin{aligned} \Omega_t^\eta &= \{(x, y) \in \Omega : (t, (x, y)) \in \Omega^\eta\} \\ &= \{(x, y) \in \Omega : D_t(x, y) > 0\} \end{aligned}$$

Let $A,\ B,\ C \in C^\infty(\Omega^\eta, \mathbb{C})$ be such that for every t,

$$LA(t, x, y) = LB(t, x, y) = LC(t, x, y) = 0 \tag{7.1}$$

Consider the linear Cauchy problem

$$\begin{cases} \dfrac{\partial u}{\partial t}(t, x, y) = A(t, x, y)\, Tu + B(t, x, y)u + C(t, x, y) \\ u(0, x, y) = \psi(x, y) \end{cases} \tag{7.2}$$

We have the following theorem:

Theorem 7.2 *Let $A,\ B,\ C \in C^\infty(\Omega^\eta, \mathbb{C})$ satisfying (7.1). Suppose that the initial data ψ satisfies $L\psi = 0$ in Ω. and that there exist positive constants p, M_A, M_B, M_C, and M_ϕ such that for every $m = (t, x, y) \in \Omega^\eta$*

$$|A(m)| \le M_A\,, \quad |B(m)|D_t(x, y) \le M_B\,, \quad |C(m)|D_t(x, y)^{p+1} \le M_C$$

and for every $(x, y) \in \Omega$,

$$|\phi(x, y)|D(x, y)^p \le M_\phi\,.$$

Then there is a positive constant K_p such that the Cauchy problem (7.2) has a unique solution in Ω^η with $Lu = 0$ for every t, provided that

$$\eta < \frac{p}{K_p M_A + M_B}$$

For the proof we need the following lemma.

Lemma 7.2 *Suppose that w satisfies $Lw = 0$ in an open set U and there are constants $c > 0$ and $p > 0$ such that*

$$|w(x, y)| < \frac{c}{\text{dist}\,((x, y), \partial U)^p}$$

Then

$$|Lw(x, y)| < \frac{c\,K_p}{\text{dist}\,((x, y), \partial U)^{p+1}} \tag{7.3}$$

for some $K_p > 0$ independent on w.

Proof This lemma is a consequence of Nagumo's Lemma and the hypocomplexity of the vector field L. Indeed from Nagumo' Lemma (see [31]), we know that if H is a holomorphic function in a domain V then

$$|H(z)| \le \frac{c}{\text{dis}(z, \partial V)^p} \Longrightarrow |\frac{\partial H(z)}{\partial z}| \le \frac{c}{\text{dis}(z, \partial V)^{p+1}}.$$

Now estimate (7.3) follows from the fact that any solution w of $Lw = 0$ in U can be expressed as $H \circ Z$ with H holomorphic in $Z(U)$. □

Proof of Theorem. First note that

$$\left|\int_0^t \frac{ds}{D_s(x, y)^{p+1}}\right| \le \int_0^{|t|} \frac{ds}{\left(D(x, y) - \frac{|s|}{\eta}\right)^{p+1}} \le \frac{\eta}{p}\frac{1}{D_t(x, y)^p} \tag{7.4}$$

Rewrite problem (7.2) as the integral equation

$$u(t, x, y) = Qu(t, x, y) = f(t, x, y) + Pu(t, x, y) \tag{7.5}$$

with

$$f(t, x, y) = \phi(x, y) + \int_0^t C(s, x, y)\,ds \quad \text{and}$$

$$Pu(t, x, y) = \int_0^t (A(s, x, y)\,Tu(s, x, y) + B(s, x, y)u(s, x, y))\,ds$$

Let $\mathcal{S}$ be the space of functions $u \in C^0(\Omega^\eta, \mathbb{C}^m)$ such that for every t, $u(t, .) \in \text{Ker}(L)$ in Ω_t^η and

$$||u|| := \sup_{\Omega^\eta} |u(t, x, y)|\, D_t(x, y)^p < \infty$$

Then $(\mathcal{S}, ||\,.\,||)$ is a Banach space. It follows from (7.4) that

$$\left|\int_0^t C(s,x,y)\,ds\right| \le M_C \left|\int_0^t \frac{ds}{D_s(x,y)^{p+1}}\right| \le \frac{\eta M_C}{pD_t(x,y)^p}.$$

This together with

$$|\phi(x,y)| \le \frac{M_\phi}{D(x,y)^p} \le \frac{M_\phi}{D_t(x,y)^p}$$

imply that $f = \phi + \int_0^t C\,ds \in \mathcal{S}$. Note that if $u \in \mathcal{S}$ then

$$|u(s,x,y)| \le \frac{||u||}{D_s(x,y)^p}$$

Hence,

$$|B(s,x,y)u(s,x,y)| \le \frac{M_B\,||u||}{D_s(x,y)^{p+1}}$$

By Lemma 7.2 we have

$$|A(s,x,y)Lu(s,x,y)| \le \frac{M_A K\,||u||}{D_s(x,y)^{p+1}}.$$

Estimate (7.4) shows that

$$|Pu(t,x,y)| \le \int_0^{|t|} \frac{(M_A K_p + M_B)\,||u||}{D_s(x,y)^{p+1}}\,ds$$

$$\le \frac{\eta(M_A K_p + M_B)}{p\,D_t(x,y)^p}||u||$$

Therefore $P : \mathcal{S} \longrightarrow \mathcal{S}$ and

$$||P|| \le \eta\,\frac{M_A K_p + M_B}{p}$$

is a contraction if

$$\eta < \frac{p}{K_p M_A + M_B}$$

and the conclusion of the theorem follows. □

References

1. Ablowitz, M.J., Clarkson, P.A.: Solitons, Nonlinear Evolution Equations and Inverse Scattering. London Mathematical Society Lecture Note, vol. 149. Cambridge University Press, New York

(1991)
2. Adwan, Z., Berhanu, S.: On microlocal analyticity and smoothness of solutions of first-order nonlinear PDEs. Math. Ann. **352**(1), 239258 (2012)
3. Baouendi, M.S., Treves, F.: A local constancy principle for the solutions of certain of certain overdetermined systems of first-order linear partial differential equations. Advances in Mathematics Supplementary Studies, vol. 7a, pp. 245–262. Academic Press, New York (1981)
4. Begehr, H.: Complex Analytic Methods for Partial Differential Equations. An Introductory Text. World Scientific Publishing, Singapore (1994)
5. Berhanu, S., Hounie, J., Santiago, P.: A similarity principle for complex vector fields and applications. Trans. Am. Math. Soc. **353**(4), 16611675 (2001)
6. Berhanu, S., Cordaro, P., Hounie, J.: An Introduction to Involutive Structures. New Mathematical Monographs, vol. 6. Cambridge University Press, Cambridge (2008)
7. Bers, L.: An outline of the theory of pseudoanalytic functions. Bull. AMS **62**, 291–331 (1956)
8. Campana, C., Meziani, A.: Boundary value problems for a class of planar complex vector fields. J. Differ. Equ. **261**(10), 56095636 (2016)
9. Campana, C., Dattori, P., Meziani, A.: Properties of solutions of a class of hypocomplex vector fields. Analysis and Geometry in Several Complex Variables. Contemporary Mathematics, vol. 2950. American Mathematical Society, Providence (2017)
10. Campana, C., Dattori, P., Meziani, A.: Riemann-Hilbert problem for a class of hypocomplex vector fields. Complex Var. Elliptic Equ. **62**(10), 16561667 (2017)
11. Freund, L.B.: Dynamic Fracture Mechanics. Cambridge University Press, New York (1990)
12. Gakhov, F.D.: Boundary Value Problems. Dover Publications Inc, New York (1990)
13. Hounie, J., Dos Santos Filho, J.R.: Well-posed Cauchy problems for complex nonlinear equations must be semilinear. Math. Ann. **294**(3), 439447 (1992)
14. Lerner, N., Morimoto, Y., Xu, C.-J.: Instability of the Cauchy-Kovalevskaya solution for a class of nonlinear systems. Am. J. Math. **132**(1), 99123 (2010)
15. Lewy, H.: An example of a smooth linear partial differential equation without solution. Ann. Math. **66**(2), 155158 (1957)
16. Meziani, A.: The Mizohata complex. Trans. Am. Math. Soc. **349**(3), 10291062 (1997)
17. Meziani, A.: On the similarity principle for planar vector fields: application to second order PDE. J. Differ. Equ. **157**(1), 119 (1999)
18. Meziani, A.: Representation of solutions of planar elliptic vector fields with degeneracies. Geometric Analysis of PDE and Several Complex Variables. Contemporary Mathematics, vol. 368, pp. 357–370. American Mathematical Society, Providence (2005)
19. Meziani, A.: On first and second order planar elliptic equations with degeneracies. Mem. Am. Math. Soc. **217**(1019) (2012)
20. Meziani, A., Ainouz, A., Boutarene, K.: The Riemann-Hilbert problem for elliptic vector fields with degeneracies. Complex Var. Elliptic Equ. **59**(6), 751768 (2014)
21. Muskhelishvili, N.I.: Singular Integral Equations. Boundary Problems of Function Theory and Their Application to Mathematical Physics. Dover Publications, Inc., New York (1992)
22. Nirenberg, L.: Lectures on Linear Partial Differential Equations. Expository Lectures, CBMS Regional Conference, vol. 17. AMS, Providence (1973)
23. Nirenberg, L., Treves, F.: Solvability of a first order linear partial differential equation. Commun. Pure Appl. Math. **16**, 331–351 (1963)
24. Nirenberg, L., Treves, F.: On local solvability of linear partial differential equations. I. Necessary conditions. Commun. Pure Appl. Math. **23**, 1–38 (1970)
25. Nirenberg, L., Treves, F.: On local solvability of linear partial differential equations. II. Sufficient conditions. Commun. Pure Appl. Math. **23**, 459–508 (1970)
26. Rodin, Y.L.: Generalized Analytic Functions on Riemann Surfaces. Lecture Notes in Mathematics, vol. 1288. Springer, Berlin (1980)
27. Rodin, Y.L.: The Riemann boundary value problem on Riemann surfaces and the inverse scattering problem for the Landau-Lifschitz equation. Physica **11D**, 90–108 (1984)
28. Treves, F.: Hypo-analytic Structures: Local Theory. Princeton Mathematical Series, vol. 40. Princeton University Press, Princeton (1992)

29. Tutschke, W.: Solution of Initial Value Problems in Classes of Generalized Analytic Functions. The Method of Scales of Banach Spaces. With German, French and Russian Summaries. Teubner-Texte zur Mathematik [Teubner Texts in Mathematics], vol. 110. BSB B. G. Teubner Verlagsgesellschaft, Leipzig (1989)
30. Vekua, I.V.: Generalized Analytic Functions. Pergamon Press, Oxford (1962)
31. Walter, W.: An Elementary Proof of the Cauchy-Kovalevsky Theorem. The American Mathematical Monthly, vol. 92(2), pp. 115–126. Mathematical Association of America, Washington (1985)

Almost-Positivity Estimates of Pseudodifferential Operators

Alberto Parmeggiani

Abstract In this paper I will give a survey on a priori estimates such as the Gårding, Sharp-Gårding, Melin, Hörmander and the Fefferman–Phong inequalities for pseudodifferential operators, discuss some generalizations and open problems in some directions. Finally, I will describe what is known at present in the case of systems of pseudodifferential operators, the latter being a still largely open and unexplored area.

Keywords Lower bound estimates · Almost-positivity

2010 Mathematics Subject Classification Primary 35B45 · Secondary 35A30 · 35S05

1 Introduction

It is well-known that a priori estimates such as the Gårding, Sharp-Gårding, Melin, Hörmander and the Fefferman–Phong inequalities play a central role in a wide range of problems such as, just to mention a few of them, the Friedrich extension (of a partial differential operator, scalar or system), spectral estimates, energy estimates for the well-posedness of the Cauchy problem for parabolic, Schrödinger or hyperbolic type equations (especially in presence of degeneracy, e.g. multiple characteristics in the hyperbolic case), location and propagation of singularities (Schrödinger and hyperbolic types), local and global solvability of PDEs, hypoellipticity and subellipticity.

In this paper I will in the first place recall the mentioned inequalities and explain the influence of the geometry of the characteristic set and of the symbol, especially

I wish to thank the organizers of the ISAAC conference 2017 held in Växjö (Sweden), and in particular Luigi Rodino and Joachim Toft

A. Parmeggiani (✉)
Department of Mathematics, University of Bologna, Piazza di Porta San Donato 5, 40126 Bologna, Italy
e-mail: alberto.parmeggiani@unibo.it

L. G. Rodino and J. Toft (eds.), *Mathematical Analysis and Applications—Plenary Lectures*, Springer Proceedings in Mathematics & Statistics 262,
https://doi.org/10.1007/978-3-030-00874-1_4

of the lower-order terms, in the estimates. I will then discuss generalizations, and open problems, in some directions (higher characteristics, relaxation of the geometric assumptions) and finally describe what is known at present about systems of pseudodifferential operators. I will focus mainly on the various generalizations in the scalar and system cases I have been working on, but I will mention the other generalizations I am aware of, addressing the reader to the bibliography for references.

As already recalled above, many areas of the geometric analysis of PDEs and its applications use as a fundamental tool almost-positivity estimates, among which the most famous are the *Gårding* and the *Sharp Gårding* inequalities, the *Melin* and the *Hörmander* inequalities, the *Fefferman–Phong* inequality. They all give, to various degrees of requirements on the operators, a first fundamental information that allows one to carry out a further and deeper study of a test inequality or of the properties of the solutions to a linear PDE or a linear system of PDEs (but, of course, also in case of nonlinear PDEs by going through the analysis of the linearized equation).

I have to establish some notation that I will be using throughout. In what follows, with $X \subset \mathbb{R}^n$ denoting an open set and $m \in \mathbb{R}$, I will write $\Psi^m_{\mathrm{cl}}(X)$ for the set of properly-supported classical pseudodifferential operators (ψdos) of order m and, when g is a Hörmander metric on $\mathbb{R}^n \times \mathbb{R}^n$ and h the corresponding Planck's function, $S(h^m, g)$ will denote the set of symbols with growth controlled by the admissible weight function h^m. The cotangent bundle of X with the zero-section removed is denoted by $T^*X \setminus 0$. In general, given $A, B > 0$ I will say that $A \lesssim B$, respectively that $A \gtrsim B$, when there exists $C > 0$ such that $A \leq CB$, respectively $A \geq CB$. I will then say that $A \approx B$ when $A \lesssim B$ and $A \gtrsim B$. The usual L^2-scalar product will be denoted by $(\cdot, \cdot)$ and the norm of the Sobolev space $H^s(\mathbb{R}^n)$ of order s will be denoted by $\|\cdot\|_s$.

Now, given a formally self-adjoint pseudodifferential operator $P = P^*$ of order m, an *almost-positivity* estimate is generically of the following kind:

$$(Pu, u) \geq -C\|u\|_s^2, \ \forall u \in C_0^\infty, \tag{1.1}$$

where the exponent $s \in \mathbb{R}$ is related to P. Note that when P has order m the case $s = m/2$ is trivial by the usual continuity of P between Sobolev spaces. Since we may rewrite (1.1) as

$$\exists s \in \mathbb{R}, \ \exists C > 0, \ \ P + C|D|^{2s} \geq 0 \ \text{ on } \ C_0^\infty, \tag{1.2}$$

it is clear that when $s < m/2$ the existence of such an almost-positivity estimate depends on the structure of the symbol of P and on the kind of "lower order terms" of the kind $|D|^{2s}$ that may be added to P to make it nonnegative. As I will show below, such an estimate is deeply connected to the characteristic set of P (equivalently, the set where P is not elliptic) and to the structure of its symbol. They influence the number s of derivatives we may control through the L^2-scalar product. The structure of this influence will in turn make (hopefully) clear the choice of some of the tools used in the generalizations.

After recalling the above mentioned estimates, I will be discussing generalizations in several possible directions mainly of the Hörmander and of the Fefferman–Phong inequalities, namely

- to the case of operators with characteristics higher than 2;
- to cases in which one may relax the main geometric hypotheses;
- to the case of systems.

As already said, the generalizations I will be describing are a selection of those I have been working on. But I will also mention other important generalizations due to other authors, referring to them in the bibliography, that I have tried to make as complete as possible; I apologize for any omissions.

2 The Main Almost-Positivity Estimates

In this section I recall the fundamental estimates I have mentioned in the Introduction. I will always assume the order $m > 0$.

Let $X \subset \mathbb{R}^n$ be open and let $P = P^* \in \Psi^m_{\mathrm{cl}}(X)$. Let p_m be the principal symbol of P. The first inequality, when P is a positive-elliptic operator (that is, its principal symbol p_m is positive elliptic) is the celebrated Gårding inequality, that was proved by Lars Gårding in [11].

Theorem 2.1 *The condition that $p_m > 0$ on $T^*X \setminus 0$ is equivalent to the inequality: For any given compact $K \subset X$ there exist constants $c_K, C_K > 0$ such that*

$$(Pu, u) \geq c_K \|u\|^2_{m/2} - C_K \|u\|^2_{(m-1)/2}, \quad \forall u \in C_0^\infty(K). \tag{2.1}$$

A consequence of (2.1) is that all coercive problems related to a positive partial differential elliptic operator with smooth coefficients can therefore be solved by Hilbert space methods. A natural question then immediately arizes: *What happens when p_m fails to be (positive) elliptic and is allowed to vanish somewhere in $T^*X \setminus 0$?* Lars Hörmander (see [15]) answered that question by proving the Sharp Gårding inequality.

Theorem 2.2 *The condition that $p_m \geq 0$ on $T^*X \setminus 0$ is equivalent to the inequality: For any given compact $K \subset X$ there exists a constant $C_K > 0$ such that*

$$(Pu, u) \geq -C_K \|u\|^2_{(m-1)/2}, \quad \forall u \in C_0^\infty(K). \tag{2.2}$$

Note that in the preceding statements, only the principal part is called into the game. It is important to remark that:

- In both results there is no condition on the characteristic set $\Sigma := p_m^{-1}(0) \subset T^*X \setminus 0$ of P and on the lower order terms in the (total) symbol (remark that $\Sigma = \emptyset$ in the elliptic case);

- The requirements $p_m > 0$, resp. $p_m \geq 0$, on $T^*X \setminus 0$ are seen to be necessary for (2.1), resp. (2.2), to hold as follows. Given any $(x_0, \xi_0) \in T^*X \setminus 0$, one proves the necessity by testing the inequality on the family $u_t(x) = e^{it^2\langle x, \xi_0\rangle} v(t(x - x_0))$, $t > 0$, $v \in C_0^\infty(X)$, and letting $t \to +\infty$.

Before studying the influence of the lower order terms of the symbol of P, we have to recall some basic symplectic invariants, in terms of which one states the geometric assumptions. Let σ be the standard symplectic form on $T^*X \setminus 0$.

- The *subprincipal symbol* of P is defined to be

$$p^s_{m-1}(x, \xi) = p_{m-1}(x, \xi) + \frac{i}{2} \sum_{j=1}^{n} \partial^2_{x_j \xi_j} p_m(x, \xi), \quad (x, \xi) \in T^*X \setminus 0.$$

It is *invariant at the zeros of order* 2 of p_m. Using the framework of 1/2-densities gives that p^s_{m-1} is actually a full invariant of the symbol, along with the principal symbol p_m, but I have preferred to avoid such a context (which is however needed when dealing with the inequalities, and PDE problems, on a smooth manifold, or when dealing with the global theory of Fourier integral operators).
- The *Hamilton vector field* H_{p_m} of P is the vector field on $T^*X \setminus 0$ defined by

$$H_{p_m} = \sum_{j=1}^{n} \Big(\frac{\partial p_m}{\partial \xi_j} \frac{\partial}{\partial x_j} - \frac{\partial p_m}{\partial x_j} \frac{\partial}{\partial \xi_j} \Big).$$

- The *Hamilton map* $F(\rho)$ of P at the characteristic points $\rho \in \Sigma$ of order 2 of p_m is the linearization of H_{p_m} at ρ and is defined by polarization by

$$\sigma(v, F(\rho)w) = \frac{1}{2} \langle \mathrm{Hess}(p_m)(\rho) v, w \rangle, \quad v, w \in T_\rho T^*X.$$

- The *positive trace* of F at $\rho \in \Sigma$ is defined to be

$$\mathrm{Tr}^+ F(\rho) = \sum_{\substack{\mu > 0 \\ i\mu \in \mathrm{Spec}(F(\rho))}} \mu.$$

- Note that when $p_m \geq 0$ on $T^*X \setminus 0$ then $p_m = 0 \Rightarrow dp_m = 0$, whence on Σ both p^s_{m-1} and F are invariantly defined, and therefore, since the spectrum of F is made in this case of the only generalized eigenvalue 0 and of the eigenvalues $\pm i\mu$, $\mu > 0$, also $\mathrm{Tr}^+ F$ is invariantly defined.

Before dealing with the next step, that is, the Melin and the Hörmander inequalities, let me remark that a general approach for lower-bound estimates is to write, in some pseudodifferential calculus, P as

$$P = \sum_k X_k^* X_k + R \tag{2.3}$$

where the X_k and R are pseudodifferential operators such that the X_k are of order $m/2$ and R is of order $m-1$, with $R \geq 0$ to which one may therefore apply the Sharp Gårding inequality.

I next recall Melin's inequality, proved by Anders Melin in [20] (see also [15]).

Theorem 2.3 *Let $p_m \geq 0$ on $T^*X \setminus 0$ and suppose*

$$p_m(\rho) = 0 \Longrightarrow p^s_{m-1}(\rho) + \mathrm{Tr}^+ F(\rho) > 0. \tag{2.4}$$

Then for all compact $K \subset X$ there exist $c_K, C_K > 0$ such that

$$(Pu, u) \geq c_K \|u\|^2_{(m-1)/2} - C_K \|u\|^2_{(m-2)/2}, \quad \forall u \in C_0^\infty(K). \tag{2.5}$$

Moreover, the condition

$$p_m(\rho) = 0 \Longrightarrow p^s_{m-1}(\rho) + \mathrm{Tr}^+ F(\rho) \geq 0 \tag{2.6}$$

yields that for all $\varepsilon > 0$, for all compact $K \subset X$ there exists $C_{K,\varepsilon} > 0$ such that

$$(Pu, u) \geq -\varepsilon \|u\|^2_{(m-1)/2} - C_{K,\varepsilon} \|u\|^2_{(m-2)/2}, \quad \forall u \in C_0^\infty(K). \tag{2.7}$$

Again, it is important to observe that:

- Also in this case *no condition* is required on the characteristic set of P, which makes the Melin inequality a very strong tool in the geometric analysis of PDEs. As we shall see in a short while, the quantity appearing in (2.4), which is symplectically invariant, has a spectral meaning. This observation will be crucial for the later extensions I will be considering. The spectral meaning of $p^s_{m-1} + \mathrm{Tr}^+ F$ was already present in Melin's paper;
- The requirements $p_m \geq 0$ on $T^*X \setminus 0$ and (2.4) are also *necessary* for (2.5) to hold. Once more, one uses the family $u_t(x) = e^{it^2\langle x, \xi_0\rangle} v(t(x - x_0))$, $v \in C_0^\infty(X)$, as $t \to +\infty$ first when $(x_0, \xi_0) \in T^*X \setminus 0$ and then when $(x_0, \xi_0) \in \Sigma$;
- Finally, the conclusion (2.7) in the theorem follows from the first one by considering $P + \varepsilon |D|^{m-1}$, $\varepsilon > 0$.

The next natural question when looking at the inequality (2.7) is: *Is it possible to take the limit $\varepsilon \to 0+$?* That is, in the case of classical ψdos, is it possible to go beyond the Melin inequality and obtain a control from below, say when the order of P is 2, by an L^2-norm? *What is the influence of the geometry of Σ and of the lower order terms in this case?*

Hörmander proved in [13] the following result (which I call the *Hörmander inequality*), that shows that an estimate of that kind still holds when there are directions in $T^*X \setminus 0$ along which the symbol of P may tend to $-\infty$ with strength $|\xi|^{m-1}$. Recall that P is *transversally elliptic* with double characteristics with respect to Σ if

$$|p_m(x,\xi)| \approx |\xi|^m \mathrm{dist}_\Sigma(x,\xi/|\xi|)^2,$$

for all $(x,\xi) \in T^*X \setminus 0$.

Theorem 2.4 *Suppose that Σ be a smooth (conic) manifold of $T^*X \setminus 0$ on which the rank of the symplectic form σ at Σ is constant, where*

$$\mathrm{rk}\,\sigma\big|_\Sigma(\rho) = \dim\Big(\frac{T_\rho\Sigma}{T_\rho\Sigma \cap T_\rho\Sigma^\sigma}\Big),\quad \rho \in \Sigma,$$

$T_\rho\Sigma^\sigma$ being the symplectic orthogonal of $T_\rho\Sigma$.

Suppose that p_m be positively transversally elliptic with double characteristics with respect to Σ, that is

$$p_m(x,\xi) \approx |\xi|^m \mathrm{dist}_\Sigma(x,\xi/|\xi|)^2,\quad \forall (x,\xi) \in T^*X \setminus 0, \tag{2.8}$$

and that the Melin condition (2.6) holds at each point $\rho \in \Sigma$. Then, for all compact $K \subset X$ there exists $C_K > 0$ such that

$$(Pu,u) \geq -C_K\|u\|^2_{(m-2)/2},\quad \forall u \in C_0^\infty(K). \tag{2.9}$$

It is important to recall the following equivalent description of the transversal ellipticity condition: p_m is transversally elliptic with double characteristics with respect to Σ iff

$$\mathrm{Ker}\, F(\rho) = T_\rho\Sigma,\quad \forall \rho \in \Sigma. \tag{2.10}$$

I wish to give a very rough idea of the way the geometric conditions in the statement are used to show that P can be microlocally written in the form (2.3), since this will be used later in one of the generalizations. The condition on the rank of $\sigma\big|_\Sigma$ yields the existence of two nonnegative integers ℓ, ν, constant throughout Σ, such that

$$\ell = \dim(T_\rho\Sigma \cap T_\rho\Sigma^\sigma),\quad 2\nu = \dim\Big(\frac{T_\rho\Sigma^\sigma}{T_\rho\Sigma \cap T_\rho\Sigma^\sigma}\Big), \tag{2.11}$$

where

$$\mathrm{codim}\,\Sigma = 2\nu + \ell,\quad \mathrm{rk}\,\sigma\big|_\Sigma = 2n - 2(\nu+\ell). \tag{2.12}$$

Hence $T\Sigma \cap T\Sigma^\sigma \to \Sigma$ is a smooth vector-bundle of rank ℓ. The nonnegativity of p_m yields that

$$\mathrm{Ker}\, F(\rho) \subset \mathrm{Ker}(F(\rho)^2) = \mathrm{Ker}(F(\rho)^3),$$

$$\mathrm{Spec}(F(\rho)) = \{0\} \cup \{\pm i\mu_j;\ \mu_j > 0,\ 1 \leq j \leq \nu\},$$

where 0 is a generalized eigenvector and the $\pm i\mu_j$ are "regular" eigenvalues (that is, they are not associated with a Jordan form). Then, using the linear algebra of the Hamilton map and the transversal ellipticity condition (2.10) gives that the complexified normal bundle of Σ ($N\Sigma = TT^*X\big|_{\Sigma}/T\Sigma \to \Sigma$ is thc normal bundle of Σ and ${}^{\mathbb{C}}N\Sigma \to \Sigma$ its complexification) is decomposed as a Whitney vector-bundle σ-orthogonal sum

$$
{}^{\mathbb{C}}N\Sigma = {}^{\mathbb{C}}\mathrm{Im}(F^2) \oplus {}^{\mathbb{C}}\Big(\frac{\mathrm{Ker}(F^2)}{\mathrm{Ker}(F)}\Big) = (V \oplus \bar{V}) \oplus {}^{\mathbb{C}}W \to \Sigma, \tag{2.13}
$$

where $W \to \Sigma$ is a real bundle of rank ℓ, $V \to \Sigma$ is a complex bundle of rank ν whose fibers $V(\rho)$ are given by

$$
V(\rho) = \bigoplus_{\substack{\mu>0 \\ i\mu\in\mathrm{Spec}(F(\rho))}} \mathrm{Ker}(F(\rho) - i\mu).
$$

This geometric information allows one to write, microlocally near Σ, P in the form (2.3), where (say $m = 2$) the X_k, R are first order with $R \geq 0$ and the Sharp Gårding estimate gives the result. In fact, microlocally near Σ, one constructs the symbols of the X_k by using sections of the bundles $V \to \Sigma$ and $W \to \Sigma$ and the Morse Lemma, and it turns out that the symbol of R in (2.3) at Σ is given by the subprincipal symbol of P plus the positive-trace of the Hamilton map, which is nonnegative by assumption, making it possible to apply the Sharp Gårding inequality (near Σ). I find this proof geometrically elegant and beautiful.

We shall see in the generalizations to the case of higher characteristics that the conditions on Σ will have to be replaced by "spectral conditions" and that writing P in the form (2.3) will have to be replaced by a similar expression that is based on the lowest eigenvalue of the localized operator and the use of the quantization of spectral projectors (Boutet de Monvel's Hermite operators [4]).

I wish now to turn to the last celebrated estimate I have listed earlier on, that is, the Fefferman–Phong inequality. I state it in the Weyl-Hörmander calculus (see [15] or [17]). The Weyl-Hörmander setting allows on the one hand using more general classes of ψdos, but on the other hand the conditions must then be put on the *total* symbol of the operator, and not on some of the homogeneous terms in the asymptotic expansion as in the case of classical ψdos. In this respect, conditions on the total symbol are more restrictive. Moreover, the characteristic set becomes, in the Weyl-Hörmander calculus, the set in phase-space where the symbol is not elliptic with respect to the admissible weight function that defines the order of the operator.

Let g be a Hörmander metric on $\mathbb{R}^n \times \mathbb{R}^n$, and let h be the corresponding Planck's function.

When a symbol $p \in S(h^{-m}, g)$ (some $m > 0$) is elliptic, that is $p \gtrsim h^{-m}$ on $\mathbb{R}^n \times \mathbb{R}^n$, one has of course the Gårding inequality in the Sobolev spaces $H(g, h^{-m/2})$ (see [17] for the definition of such spaces).

More interesting are the cases when $p \geq 0$ only. As customary, I will write $p^{\mathrm{w}}(x, D)$ for the ψdo obtained by Weyl-quantizing the symbol $p \in S(h^{-m}, g)$.

In this general setting one has the (scalar) Sharp Gårding inequality, due to Hörmander (see [15]).

Theorem 2.5 *Let $0 \leq p \in S(h^{-1}, g)$. Then there exists $C > 0$ such that*

$$(p^{\mathrm{w}}(x, D)u, u) \geq -C\|u\|_0^2, \quad \forall u \in \mathscr{S}(\mathbb{R}^n). \tag{2.14}$$

Of course, the theorem holds just by assuming that $p \geq -c$ for some $c \in \mathbb{R}$.

It was then shown by Charlie Fefferman and Duong Phong [8] (see also [15]) that (still in the scalar case) one may obtain a radically better estimate.

Theorem 2.6 *Let $0 \leq p \in S(h^{-2}, g)$. Then there exists $C > 0$ such that*

$$(p^{\mathrm{w}}(x, D)u, u) \geq -C\|u\|_0^2, \quad \forall u \in \mathscr{S}(\mathbb{R}^n). \tag{2.15}$$

Of course, once more, the theorem holds just by assuming that $p \geq -c$ for some $c \in \mathbb{R}$.

It is important to note that in Theorem 2.5 the operator is *first-order* in the calculus, whereas in Theorem 2.6 the operator is *second-order* in the calculus. Therefore, the Sharp Gårding inequality allows a control from below in the L^2-norm of a *first-order* operator, whereas the Fefferman–Phong inequality allows a control from below in the L^2-norm of a *second-order* operator.

It may also be useful to restate the above theorems in case of $0 \leq p \in S(m, g)$, m being an admissible g-weight (see [15] or [17]). One has the following statements.

Theorem 2.7 *Let m be an admissible g-weight.*

(i) As for the Sharp Gårding inequality one has that if $0 \leq p \in S(m, g)$ then there exists $q \in S(mh, g)$ such that $p^{\mathrm{w}}(x, D) \geq q^{\mathrm{w}}(x, D)$ (on $\mathscr{S}(\mathbb{R}^n)$).
(ii) As for the Fefferman–Phong one has that if $0 \leq p \in S(m, g)$ then there exists $q \in S(mh^2, g)$ such that $p^{\mathrm{w}}(x, D) \geq q^{\mathrm{w}}(x, D)$.

The order function h^{-2} in the Fefferman–Phong inequality is sharp, in the sense that in the Weyl-Hörmander calculus one cannot improve the exponent –2 (see [15]).

Note that when dealing with the total symbol in the Weyl-Hörmander setting, the characteristic set has disappeared, to be replaced by the set where P is not elliptic. Loosely speaking, the proof of the Fefferman–Phong inequality goes along the following line: after a Calderón–Zygmund microlocalization, one may still write P on blocks on which the symbol of P is nonelliptic-nondegenerate (the most difficult case) in the form (2.3) (with only one X_k, that is $P = X^*X + R$, $R \geq 0$), but the point now is to use an induction on the number of variables. More precisely, the Calderón–Zygmund microlocalization reduces matters to blocks in the phase-space $\mathbb{R}^n \times \mathbb{R}^n$ on which either the symbol is elliptic, or it has a Hessian which is nondegenerate in some direction (whence the symbol may be written on that block as a square of a first-order symbol plus a nonnegative remainder), or it is simply bounded. The (microlocal) lower bound estimate in the first case is clear, in the second case is obtained through the induction on the number of variables, and in the third just by L^2-continuity. In the end the lower bound estimate is achieved by summing together the individual (microlocal) pieces by a Cotlar–Stein argument. In my opinion, also this proof is remarkably elegant and beautiful, the geometry here being highlighted by the Calderón–Zygmund localization.

In closing this section, it is interesting to observe that when P is a second-order ψdo the Fefferman–Phong inequality (2.15) and the Hörmander inequality (2.9) give the same lower bound with the L^2-norm. (The statement on C_0^∞ versus the statement on $\mathscr{S}$ is just due to the seminorms one endows the symbol spaces with.) Where the Fefferman–Phong inequality allows one to deal with more general pseudodifferential classes, but requires a more demanding control on the total symbol of the operator (regrettably enough, the total symbol has no invariant meaning on a manifold), the Hörmander inequality deals with the more restrictive class of classical ψdos for which however it makes sense to introduce geometrical assumptions on the characteristic manifold and on the quantities related to it (as in the (2.6) condition), but then it allows one to consider symbols that are definitely not necessarily bounded from below.

3 Generalizations of the Hörmander Inequality

In this section I will describe some extensions [23, 24, 30, 33] of the Hörmander inequality (2.9). I will describe three main directions of generalization:

(i) one which allows operators with higher characteristics, keeping the geometric hypotheses on Σ, i.e. smoothness and rank condition of $\sigma|_\Sigma$, and the transversal ellipticity;
(ii) one which relaxes the condition on the rank of $\sigma|_\Sigma$ in the case of transversally elliptic operator with double characteristics;
(iii) one which relaxes the transversal ellipticity assumption.

Before embarking in all that, I wish first to recall another fundamental tool in the study of a priori estimates: *the localized operator*. The concept, which is already present in Melin's proof of inequality (2.5), is due to Louis Boutet de Monvel and Johannes Sjöstrand (see [4, 45]) in their pioneering study of hypoelliptic operators. It was successively developed by Boutet de Monvel, A. Grigis and B. Helffer in their study of hypoellipticity in the case of operators with characteristics of order higher than 2, and finally exploited by C. Parenti and myself in the study of almost-positivity estimates for operators with multiple characteristics and in the study of operators which are hypoelliptic but lose many derivatives, and in the study of semi-global solvability theorems in higher characteristics (see [41]).

Roughly speaking, one may think of the localized operator as the quantization of the *localized polynomial*, which is the "meaningful" part of the Taylor expansion of the symbol of P at points of Σ in directions normal to Σ. A natural class of ψdos in this case is the Sjöstrand class $\mathrm{OPN}^{m,k}_{\mathrm{cl}}(X, \Sigma)$, where $k \geq 2$ is an integer and $\Sigma \subset T^*X \setminus 0$ is a smooth conic submanifold. A ψdo P belongs to such a class if P is a classical (properly supported) ψdo of order m and if the homogeneous terms p_{m-j}, $0 \le j \le [k/2]$ ($[k/2]$ being the integer part of $k/2$) in the expansion of its symbol $p \sim \sum_{j\ge 0} p_{m-j}$ vanish to order $k-2j$ at Σ, that is,

- $|p_{m-j}(x,\xi)| \lesssim |\xi|^{m-j}\mathrm{dist}_\Sigma(x,\xi/|\xi|)^{(k-2j)_+}$, for $0 \le j \le [k/2]$, where $(k-2j)_+ = \max\{k-2j, 0\}$.

Therefore, when $P \in \mathrm{OPN}^{m,2}_{\mathrm{cl}}(X,\Sigma)$ (i.e. one has double characteristics), there is no vanishing condition at Σ on the $m-1$st-term. Furthermore, when the operator $P \in \mathrm{OPN}^{m,k}_{\mathrm{cl}}(X,\Sigma)$ is also *transversally elliptic* with characteristics of order k with respect to Σ, we therefore have the conditions

$$|p_m(x,\xi)| \approx |\xi|^m \mathrm{dist}_\Sigma(x,\xi/|\xi|)^k \text{ and } |p_{m-j}(x,\xi)| \lesssim |\xi|^{m-j}\mathrm{dist}_\Sigma(x,\xi/|\xi|)^{(k-2j)_+},$$

for $0 \le j \le [k/2]$.

Let $p_{\mathrm{w}} \sim e^{-i\langle D_x, D_\xi\rangle/2}p$ be the Weyl-symbol of P.

Definition 3.1 Let $\rho \in \Sigma$. Let $P \in \mathrm{OPN}^{m,k}_{\mathrm{cl}}(X,\Sigma)$. The **localized polynomial** of P at ρ is the map

$$T_\rho T^*X \ni v \longmapsto p^{(k)}(\rho, v) := \sum_{j=0}^{[k/2]} \frac{1}{(k-2j)!}\Big(V^{k-2j} p_{\mathrm{w},m-j}\Big)(\rho), \quad v = V(\rho), \tag{3.1}$$

V being a (smooth) section of $TT^*X \to T^*X$ defined near ρ and such that $V(\rho) = v$.

Suppose now (for simplicity) that Σ is symplectic, that is, that $T_\rho\Sigma \cap T_\rho\Sigma^\sigma = \{0\}$ for all $\rho \in \Sigma$. The **localized operator** P_ρ of P at $\rho \in \Sigma$ is given by the Weyl quantization of $p^{(k)}(\rho,\cdot)$ (which may be defined since $T_\rho\Sigma^\sigma$ is a symplectic vector space)

$$P_\rho = \mathrm{Op}^{\mathrm{w}}(p^{(k)}(\rho,\cdot)).$$

(See [4, 45] or [29] for more general cases.)

It turns out that the definition is independent of V and that the map

$$p^{(k)}\colon TT^*X\big|_\Sigma \ni (\rho, v) \longmapsto p^{(k)}(\rho, v) \in \mathbb{C}$$

is smooth and a *polynomial* of degree $\le k$ in the fibers v. Moreover, if

$$\Lambda(p^{(k)}(\rho,\cdot)) := \{w \in T_\rho T^*X;\ p^{(k)}(\rho, v+tw) = p^{(k)}(\rho, v),\ \forall v \in T_\rho T^*X,\ \forall t \in \mathbb{R}\}$$

is the lineality of $p^{(k)}(\rho,\cdot)$, one has that $T_\rho\Sigma \subset \Lambda(p^{(k)}(\rho,\cdot))$ for all $\rho \in \Sigma$, with equality holding iff P is transversally elliptic (that is, roughly speaking, $p^{(k)}(\rho,\cdot)$ is a polynomial which depends only on the normal variables in $N_\rho\Sigma = T_\rho T^*X/T_\rho\Sigma$).

Furthermore, when Σ is *symplectic*, since we may always find a local system of coordinates (a "canonical flattening") for which $\Sigma = \{t = \tau = 0\}$, writing $x = (t, y)$, $\xi = (\tau, \eta)$ and $\rho \simeq (y, \eta)$, we then have

$$P_{(y,\eta)} = \mathrm{Op}^{\mathrm{w}}\Big(\sum_{\substack{0 \le j \le k/2 \\ |\alpha|+|\beta|=k-2j}} \frac{1}{\alpha!\beta!}\Big(\partial_t^\alpha \partial_\tau^\beta p_{\mathrm{w},m-j}\Big)(t=0, y, \tau=0, \eta) t^\alpha \tau^\beta \Big).$$

It turns out (see for example [29]) that

If Σ is symplectic, the spectrum of P_ρ, as an unbounded operator on $L^2(\mathbb{R}^\nu)$ realized as a maximal operator (its domain is the Shubin space $B^k(\mathbb{R}^\nu)$), is invariant under the different choices of local "flattenings" of Σ near ρ.

3.1 Extension to Higher Transversally Elliptic Characteristics

In [30] Parenti and I extended Hörmander's inequality (2.9) to operators $P = P^* \in \mathrm{OPN}^{m,k}_{\mathrm{cl}}(X, \Sigma)$ that are *positively transversally elliptic* with characteristics of order $k \ge 2$, that is, $p_m(x, \xi) \approx |\xi|^m \mathrm{dist}_\Sigma(x, \xi/|\xi|)^k$ where *necessarily k must be an even integer*, where Σ is either *symplectic* or *involutive* (recall that the latter condition means that $T_\rho\Sigma^\sigma \subset T_\rho\Sigma$ for all $\rho \in \Sigma$). In the symplectic case, $P_\rho = P_\rho^*$ for all $\rho \in \Sigma$ has a discrete spectrum, and one may consider the map $\lambda_0 \colon \Sigma \ni \rho \longmapsto \lambda_0(\rho) \in \mathbb{R}$, where $\lambda_0(\rho)$ is the *lowest eigenvalue* of P_ρ, $\rho \in \Sigma$. We proved the following theorem.

Theorem 3.2 *(i) Suppose Σ symplectic (codim $\Sigma = 2\nu$) and suppose that $P_\rho \ge 0$ for all $\rho \in \Sigma$. Suppose that for any given $\rho_0 \in \Sigma$ such that $\lambda_0(\rho_0) = 0$ there exists a conic neighborhood $\Gamma_{\rho_0} \subset \Sigma$ of ρ_0 such that for all $\rho \in \Gamma_{\rho_0}$ the spaces $W_\rho = \mathrm{Ker}(P_\rho - \lambda_0(\rho)I)(\subset \mathscr{S}(\mathbb{R}^\nu))$ have finite dimension $d(\rho)$ which is **constant** and that one may find a basis of W_ρ made of eigenfunctions of P_ρ that are either **all even** or **all odd**. Then one has that for any given compact $K \subset X$ there exists $C_K > 0$ such that*

$$(Pu, u) \ge -C_K \|u\|^2_{m/2-(k+2)/4}, \quad \forall u \in C_0^\infty(K). \tag{3.2}$$

(ii) Suppose Σ involutive and that $p^{(k)} \ge 0$ on $N\Sigma = TT^*X\big|_\Sigma / T\Sigma$. Suppose furthermore that $p^{(k)}(\rho, v) = 0 \Longrightarrow v = 0$ and that

$$p^{(k)}(\rho, 0) = 0 \Longrightarrow (\partial^2 p^{(k)}/\partial v^2)(\rho, 0) \quad \textit{is invertible.}$$

Then inequality (3.2) above holds.

One may slightly generalize the theorem (when Σ is symplectic) and consider a case in which the kernel of P_ρ at each $\rho_0 \in \lambda_0^{-1}(0)$ is formed for ρ near ρ_0 by a fixed locally constant number of "low lying" eigenvalues of P_ρ that all vanish at ρ_0 (see [31]), the parity condition being required throughout.

Remark 3.3 It is important to see what happens when $k = 2$. In this case the dimensions of the W_ρ are automatically all constant with $d(\rho) = 1$, so that also the parity assumption on the eigenfunction is **automatically** satisfied. In this case

$$\lambda_0(\rho) = p^s_{m-1}(\rho) + \mathrm{Tr}^+ F(\rho)$$

and one recovers Hörmander's inequality (2.9). Theorem 3.2 gives therefore a different proof of it when P has double characteristics and Σ is either symplectic or involutive.

When the parity condition does not hold, one is no longer able to obtain (3.2) and the technique gives only (see [30]) the weaker lower bound

$$(Pu, u) \geq -C_K \|u\|^2_{m/2-(k+1)/4}, \quad \forall u \in C_0^\infty(K). \tag{3.3}$$

M. Mughetti and F. Nicola showed in [22] the *necessity* of the parity assumption in order to obtain (3.2), and hence that the result of [30] is the correct analog of the Hörmander inequality (2.9) for operators with higher characteristics. Indeed, they constructed an example of operator P for which the parity assumption is not fulfilled and for which only (3.3) may hold. The same authors then generalized in [23] inequality (3.2) to the case in which Σ is such that the rank of $\sigma\big|_\Sigma$ is constant, the canonical 1-form ξdx does not vanish on $T_\rho\Sigma$ for all $\rho \in \Sigma$ and Σ is neither symplectic nor involutive.

Before passing to the next kind of generalization, I wish to give and example of operator with multiple symplectic characteristics for which Theorem 3.2 holds. Let Σ be symplectic. Consider ψdos $P_j = P_j^* \in \mathrm{OPN}^{2,2}_{\mathrm{cl}}(X, \Sigma)$, $1 \leq j \leq N$, such that they are positively transversally elliptic with respect to Σ and such that, F_j being the fundamental matrix of P_j, one has $[F_j(\rho), F_k(\rho)] = 0$ for all $\rho \in \Sigma$. Write, with $2\nu = \mathrm{codim}\,\Sigma$,

$$\mathrm{Spec}(F_j(\rho)) = \{0\} \cup \{\pm i\mu_{j,k}(\rho),\ 1 \leq k \leq \nu\},$$

where, for each j, the $\mu_{j,k}(\rho) > 0$ are repeated according to multiplicity. Consider the operator

$$\mathsf{P} = \sum_{|\alpha| \leq m} A_\alpha P_1^{\alpha_1} P_2^{\alpha_2} \ldots P_N^{\alpha_N},$$

where the $A_\alpha = A_\alpha^*$ are classical ψdos of order 0 with principal symbol a_α. Suppose that for any given conic set $\Gamma \subset T^*X \setminus 0$ with compact base there exists $C_\Gamma > 0$ and a $\gamma \in \mathbb{Z}_+^N$ with $|\gamma| = m$ such that

$$\sum_{|\alpha|=m} a_\alpha(x, \xi)\zeta^\alpha \geq C_\Gamma \zeta^\gamma, \ \forall (x, \xi) \in \Gamma, \ \forall \zeta \in [0, +\infty)^N.$$

One then has that $\mathsf{P} \in \mathrm{OPN}^{2m,2m}_{\mathrm{cl}}(X, \Sigma)$ is positively transversally elliptic and that

$$\mathsf{P}_\rho = \sum_{|\alpha|=m} a_\alpha(\rho) P^{\alpha_1}_{1,\rho} P^{\alpha_2}_{2,\rho} \dots P^{\alpha_N}_{N,\rho}, \ \rho \in \Sigma.$$

The operator P_ρ is self-adjoint because the $P_{j,\rho}$ are self-adjoint and commute. Define next

$$\zeta_j(\rho;\beta) = \sum_{k=1}^{\nu} \mu_{jk}(\rho)(2\beta_k + 1) + p_1^s(\rho), \ \rho \in \Sigma, \ \beta \in \mathbb{Z}^\nu_+, \ 1 \le j \le N.$$

Then $\mathsf{P}_\rho \ge 0$ on $\mathscr{S}(\mathbb{R}^\nu)$ iff

$$Q(\rho;\beta) = \sum_{|\alpha|=m} a_\alpha(\rho) \prod_{j=1}^{N} \zeta_j(\rho;\beta)^{\alpha_j} \ge 0, \ \forall \beta \in \mathbb{Z}^\nu_+.$$

We have

$$\lambda_0(\rho) = \min_{\beta \in \mathbb{Z}^\nu_+} Q(\rho;\beta)$$

and the dimension of the eigenspace corresponding to $\lambda_0(\rho)$ is $\mathrm{card}(J(\rho))$ where

$$J(\rho) = \{\beta \in \mathbb{Z}^\nu_+; \ Q(\rho;\beta) = \lambda_0(\rho)\}.$$

The conditions required by Theorem 3.2 in this case are therefore that

$$\mathrm{card}(J(\rho)) = \text{constant in some neighborhood } V_{\rho_0} \text{of } \rho_0, \ \forall \rho_0 \in \lambda_0^{-1}(0),$$

and that the parity of the corresponding eigenfunctions be constant as well, that is,

$$(-1)^{|\beta|} = \text{constant } \forall \beta \in J(\rho), \ \forall \rho \in V_{\rho_0}, \ \forall \rho_0 \in \lambda_0^{-1}(0).$$

For more examples see [30].

3.2 *Relaxing the Hypothesis on the Rank of* $\sigma\big|_\Sigma$

I next consider a result proved by Parenti and myself in [33] in which we relax the hypothesis on the rank of $\sigma\big|_\Sigma$, *keeping the transversal ellipticity assumption*. Given a conic set $\Omega \subset T^*X \setminus 0$, denote by

$$\mathbb{S}^*(\Omega) = \{(x,\xi) \in \Omega; \ |\xi| = 1\}$$

the cosphere of the set Ω. We proved the following result.

Theorem 3.4 *Let $P = P^*$ be positively transversally elliptic with double characteristics with respect to Σ and suppose Melin's condition (2.6) be fulfilled. Suppose that there exists a closed conic subset $\Sigma' \subset \Sigma$ such that $\overline{\Sigma \setminus \Sigma'} = \Sigma$ on which*

$$\begin{cases} \operatorname{rk} \sigma\big|_{\Sigma}(\rho) = 2n - 2(\nu + \ell), \ \forall \rho \in \Sigma \setminus \Sigma', \\ \operatorname{rk} \sigma\big|_{\Sigma}(\rho) < 2n - 2(\nu + \ell), \ \forall \rho \in \Sigma', \end{cases} \tag{3.4}$$

where ℓ and ν are defined in (2.11) and (2.12). Suppose finally that the map

$$\mathbb{S}^*(\Sigma \setminus \Sigma') \ni \rho \longmapsto \Pi_+(\rho) := F(\rho)\Big(\frac{1}{2\pi i}\int_{\gamma_+} (\zeta - F(\rho))^{-1} d\zeta\Big) \tag{3.5}$$

can be smoothly extended to the whole $\mathbb{S}^*(\Sigma)$ (where $\gamma_+ \subset \{z \in \mathbb{C};\ \operatorname{Im} z > 0\}$ is a counter-clockwise oriented curve enclosing all the eigenvalues $i\mu$, $\mu > 0$, of $F(\rho)$). Then $\Sigma \ni \rho \longmapsto \operatorname{Tr}^+ F(\rho)$ is C^∞ and the Hörmander inequality (2.9) holds.

Note that the map

$$\rho \longmapsto \frac{1}{2\pi i}\int_{\gamma_+} (\zeta - F(\rho))^{-1} d\zeta$$

is the projector onto

$$\bigoplus_{\substack{\mu>0 \\ i\mu \in \operatorname{Spec} F(\rho)}} \operatorname{Ker}(F(\rho) - i\mu),$$

and that it is easier extending Π_+ than extending the projector itself. Since the map $F\colon {}^{\mathbb{C}}N\Sigma \longrightarrow {}^{\mathbb{C}}\operatorname{Im} F$ is a vector bundle isomorphism, a good decomposition (2.13) of ${}^{\mathbb{C}}N\Sigma$ is induced from a good decomposition of ${}^{\mathbb{C}}\operatorname{Im} F$, and for that, roughly speaking, one needs (see [33]) that the eigenvalues in $i\mathbb{R}_+$ of the Hamilton map that vanish must "bounce back" and not go across the generalized eigenvalue 0. One then may proceed as in the proof of the Hörmander inequality and find sections so that P can be written in the form (2.3), thus obtaining the result.

At this point I give an example of operator for which the above theorem holds. Let $P = P^*$ be second order with principal symbol $p_2(x, \xi) = \xi_1^2 + (f(x_1) + x_2)^2 \xi_2^2$ in $T^*\mathbb{R}^2$, with $f \in C^\infty(\mathbb{R}^2; \mathbb{R})$. In this case

- $\Sigma = \{\xi_1 = 0,\ x_2 = -f(x_1),\ \xi_2 \neq 0\}$ is a smooth manifold;
- p_2 is transversally elliptic with respect to Σ;
- $f'(x_1) \neq 0 \Longrightarrow \operatorname{rk} \sigma\big|_{\Sigma} = 2$ (where $\ell = 0$, $\nu = 1$) and $f'(x_1) = 0 \Longrightarrow \operatorname{rk} \sigma\big|_{\Sigma} = 0$ (where $\ell = 2$, $\nu = 0$). This corresponds to a transition of the Hessian $\sigma(v, F(\rho)v)$ of $p_2/2$, $\rho \in \mathbb{S}^*\Sigma$, from the quadratic form

$$\delta\xi_1^2 + (f'(x_1)\delta x_1 + \delta x_2)^2 \quad \text{to the quadratic form} \quad \delta\xi_1^2 + \delta x_2^2,$$

where $(\delta x, \delta\xi) = (\delta x_1, \delta x_2, \delta\xi_1, \delta\xi_2)$ are coordinates in $T_\rho T^*\mathbb{R}^2$.

- The Hamilton map acts as

$$F(\rho)\begin{bmatrix}\delta x\\ \delta\xi\end{bmatrix}=\begin{bmatrix}\delta\xi_1\\ 0\\ -f'(x_1)(f'(x_1)\delta x_1+\delta x_2)\\ -(f'(x_1)\delta x_1+\delta x_2)\end{bmatrix},\quad \begin{bmatrix}\delta x\\ \delta\xi\end{bmatrix}\in T_\rho T^*\mathbb{R}^2,$$

and the eigenvalue equation $F(\rho)v=\lambda v$ yields that there is no $\lambda\neq 0$ unless $f'(x_1)\neq 0$, in which case $i|f'(x_1)|$ is the only eigenvalue in $i\mathbb{R}_+$ with corresponding eigenvector

$$v=\begin{bmatrix}1\\ 0\\ i|f'(x_1)|\\ i\,\mathrm{sgn}(f'(x_1))\end{bmatrix}.$$

- Therefore the extendibility of Π_+ holds iff f' *has a constant sign near any given characteristic point*, that is, f has to be *locally monotonic* there. In such a case $\mathrm{Tr}^+F(x,\xi)=|f'(x_1)||\xi_2|$, which is therefore smooth.

Note that a case in which $f(x_1)=x_1^{2k}$ does not satisfy the extendibility assumption. However, inequality (2.9) may still hold, for one may choose p_1^s such that on Σ one has Melin's strict condition $p_1^s+\mathrm{Tr}^+F>0$. Of course, once more, the *delicate* point is when one only has that $p_1^s+\mathrm{Tr}^+F\geq 0$ with points ρ where $p_1^s(\rho)+\mathrm{Tr}^+F(\rho)=0$ at which therefore the strong Melin inequality (2.5) may not a priori hold.

3.3 *Relaxing the Hypothesis on the Transversal Ellipticity*

Relaxing the transversal ellipticity assumption, *keeping the rank of* $\sigma|_\Sigma$ *constant*, is very complicated. Saying that p_m with double characteristics is not transversally elliptic means that there are characteristic points $\rho\in\Sigma$ such that

$$T_\rho\Sigma\subsetneq \mathrm{Ker}\,F(\rho).$$

In [24], Mughetti, Parenti and myself considered two kinds of non-transversal ellipticity, a *weak* and a *strong* failure of transversal ellipticity, and I will consider here only the strong one:

$$T_\rho\Sigma\subsetneq \mathrm{Ker}\,F(\rho),\quad \forall\rho\in\Sigma.$$

The assumption on Σ is that it is *symplectic*, *transversal* intersection of *involutive* manifolds Σ_1 and Σ_2 with codim $\Sigma_j=\nu$, $j=1,2$. Suppose also that the principal and subprincipal symbols of P satisfy the estimates

$$|p_m(x,\xi)| \lesssim |\xi|^m\Big(\mathrm{dist}_{\Sigma_1}(x,\xi/|\xi|)^{2h} + \mathrm{dist}_{\Sigma_2}(x,\xi/|\xi|)^2\Big),$$

$$|p^s_{m-1}(x,\xi)| \lesssim |\xi|^{m-1}\Big(\mathrm{dist}_{\Sigma_1}(x,\xi/|\xi|)^{h-1} + \mathrm{dist}_{\Sigma_2}(x,\xi/|\xi|)\Big),$$

where the "anisotropy" index h is such that $h \geq 2$.

Note that in this case

$$\mathrm{Ker}\, F(\rho) = T_\rho\Sigma_2 \quad \text{and} \quad \mathrm{Tr}^+ F = 0,$$

whence the necessary condition (2.6) is fulfilled.

Note also that if p^s_{m-1} had the exponent h in the vanishing behavior, then the Fefferman–Phong inequality (2.15) could be applied (provided of course that $p_m \geq 0$) yielding immediately Hörmander's inequality (2.9).

It turns out (see [24]) that when p_m is *positively anisotropically transversally elliptic*, that is,

$$p_m(x,\xi) \gtrsim |\xi|^m\Big(\mathrm{dist}_{\Sigma_1}(x,\xi/|\xi|)^{2h} + \mathrm{dist}_{\Sigma_2}(x,\xi/|\xi|)^2\Big),$$

if the Hörmander inequality (2.9) holds then necessarily one has

$$(P_\rho f, f)_{L^2(\mathbb{R}^\nu)} \geq 0, \quad \forall f \in \mathscr{S}(\mathbb{R}^\nu), \ \forall \rho \in \Sigma,$$

where P_ρ is the *anisotropic localized* operator of P at $\rho \in \Sigma$ (see [24]). It turns out from the discussion in [24] that, because of the fact that P_ρ preserves parity when h is odd, there is a difference when h is even or odd which reduces matters to considering a case in which h is odd.

One has the following theorem.

Theorem 3.5 *Suppose that the above conditions on Σ, p_m and p^s_{m-1} hold and that p_m is positively anisotropically transversally elliptic. Suppose furthermore that $P_\rho \geq 0$ for all $\rho \in \Sigma$. Let*

$$\lambda_0(\rho) = \inf \mathrm{Spec}(P_\rho), \quad \rho \in \Sigma.$$

Suppose that λ_0 vanishes in Σ and that, when $\nu > 1$, the principal symbol may be written as $p_m = q_1 + q_2$, where

$$q_1(x,\xi) \approx |\xi|^2 \mathrm{dist}_{\Sigma_1}(x,\xi/|\xi|)^{2h}, \quad q_2(x,\xi) \approx |\xi|^2 \mathrm{dist}_{\Sigma_2}(x,\xi/|\xi|)^2.$$

Then inequality (2.9) holds.

The "strange" condition $p_m = q_1 + q_2$ when $\nu > 1$ is necessary to ensure the needed parity of the eigenfunctions of P_ρ when ρ is a zero of λ_0 (i.e. the eigenfunctions must be *either all even or all odd*). If p_m does not have the required form then

P_ρ is a "magnetic" Schrödinger operator and $\lambda_0(\rho)$ may have a multiplicity greater than 1, and hence may have eigenfunctions of different parities and therefore give an obstruction to the desiderd inequality (as in the case of the Hörmander inequality for higher transversally elliptic symplectic characteristics).

A typical model of the above anisotropic situation is the following. Let $x = (x', x'') \in \mathbb{R}^\nu \times \mathbb{R}$, $\Sigma_1 = \{x' = 0\}$, $\Sigma_2 = \{\xi' = 0\}$, and

$$P = P^* = \sum_{j=1}^{\nu} D_{x'_j}^2 + \sum_{|\alpha|=2h} a_\alpha x'^\alpha D_{x''}^2 + \sum_{j=1}^{\nu} b_j(x) D_{x'_j} + c(x) D_{x''} + e(x),$$

smooth coefficients, where

$$\sum_{|\alpha|=2h} a_\alpha x'^\alpha > 0, \ \forall x' \neq 0.$$

Since on $\Sigma = \{(0, x'', 0, \xi'');\ \xi'' \neq 0\}$ one has $p_1^s(\rho) = c(0, x'')\xi''$ and $\mathrm{Tr}^+ F(\rho) = 0$, Melin's condition amounts to requiring that $c(0, x'') = 0$ for all x'' and the vanishing condition of the subprincipal part p_1^s is therefore $|c(x)| \lesssim |x'|^{h-1}$, for x near 0. In this case, the localized operator P_ρ, through which the conditions of Theorem 3.5 are given, is

$$P_\rho = \sum_{j=1}^{\nu} D_{x'_j}^2 + \sum_{|\alpha|=2h} a_\alpha x'^\alpha |\xi''|^2 + \sum_{|\beta|=h-1} \frac{1}{\beta!} (\partial_{x'}^\beta c)(0, x'') x'^\beta \xi''.$$

I address the reader to the paper [24] for more details on this.

4 Other Generalizations

In this section, I wish to briefly mention and give some references to important generalizations of the Sharp Gårding, the Melin, the Hörmander and the Fefferman–Phong inequalities in other directions.

4.1 The Sharp Gårding Inequality

The Sharp Gårding inequality was generalized to classes of low-regularity symbols by D. Tataru in [47], using FBI methods.

There are also very interesting generalizations by M. Ruzhansky and V. Turunen [44] to che case of compact Lie groups, and by V. Fischer and M. Ruzhansky [9] to the case of graded nilpotent Lie groups.

4.2 The Melin Inequality

Melin's inequality was generalized to operators in the class $\mathrm{OPN}^{m,k}_{\mathrm{cl}}(X,\Sigma)$ by A. Mohamed [21]. It takes the following form (see [29], page 197): *Let Σ be smooth. If $P = P^* \in \mathrm{OPN}^{m,k}_{\mathrm{cl}}(X,\Sigma)$ is transversally elliptic and $P_\rho \geq 0$ and injective on* $\mathscr{S}(\mathbb{R}^n)$ for all $\rho \in \Sigma$, *then for all $\varepsilon > 0$, for all $\mu < m/2 - k/4$ and for all compact $K \subset X$ there exists $C > 0$ such that*

$$(Pu, u) \geq -\varepsilon \|u\|^2_{m/2-k/4} - C\|u\|^2_{\mu}, \quad \forall u \in C_0^\infty(K).$$

Note that when $k = 2$ the above result does not fully recover the result by Melin, for in the present case one requires P to be *transversally elliptic* and Σ to be *smooth*.

A very interesting generalization was obtained by J. Toft [48] in the framework of the Weyl calculus in the case of minimal regularity assumptions on the symbol, by using a regularization and decomposition method.

Another very interesting generalization of Melin's inequality is due to F. Hérau [12] in the case of a Wiener type pseudodifferential algebra. I address the reader to his paper for more details.

4.3 The Hörmander Inequality

Hörmander's inequality (2.9) was generalized by F. Nicola [25] to the case of (transversally elliptic) anisotropic pseudodifferential operators, and by F. Nicola and L. Rodino [27] to a very interesting class of (transversally elliptic) operators for which the nonnegativity of invariants obtained by the Taylor expansion of the subprincipal symbol on Σ gives information which is also seen to be sharp.

It is also interesting to mention that in [32] Parenti and myself proved a "strong" Hörmander inequality for operators with positively transversal multiple symplectic characteristics of the kind: For all compact $K \subset X$ there are *positive* constants c_K, C_K such that

$$(Pu, u) \geq c_K \|u\|^2_{m/2-(k+2)/4} - C_K \|u\|^2_{m/2-(k+4)/4}, \quad \forall u \in C_0^\infty(K),$$

in terms of a strong Melin inequality for the Schur reduction (which is of order $m - k/2$) on Σ. See [32], Theorem 5.3.

4.4 The Fefferman–Phong Inequality

Fefferman–Phong's inequality has been generalized in the aforementioned papers by F. Hérau, D. Tataru and J. Toft. There is also a remarkable generalization by

J.-M. Bony [2], where he shows that nonnegativity of the second order symbol and the control of derivatives of order greater then or equal to 4 suffice to get the Fefferman–Phong inequality (2.15).

Subsequently, N. Lerner and Y. Morimoto [18] established the Fefferman–Phong inequality in the standard metric of the $S^2_{1,0}$ calculus, when the derivatives of the symbol up to order $2n + 4 + \varepsilon$ (n the dimension and $\varepsilon > 0$) are controlled, and afterwards A. Boulkhemair [3] generalized their result by requiring a control only up to order $n/2 + 4 + \varepsilon$. This kind of result is fundamental for nonlinear problems, especially nonlinear hyperbolic equations.

In [19], N. Lerner and J. Nourrigat obtained a remarkable generalization of the Fefferman–Phong inequality in the framework of the Weyl calculus for one-dimensional ψdos when the (second order) symbol has a nonnegative average over suitable symplectic blocks. In the same paper they also establish nonnegativity results for a class of polynomial-coefficient magnetic Schrödinger operators, a very useful result that, as we have already seen earlier, may be used to study the anisotropic Hörmander inequality.

Finally, in [7], F. Colombini, D. Del Santo and C. Zuily studied the case of the Fefferman–Phong inequality in the framework of Hörmander's metrics which are *not* temperate but only *locally* temperate. Such a result is meaningful for some problems coming from spectral theory and from the propagation of singularities in nonlinear hyperbolic problems.

4.5 Open Problems

In closing this short Sect. 4, I wish to mention a couple of very interesting (in my opinion) problems:

(i) The first one is establishing any of the inequalities I have illustrated above in the case of ψdos on manifolds with singularities. A very interesting initial step could be to obtain the Sharp Gårding inequality for operators on manifolds with a conic singularity, in the framework of the Melrose or Schulze calculi.

(ii) The second one is to obtain the Hörmander and the Fefferman–Phong inequalities in the cases, respectively, of compact and nilpotent Lie groups (as regards this problem, I have some work in progress).

5 Generalizations to Systems

The case of systems is widely open. In this section I start by briefly recalling the main results on the various inequalities we have been considering here except for the Fefferman–Phong inequality, that I will afterwards pass to discussing in more details. I will always assume the order $m > 0$.

In the first place one has, as is well-known, that the Gårding inequality holds true for all positive-elliptic systems of mth-order ψdos.

There are extensions of the Sharp Gårding inequality to the case of $N \times N$ systems of ψdos due to P. Lax and L. Nirenberg [16] and to R. Beals [1]. The most general statement is due to L. Hörmander [15] and may be stated as follows.

Theorem 5.1 *Let $0 \leq p \in S(h^{-1}, g; \mathscr{L}(H, H))$ be a self-adjoint and nonnegative symbol of order* 1 *with values in the Hermitian bounded operators on some Hilbert space H. Then there exists $C > 0$ such that*

$$(p^{\mathrm{w}}(x, D)u, u) \geq -C\|u\|_0^2, \quad \forall u \in \mathscr{S}(\mathbb{R}^n; H). \tag{5.1}$$

The proof is very nice and goes by a semiclassical approximation and comparison of the operator with a system of harmonic oscillators.

But the next question is: *How about the Melin, the Hörmander and the Fefferman–Phong inequalities? Can they be generalized to systems?*

In the case of Melin's inequality, the best result (to my knowledge of course) for $N \times N$ systems of mth-order ψdos is due to R. Brummelhuis and J. Nourrigat in the paper [6], where they define a family of localized test-systems (through sequences of points associated with characteristic points) with polynomial coefficients from whose nonnegativity properties (on Schwartz functions) one may obtain the Melin inequality (2.5).

In the case of the Hörmander inequality, there are the result by L. Hörmander [13] and that by C. Parenti and myself [31] for $N \times N$ systems of mth-order ψdos with double characteristics. By that I mean that the characteristic set

$$\Sigma = \{(x, \xi) \in T^*X \setminus 0;\ \det p_m(x, \xi) = 0\}$$

is a smooth conic manifold on which $\operatorname{rk} \sigma\big|_{\Sigma}$ is constant and for which one has the conditions

$$\begin{cases} \det p_m(x, \xi) \approx |\xi|^{mN} \operatorname{dist}_{\Sigma}(x, \xi/|\xi|)^{2\ell}, \\ \dim \operatorname{Ker}(p_m(\rho))\big|_{\Sigma} = \ell,\ \forall \rho \in \Sigma. \end{cases} \tag{5.2}$$

The condition roughly says that the vanishing eigenvalues of the principal symbol *all* vanish to second order at Σ.

In [13] Hörmander considered the case Σ general but with $\ell = 1$ whereas Parenti and I in [31] treated the case where Σ is symplectic with general $1 \leq \ell \leq N$. We defined a localized system (one does not have to consider all possible limits at the characteristic points, as Brummelhuis and Nourrigat had to, because of the regular geometry of Σ and the transversal ellipticity assumption), and through the behavior of its lowest eigenvalue (that turns out to be a continuous function on Σ) and assuming the *parity* of the eigenfunctions corresponding to the low-lying eigenvalues that collapse to zero, near points where the lowest eigenvalue vanishes (that is, the localized system must be *regular* in the terminology of [31]), we could prove the Hörmander inequality for such systems. The approach uses Boutet's Hermite operators and is a generalization of that in [30]. It is intercsting to stress once more that without the parity assumption on the eigenfunctions, one obtains (see [31]) only the weaker estimate (weak Hörmander's inequality)

$$(Pu, u) \geq -C_K \|u\|_{(m-3/2)/2}, \quad \forall u \in C_0^\infty(K; \mathbb{C}^N), \tag{5.3}$$

in place of (2.9) (which instead has the Sobolev exponent $(m-2)/2$). Afterwards, F. Nicola in [26] was able to show that for $N \times N$ systems with double and symplectic characteristics the *sole nonnegativity condition* of the lowest eigenvalue of the localized system yields indeed inequality (5.3).

At this point it is worth mentioning that the study of the spectrum of systems with polynomial coefficients (localized systems are an instance of these) is still largely open. A few results on that kind of problems (for systems) are due to M. Wakayama and myself (see, in particular, [37, 40, 42] and references therein).

From now on I will focus on the Fefferman–Phong inequality in the framework of the Weyl calculus, for $N \times N$ systems whose (total) symbol is Hermitian nonnegative. Hence the context is fixed by second order symbols A such that

$$A(X) = A(X)^* \geq 0, \quad \forall X = (x, \xi) \in \mathbb{R}^n \times \mathbb{R}^n. \tag{5.4}$$

We have immediately a surprise when N $\geq$ 2: *The Fefferman–Phong inequality*

$$\exists C > 0, \quad (A^{\mathrm{w}}(x, D)u, u) \geq -C\|u\|_0^2, \quad \forall u \in \mathscr{S}(\mathbb{R}^n; \mathbb{C}^N),$$

does not hold in the full generality of condition (5.4) *above.*

The first one who wrote down a counterexample was R. Brummelhuis [5], followed by a series of counterexamples that I gave in [34] that are geometrically characterized (including also some anisotropic cases), and that are "robust" under lower order perturbations. In the same paper [34] I exploited the tool of the localized systems and clarified the relation between Brummelhuis' counterexample and a fundamental example of Hörmander in his paper on the Weyl calculus [14], of a 2×2 example of a Hermitian nonnegative symbol in dimension $n = 1$ whose Weyl quantization is not nonnegative. Namely, the 2×2 Hermitian nonnegative symbol

$$A_H(X) = \begin{bmatrix} \xi^2 & x\xi \\ x\xi & x^2 \end{bmatrix} = \begin{bmatrix} \xi \\ x \end{bmatrix}^* \otimes \begin{bmatrix} \xi \\ x \end{bmatrix}, \quad X \in \mathbb{R} \times \mathbb{R},$$

has Weyl quantization $A_H^{\mathrm{w}}(x, D)$ which may be not nonnegative: there exists $u_0 \in \mathscr{S}(\mathbb{R}; \mathbb{C}^2)$ such that $(A_H^{\mathrm{w}}(x, D)u_0, u_0) < 0$.

I next give, using system A_H, the simplest example I am aware of (see [39]).

5.1 *The Simplest Counterexample to the Fefferman–Phong Inequality in the Case of Systems*

I will be working with the (semiclassical) Hörmander metric $g = |dX|^2/M$, $M \gg 1$ being a large parameter. The Planck function is therefore $h(X) = 1/M$. Let $\chi_0 \in C_0^\infty(\mathbb{R} \times \mathbb{R})$ be supported in $\{|t| < 2\}$, with $0 \leq \chi_0 \leq 1$ and such that $\chi \equiv 1$ on $|t| \leq 1$. Let

$$0 \leq p_M(X) = p_M(X)^* = M\chi_0(\frac{X}{M^{1/2}})A_H(X) \in S(h^{-2}, g; \mathsf{M}_2), \tag{5.5}$$

where M_N denotes the set of $N \times N$ complex matrices.

One has the following result (see [39]; I recall here the easy proof; it will be useful in the following).

Lemma 5.2 *The system $p_M^{\mathrm{w}}(x, D)$ cannot satisfy the Fefferman–Phong inequality. That is, there are no constants $C \in \mathbb{R}$ and $M_0 \geq 1$ such that $p_M^{\mathrm{w}}(x, D) \geq -C$ for all $M \geq M_0$.*

Proof Suppose the Fefferman–Phong inequality holds, i.e. that there exists $C \in \mathbb{R}$ and $M_0 \geq 1$ such that (2.15) holds, that is, $p_M^{\mathrm{w}}(x, D) \geq -C$ for all $M \geq M_0$. Pick $u_0 \in \mathscr{S}(\mathbb{R}; \mathbb{C}^2)$ such that $(A_H^{\mathrm{w}}(x, D)u_0, u_0) < 0$. Since

$$\mathscr{S}' \ni a \longmapsto (a^{\mathrm{w}}(x, D)u, v) \quad \text{is continuous for all } u, v \in \mathscr{S},$$

we have that

$$(p_M^{\mathrm{w}}(x, D)u_0, u_0)/M \longrightarrow (A_H^{\mathrm{w}}(x, D)u_0, u_0), \quad \text{as } M \to +\infty,$$

whence

$$0 \leq \lim_{M \to +\infty} (p_M^{\mathrm{w}}(x, D)u_0, u_0)/M = (A_H^{\mathrm{w}}(x, D)u_0, u_0) < 0,$$

a contradiction. □

This is discouragingly simple! However, that is not yet the end of the story: there are positive results nevertheless, and it is still an open (and challenging) problem that of characterizing those systems, satisfying (5.4), for which the Fefferman–Phong inequality holds.

In the way of a "philosophical" reflection, loosely speaking in my opinion systems of ψdos are nonlinear objects: in $Pu = f$ the energy in phase-space of f is distributed among the various components of u through the matrix symbol of P which has, starting with its determinant, a "nonlinear" phase-space geometry.

5.2 Positive Results

I wish next to describe some positive results that one does have, which shed some light on what the actual phase-space geometry of a system may be.

Theorem 5.3 *Consider for $X \in \mathbb{R}^n \times \mathbb{R}^n$ the $N \times N$ Hermitian system*

$$p(X) = p(X)^* = A(x)e(\xi) + \sum_{j=1}^{N} B_j(x)\xi_j + C(x), \tag{5.6}$$

where $e \in S^2_{1,0}$ is a positive elliptic quadratic form in ξ and $A, B_j, C \in C^\infty_{L^\infty}(\mathbb{R}^n; \mathsf{M}_N)$. Suppose $p(X) \geq -c_0$. Then the Fefferman–Phong inequality holds for $p^{\mathrm{w}}(x, D)$.

Theorem 5.3 was first proved by L.-Y. Sung [46] in dimension $n = 1$ using a difficult expansion in Hermite eigenfunctions, and then by myself [35, 36] for general dimension n using a Calderón–Zygmund decomposition of phase-space in the spirit of Fefferman–Phong, that allowed for a reduction procedure to an $(N-1) \times (N-1)$ system and hence a finite induction on the size of the system.

Theorem 5.3 covers a nice class of systems (essentially generalizations of systems of ODEs), but Brummelhuis' counterexample and example (5.5) are not of the form (5.6). One may then naturally ask whether there are remedies to the counterexamples seen above. The symbol of Brummelhuis' counterexamples is

$$A_B(X) = \begin{bmatrix} \xi_1^2 & -ix_1\xi_1\xi_2 \\ ix_1\xi_1\xi_2 & x_1^2\xi_2^2 \end{bmatrix}, \quad X \in \mathbb{R}^2 \times \mathbb{R}^2.$$

While $A_B^{\mathrm{w}}(x, D)$ cannot satisfy the Fefferman–Phong inequality, one readily sees that the system

$$A_0(X) = A_B(X) + \xi_2^2 \begin{bmatrix} 0 & 0 \\ 0 & 1 \end{bmatrix}$$

does satisfy the Fefferman–Phong inequality. The main difference between A_B and A_0 is that the latter has an *elliptic* matrix-trace (that is, elliptic in the pseudodifferential calculus). By noting that, I then was able to prove the following result [38] (which is the best 2×2 result I am aware of).

Theorem 5.4 *Let g be a Hörmander metric. Let $a, b, c \in S(h^{-2}, g)$ be such that*

$$p(X) = \begin{bmatrix} a(X) & \overline{c(X)} \\ c(X) & b(X) \end{bmatrix} = p(X)^* \geq 0, \quad X = (x, \xi) \in \mathbb{R}^n \times \mathbb{R}^n.$$

Suppose the following ellipticity assumption on the matrix-trace of $p(X)$

$$\mathsf{Tr}(p(X)) \approx h(X)^{-2}, \tag{5.7}$$

and that

$$e_1 := a\{c, \bar{c}\} - 2i\,\mathsf{Im}(c\{a, \bar{c}\}),\ e_2 := b\{c, \bar{c}\} - 2i\,\mathsf{Im}(c\{b, \bar{c}\}) \in S(h^{-4}, g). \quad (5.8)$$

Then the Fefferman–Phong inequality (2.15) holds for $p^{\mathrm{w}}(x, D)$.

One may find in [39] (Corollary 14.3) an extension of the above theorem to the case of a Hermitian nonnegative system $p \in S(m, g; \mathsf{M}_2)$, where m is an admissible g-weight.

The next remark is in order, to clarify the role of the seemingly strange condition (5.8).

Remark 5.5 Note that a priori the symbols $e_1, e_2 \in S(h^{-5}, g)$. Hence condition (5.8) amounts to requiring that they be at most of order 4. When c is either real or purely imaginary the condition is fulfilled, and therefore such condition is fulfilled in the counterexamples A_B and p_M (see (5.5)). Note, however, that neither A_B nor p_M satisfy condition (5.7).

It is finally important to remark that (see [39]) conditions (5.7) and (5.8) are both symplectically invariant and invariant under conjugation by constant unitary matrices.

Following [38], one has an example of system for which (5.7) is fulfilled while (5.8) is not, and for which the Fefferman–Phong inequality does not hold. Indeed, one may consider the system

$$p(X) = \begin{bmatrix} \xi_1^2 & (1 + ix_1)\xi_1\xi_2 \\ (1 - ix_1)\xi_1\xi_2 & (1 + x_1^2)\xi_2^2 \end{bmatrix}.$$

Hence (5.7) is fulfilled while (5.8) is not. One may construct a 2×2 elliptic zeroth-order ψdo B^{w} acting on a vector $f = \begin{bmatrix} f_1 \\ f_2 \end{bmatrix}$ as $B^{\mathrm{w}} f = B_-^{\mathrm{w}} f_1 + B_+^{\mathrm{w}} f_2$ such that

$$(B^{\mathrm{w}})^* p^{\mathrm{w}} B^{\mathrm{w}} = \begin{bmatrix} (B_-^{\mathrm{w}})^* \Lambda_-^{\mathrm{w}} B_-^{\mathrm{w}} & O_{L^2 \to L^2}(1) \\ O_{L^2 \to L^2}(1) & (B_+^{\mathrm{w}})^* \Lambda_+^{\mathrm{w}} B_+^{\mathrm{w}} \end{bmatrix},$$

where the principal symbols $\lambda_\pm$ of $\Lambda_\pm^{\mathrm{w}}$ are respectively given by

$$\lambda_+(x, \xi) = \mathsf{Tr}\, p(x, \xi), \quad \lambda_-(x, \xi) = \xi_1 \xi_2^2 / \lambda_+(x, \xi).$$

Therefore the Fefferman–Phong inequality holds for p^{w} iff the Sharp Gårding inequality holds for $(B_-^{\mathrm{w}})^* \Lambda_-^{\mathrm{w}} B_-^{\mathrm{w}}$. But since λ_- may change sign, the Sharp Gårding cannot hold, and hence, in turn, the Fefferman–Phong for p^{w} may not either. This shows that condition (5.8) is necessary in order for the Fefferman–Phong inequality (2.15) to hold.

Recall next that a symbol t is *hypoelliptic in the calculus* if $1 \lesssim t \in S(t, g)$. The use of the *proper metric*, introduced by Lerner and Nourrigat in [19], yields the following corollary of Theorem 5.4.

Corollary 5.6 *Suppose G is a Hörmander metric and let H be the corresponding Planck's function. Let $0 \le p = p^* \in S(H^{-2}, G; \mathsf{M}_2)$. Suppose that $t := \mathrm{Tr}(p)$ is hypoelliptic and that condition (5.8) is fulfilled with respect to the weight $t^{1/2}$, that is, $e_1, e_2 \in S(t^2, G)$. Then p^{w} satisfies the Fefferman–Phong inequality.*

Proof Consider the associated proper metric

$$g_X := \frac{h(X)}{H(X)} G_X, \quad X \in \mathbb{R}^n \times \mathbb{R}^n,$$

where

$$h(X)^{-2} := 1 + t(X) + |t|_1^G(X)^{4/3} H(X)^{2/3} + |t|_2^G(X)^2 H(X)^2 + |t|_3^G(X)^4 H(X)^6$$

($|t|_k^G(X)$ being the norm of the kth-differential of t at X with respect to the metric G, see [15], or [17], or else [19]). Then one has that $G_X \lesssim g_X$ for all X and, because t is hypoelliptic, $h(X)^{-2} \approx t(X)$. The nonnegativity of p yields therefore that $p \in S(h^{-2}, g; \mathsf{M}_2)$ whence Theorem 5.4 gives the result. □

It is natural, at this point, to ask whether having a "subelliptic" trace $t(X)$ gives the same result. After all, important classes of subelliptic operators do have an associated Weyl calculus. However, the answer is in general no, as the following example shows.

Let χ_0 be the cut-off function considered in Sect. 5.1. Let $\alpha > 0$ be fixed and let $M \gg 1$ be a (large) parameter. Again, we consider the metric $g = |dX|^2/M$ on $\mathbb{R} \times \mathbb{R}$.

Lemma 5.7 *Consider $p \in S(h^{-2}, g; \mathsf{M}_2)$ given by*

$$p(X) = M \chi_0\Big(\frac{X}{M^{1/2}}\Big) \underbrace{\begin{bmatrix} \xi^2 & -ix\xi \\ ix\xi & x^2 + \alpha \end{bmatrix}}_{=:A(X)}, \quad X = (x, \xi) \in \mathbb{R} \times \mathbb{R}.$$

Then $p^{\mathrm{w}}(x, D)$ cannot satisfy the Fefferman–Phong inequality (2.15).

Proof One may find $u_0 \in \mathscr{S}(\mathbb{R}; \mathbb{C}^2)$ such that $(A^{\mathrm{w}}(x, D)u_0, u_0) < 0$. In fact, with $u_0 = \begin{bmatrix} u_1 \\ u_2 \end{bmatrix}$, one has

$$(A^{\mathrm{w}}(x, D)u_0, u_0) = \|Du_1 - ixu_2\|_0^2 + \mathrm{Re}\,(u_2, u_1) + \alpha \|u_2\|_0^2.$$

Let then $\varphi(x) = e^{-\lambda x^2/2}$, $\lambda > 0$ to be picked, and let $u_1 = \varphi - x\varphi'$, $u_2 = \varphi''$. It follows that $Du_1 - ixu_2 = 0$ and

$$(A^{\mathrm{w}}(x, D)u_0, u_0) = \mathrm{Re}\,(u_2, u_1) + \alpha \|u_2\|_0^2 = -\frac{1}{2}\|\varphi'\|_0^2 + \alpha \|\varphi''\|_0^2$$

$$= -\frac{\sqrt{\lambda\pi}}{4} + \frac{3}{4}\alpha\lambda^{3/2}\sqrt{\pi} = \frac{\sqrt{\lambda\pi}}{4}(3\alpha\lambda - 1) < 0$$

whenever λ is picked sufficiently small. The proof then proceeds as in Lemma 5.2 by noting that

$$(p^{\mathrm{w}}(x, D)u_0, u_0)/M \longrightarrow (A^{\mathrm{w}}(x, D)u_0, u_0) \quad \text{as } M \to +\infty.$$

This ends the proof. □

Notice that p is a semiclassical localization near $\{x_1 = \xi_1 = 0\}$ of the system

$$\begin{bmatrix} \xi_1^2 & -ix_1\xi_1\xi_2 \\ ix_1\xi_1\xi_2 & x_1^2\xi_2^2 + \alpha|\xi_2| \end{bmatrix} = A_B(X) + \alpha|\xi_2| \begin{bmatrix} 0 & 0 \\ 0 & 1 \end{bmatrix}$$

when $|\xi_2| \approx M$, whose matrix-trace is $\xi_1^2 + x_1^2\xi_2^2 + \alpha|\xi_2|$ which is a subelliptic operator (more precisely, a hypoelliptic operator with a loss of one derivative). This shows that the only way to "correct" Brummelhuis' counterexample (equivalently, counterexample p_M in (5.5)) is making its matrix trace *elliptic* (in the calculus).

Remark 5.8 In their study of the well-posedness of the Cauchy problem for (effectively) hyperbolic operators with triple characteristics [28], T. Nishitani and V. Petkov have to use an almost-positivity estimate resembling the Fefferman–Phong inequality (Lemma 3.4 of [28]). Although very much dependent on the particular system under study (a system which is a reduction of a scalar hyperbolic operator), the inequality is very interesting and new in nature, and could be a "natural" one in the sense that it holds under pretty general assumptions.

This, however, does not reduce the interest in obtaining the optimal class of pseudodifferential systems for which the "classical" Fefferman–Phong inequality holds.

5.3 Problems

I wish to conclude this survey by listing a few problems related to the Fefferman–Phong and the other inequalities we have seen for general systems.

- Find sufficient conditions for the Fefferman–Phong inequality to hold in the case of $N \times N$ systems. I have some work in progress in the case $N = 2$ when the matrix-trace is just nonnegative. I hope to complete the 2×2 case in the near future. This should give, at least in the 2×2 case, the optimal class (in the Weyl calculus) of pseudodifferential systems with nonnegative symbol for which the Fefferman–Phong inequality holds.
- Study the Melin, the Hörmander and the Fefferman–Phong inequalities for systems of ψdos in the case of a compact Lie group and, more generally, a nilpotent

Lie group. In the case of the Melin and the Hörmander inequality, in the first place one has to understand what is playing the invariant role of subprincipal symbol and Hamilton map, and in the case of the Fefferman–Phong inequality understand a semiclassical calculus on a Lie group. The pseudodifferential calculus of Ruzhanzky and Turunen (compact Lie group case) [43] and that of Fischer and Ruzhansky (nilpotent case) [10] are crucial to approach the problem. Also in this case I have some work in progress.

- Explore, in the case of compact Lie groups, the relation between local and global estimates to see how the geometry of the group influences the rigidity of the estimates.

References

1. Beals, R.: Square roots of nonnegative systems and the sharp Gårding inequality. J. Differ. Equ. **24**, 235–239 (1977)
2. Bony, J.M.: Sur l'inégalit de Fefferman-Phong. Seminaire: quations aux Drives Partielles, 1998–1999, Exp. No. III, 16 pp. Sémin. Équ. Dériv. Partielles, École Polytech. Palaiseau (1999)
3. Boulkhemair, A.: On the Fefferman-Phong inequality. Ann. Inst. Fourier **58**, 1093–1115 (2008)
4. Boutet de Monvel, L.: Hypoelliptic operators with double characteristics and related pseudo-differential operators. Commun. Pure Appl. Math. **27**, 585–639 (1974)
5. Brummelhuis, R.: Sur les inégalités de Gå rding pour les systm̀es d'opérateurs pseudo-différentiels. C. R. Acad. Sci. Paris **315** (Serie I), 149–152 (1992)
6. Brummelhuis, R., Nourrigat, J.: A necessary and sufficient condition for Melin's inequality for a class of systems. J. Anal. Math. **85**, 195–211 (2001)
7. Colombini, F., Del Santo, D., Zuily, C.: The Fefferman-Phong inequality in the locally temperate Weyl calculus. Osaka J. Math. **33**, 847–861 (1996)
8. Fefferman, C., Phong, D.H.: On positivity of pseudo-differential operators. Proc. Natl. Acad. Sci. U.S.A. **75**, 4673–4674 (1978)
9. Fischer, V., Ruzhansky, M.: Lower bounds for operators on graded Lie groups. C. R. Math. Acad. Sci. Paris **351**, 13–18 (2013)
10. Fischer, V., Ruzhansky, M.: Quantization on Nilpotent Lie Groups. Progress in Mathematics, vol. 314, p. xiii+557. Birkhäuser, Basel (2016)
11. Gårding, L.: Dirichlet's problem for linear elliptic partial differential equations. Math. Scand. **1**, 55–72 (1953)
12. Herau, F.: Melin inequality for paradifferential operators and applications. Commun. Partial Differ. Equ. **27**, 1659–1680 (2002)
13. Hörmander, L.: The Cauchy problem for differential equations with double characteristics. J. Anal. Math. **32**, 118–196 (1977)
14. Hörmander, L.: The Weyl calculus of pseudodifferential operators. Commun. Pure Appl. Math. **32**, 360–444 (1979)
15. Hörmander, L.: The Analysis of Linear Partial Differential Operators. III. Pseudodifferential Operators. Grundlehren der Mathematischen Wissenschaften, vol. 274. Springer, Berlin (1985)
16. Lax, P.D., Nirenberg, L.: On stability for difference schemes: a sharp form of Gårding's inequality. Commun. Pure Appl. Math. **19**, 473–492 (1966)
17. Lerner, N.: Metrics on the Phase Space and Non-Selfadjoint Pseudo-Differential Operators. Pseudo-Differential Operators Theory and Applications, vol. 3, p. xii+397. Birkhäuser, Basel (2010)

18. Lerner, N., Morimoto, Y.: On the Fefferman-Phong inequality and a Wiener-type algebra of pseudodifferential operators. Publ. Res. Inst. Math. Sci. **43**, 329–371 (2007)
19. Lerner, N., Nourrigat, J.: Lower bounds for pseudo-differential operators. Ann. Inst. Fourier **40**, 657–682 (1990)
20. Melin, A.: Lower bounds for pseudo-differential operators. Arkiv för Matematik **9**, 117–140 (1971)
21. Mohamed, A.: Étude spectrale d'opérateurs hypoelliptiques à caractéristiques multiples II. Commun. Partial Differ. Equ. **8**, 247–316 (1983)
22. Mughetti, M., Nicola, F.: A counterexample to a lower bound for a class of pseudodifferential operators. Proc. Am. Math. Soc. **132**, 3299–3303 (2004)
23. Mughetti, M., Nicola, F.: On the generalization of Hörmander's inequality. Commun. Partial Differ. Equ. **30**, 509–537 (2005)
24. Mughetti, M., Parenti, C., Parmeggiani, A.: Lower bound estimates without transversal ellipticity. Commun. Partial Differ. Equ. **32**, 1399–1438 (2007)
25. Nicola, F.: Hörmander's inequality for anisotropic pseudo-differential operators. J. Partial Differ. Equ. **15**, 49–64 (2002)
26. Nicola, F.: A lower bound for systems with double characteristics. J. Anal. Math. **96**, 297–311 (2005)
27. Nicola, F., Rodino, L.: Remarks on lower bounds for pseudo-differential operators. J. Math. Pures Appl. **83**, 1067–1073 (2004)
28. Nishitani, T., Petkov, V.: Cauchy problem for effectively hyperbolic operators with triple characteristics. 25. arXiv:1706.05965
29. Parenti, C., Parmeggiani, A.: Some remarks on almost-positivity of ψdo's. Boll. Unione Mat. Ital. Sez. B Artic. Ric. Mat. (8) **1**(1), 187–215 (1998)
30. Parenti, C., Parmeggiani, A.: A generalization of Hörmander's inequality I. Commun. Partial Differ. Equ. **25**, 457–506 (2000)
31. Parenti, C., Parmeggiani, A.: Lower bounds for systems with double characteristics. J. Anal. Math. **86**, 49–91 (2002)
32. Parenti, C., Parmeggiani, A.: On hypoellipticity with a big loss of derivatives. Kyushu J. Math. **59**, 155–230 (2005)
33. Parenti, C., Parmeggiani, A.: A remark on the Hörmander inequality. Commun. Partial Differ. Equ. **31**, 1071–1084 (2006)
34. Parmeggiani, A.: A class of counterexamples to the Fefferman-Phong inequality for systems. Commun. Partial Differ. Equ. **29**, 1281–1303 (2004)
35. Parmeggiani, A.: On the Fefferman-Phong inequality for systems of PDEs. Phase Space Analysis of Partial Differential Equations. Progress in Nonlinear Differential Equations and Their Applications, vol. 69, pp. 247–266. Birkhäuser, Boston (2006)
36. Parmeggiani, A.: On positivity of certain systems of partial differential equations. Proc. Natl. Acad. Sci. U.S.A. **104**, 723–726 (2007)
37. Parmeggiani, A.: Spectral Theory of Non-Commutative Harmonic Oscillators: An Introduction. Lecture Notes in Mathematics, vol. 1992. Springer, Berlin (2010)
38. Parmeggiani, A.: A remark on the Fefferman-Phong inequality for 2×2 systems. Pure Appl. Math. Q. **6**, 1081–1103 (2010). Special Issue: In honor of Joseph J. Kohn. Part 2
39. Parmeggiani, A.: On the problem of positivity of pseudodifferential systems. Studies in Phase Space Analysis with Applications to PDEs. Progress in Nonlinear Differential Equations and Their Applications, vol. 84, pp. 313–335. Springer, New York (2013)
40. Parmeggiani, A.: Non-commutative harmonic oscillators and related problems. Milan J. Math. **82**, 343–387 (2014)
41. Parmeggiani, A.: On the solvability of certain degenerate partial differential operators. Shocks, Singularities and Oscillations in Nonlinear Optics and Fluid Mechanics. Springer INdAM Series, vol. 17, pp. 151–179. Springer, Cham (2017)
42. Parmeggiani, A., Wakayama, M.: Non-commutative harmonic oscillators-I, -II. Forum Mathematicum **14** (2002), 539–604 ibid. 669–690

43. Ruzhansky, M., Turunen, V.: Pseudo-Differential Operators and Symmetries. Background Analysis and Advanced Topics. Pseudo-Differential Operators Theory and Applications, vol. 2, p. xiv+709. Birkhäuser, Basel (2010)
44. Ruzhansky, M., Turunen, V.: Sharp Gårding inequality on compact Lie groups. J. Funct. Anal. **260**, 2881–2901 (2011)
45. Sjöstrand, J.: Parametrices for Pseudodifferential Operators with Multiple Characteristics. Arkiv för Matematik **12**, 85–130 (1974)
46. Sung, L.-Y.: Semiboundedness of systems of differential operators. J. Differ. Equ. **65**, 427–434 (1986)
47. Tataru, D.: On the Fefferman-Phong inequality and related problems. Commun. Partial Differ. Equ. **27**, 2101–2138 (2002)
48. Toft, J.: Regularizations, decompositions and lower bound problems in the Weyl calculus. Commun. Partial Differ. Equ. **25**, 1201–1234 (2000)

Morrey Spaces from Various Points of View

Yoshihiro Sawano

Abstract This article overviews recent topic on Morrey spaces as well as the fundamental properties of Morrey spaces. The topics to be covered are:

(1) The origin of Morrey spaces–An observation made by C. Morrey
(2) Examples of the functions in Morrey spaces
(3) Morrey spaes from the viewpoint of functional analysis
(4) Interpolation of Morrey spaces
(5) Weighted Morrey spaces
(6) Various related spaces.

1 Morrey Spaces

In 1938, C. Morrey observed the following fact on functional analysis and applied it to partial differential equations [18]. Let $p > n$. If

$$\sup_{x \in \mathbb{R}^n,\, r>0} |B(x,r)|^{\frac{1}{p}-1}(\|f\|_{L^1(B(x,r))} + \|\nabla f\|_{L^1(B(x,r))}) < \infty,$$

then $f \in \mathrm{Lip}^{1-n/p}(\mathbb{R}^n)$. His main claim is that: we do not have to assume

$$\|f\|_p + \|\nabla f\|_p < \infty.$$

Based on this observation Peetre [28] defined the space $\mathcal{L}^q_\lambda$. In fact he defined

$$\|f\|_{\mathcal{L}^q_\lambda} \equiv \sup_{x \in \mathbb{R}^n,\, r>0} \left(\frac{1}{r^\lambda}\int_{B(x,r)} |f(y)|^q \, \mathrm{d}y\right)^{\frac{1}{q}}. \tag{1}$$

Here $1 \le q < \infty$ and $0 < \lambda < n$.

Y. Sawano (✉)
Department of Mathematics, Tokyo Metropolitan University, Minami-Ohsawa 1-1, Hachioji Tokyo 192-0397, Japan
e-mail: ysawano@tmu.ac.jp

L. G. Rodino and J. Toft (eds.), *Mathematical Analysis and Applications—Plenary Lectures*, Springer Proceedings in Mathematics & Statistics 262,
https://doi.org/10.1007/978-3-030-00874-1_5

1.1 Our Morrey Norm

Modulo the change of notation, we consider the following norm:

Let $1 \le q \le p < \infty$. For an $L^q_{\rm loc}(\mathbb{R}^n)$-function f its Morrey norm is defined by:

$$\|f\|_{\mathcal{M}^p_q} = \|f\|_{\mathcal{M}^p_q(\text{ball})} \equiv \sup_{x\in\mathbb{R}^n,\, r>0} |B(x,r)|^{\frac{1}{p}-\frac{1}{q}} \|f\|_{L^q(B(x,r))}.$$

Here $B(x,r)$ denotes the open ball centered at $x \in \mathbb{R}^n$ and the radius $r > 0$. The Morrey space $\mathcal{M}^p_q(\mathbb{R}^n)$ is the set of all $L^q(\mathbb{R}^n)$-locally integrable functions f for which the norm $\|f\|_{\mathcal{M}^p_q}$ is finite

One may replace balls by cubes.

$$\|f\|_{\mathcal{M}^p_q(\text{cube})} \equiv \sup_{x\in\mathbb{R}^n,\, r>0} |Q(x,r)|^{\frac{1}{p}-\frac{1}{q}} \|f\|_{L^q(Q(x,r))}. \tag{2}$$

Here $Q(x,r)$ denotes the open ball centered at $x \in \mathbb{R}^n$ and having volume $(2r)^n$.

Then there exists a constant $c = c_{n,p,q} > 0$ depending on n such that

$$c_{n,p,q}{}^{-1} \|f\|_{\mathcal{M}^p_q} \le \|f\|_{\mathcal{M}^p_q(\text{cube})} \le c_{n,p,q} \|f\|_{\mathcal{M}^p_q}$$

for all measurable functions f.

1.2 The Diagonal Case $p = q$

Theorem 1 *For all* $0 < q \le p < \infty$, $\mathcal{M}^p_p(\mathbb{R}^n) = L^p(\mathbb{R}^n)$ *with coincidence of norms.*

We write down the definition of the norm $\|f\|_{\mathcal{M}^p_p}$;

$$\|f\|_{\mathcal{M}^p_p} \equiv \sup_{\substack{x\in\mathbb{R}^n\\ r>0}} |B(x,r)|^{\frac{1}{p}-\frac{1}{p}} \|f\|_{L^p(B(x,r))} = \sup_{\substack{x\in\mathbb{R}^n\\ r>0}} \|f\|_{L^p(B(x,r))}.$$

So, since $B(x,r) \subset \mathbb{R}^n$, we have

$$\|f\|_{\mathcal{M}^p_p} \le \left(\int_{\mathbb{R}^n} |f(y)|^p \, dy \right)^{\frac{1}{p}} = \|f\|_{L^p}. \tag{3}$$

Meanwhile, by the monotone convergence theorem, we have

$$\left(\int_{\mathbb{R}^n} |f(y)|^p \, dy\right)^{\frac{1}{p}} = \lim_{m\to\infty} \left(\int_{B(x,m)} |f(y)|^p \, dy\right)^{\frac{1}{p}}$$
$$\leq \sup_{r>0} \left(\int_{B(x,r)} |f(y)|^p \, dy\right)^{\frac{1}{p}}$$
$$= \|f\|_{\mathcal{M}^p_p}. \tag{4}$$

Here $x \in \mathbb{R}^n$ is a fixed point. Combining these two observations, we obtain the desired result.

Write out the definition of the norm $\|\cdot\|_{\mathcal{M}^p_p}$. We thus notice $L^p = \mathcal{M}^p_p$.

We are thus interested in the case $q < p$.

1.3 Extermal Function

The Morrey space $\mathcal{M}^{n/\alpha}_q(\mathbb{R}^n)$ with $1 < q < n/\alpha$ contains $|x|^{-\alpha}$. To check this, we observe that for $f(x) = |x|^{-\alpha}$,

$$\sup_{x\in\mathbb{R}^n} |B(x,r)|^{\frac{1}{p}-\frac{1}{q}} \left(\int_{B(x,r)} |f(y)|^q \, dy\right)^{\frac{1}{q}} = |B(r)|^{\frac{1}{p}-\frac{1}{q}} \left(\int_{B(r)} |f(y)|^q \, dy\right)^{\frac{1}{q}} \sim 1, \tag{5}$$

where $B(r)$ is an abbreviation of $B(x,r)$ with x the origin.

1.4 L^p Versus $\mathcal{M}^p_q$

About the norm

$$\|f\|_{\mathcal{M}^p_q} = \sup_{x\in\mathbb{R}^n,\, r>0} |B(x,r)|^{\frac{1}{p}-\frac{1}{q}} \|f\|_{L^q(B(x,r))}$$

we have the following embedding using the Hölder inequality:

$$\mathcal{M}^p_q \hookleftarrow L^p.$$

1.5 Why Do We Need to Introduce the Parameter q?

From the definition of the norm,

$$\|f\|_{\mathcal{M}^p_q} = \|f\|_{\mathcal{M}^p_q(\text{ball})} \equiv \sup_{x\in\mathbb{R}^n,\, r>0} |B(x,r)|^{\frac{1}{p}-\frac{1}{q}} \|f\|_{L^q(B(x,r))}.$$

We have the nesting property:

Proposition 2 *For all measurable functions f,*

$$\|f\|_{\mathcal{M}^p_{q_1}} \le \|f\|_{\mathcal{M}^p_{q_0}}, \tag{6}$$

where $1 \le q_1 \le q_0 < p$. In particular, $\mathcal{M}^p_{q_0}(\mathbb{R}^n) \hookrightarrow \mathcal{M}^p_{q_1}(\mathbb{R}^n)$.

Proof We write out the norms in full:

$$\|f\|_{\mathcal{M}^p_{q_1}} \equiv \sup_{x\in\mathbb{R}^n,\, r>0} |B(x,r)|^{\frac{1}{p}-\frac{1}{q_1}} \left(\int_{B(x,r)} |f(y)|^{q_1}\, dy\right)^{\frac{1}{q_1}} \tag{7}$$

$$\|f\|_{\mathcal{M}^p_{q_0}} = \sup_{x\in\mathbb{R}^n,\, r>0} |B(x,r)|^{\frac{1}{p}-\frac{1}{q_0}} \left(\int_{B(x,r)} |f(y)|^{q_0}\, dy\right)^{\frac{1}{q_0}}. \tag{8}$$

By the Hölder inequality (for probability measure), we have

$$\left(\frac{1}{|B(x,r)|}\int_{B(x,r)} |f(y)|^{q_1}\, dy\right)^{\frac{1}{q_1}} \le \left(\frac{1}{|B(x,r)|}\int_{B(x,r)} |f(y)|^{q_0}\, dy\right)^{\frac{1}{q_0}}. \tag{9}$$

Inserting (7) and (8) to inequality (9), we have $\|f\|_{\mathcal{M}^p_{q_1}} \le \|f\|_{\mathcal{M}^p_{q_0}}$, □

So far, we proved $\mathcal{M}^p_{q_1} \supset \mathcal{M}^p_{q_2} \supsetneq L^p = \mathcal{M}^p_p$ when $p < q_2 < q_1$. Thus, we may ask ourselves whether the first inclusion is strict. Actually, we construct a set E such that

$$\chi_E \in \mathcal{M}^p_u \setminus \mathcal{M}^p_q$$

when $p > u > q > 1$ [30].

Let $p > q > 1$ and $R > 1$ be fixed so that

$$(R+1)^{-\frac{1}{p}} = 2^{\frac{1}{q}}(1+R)^{-\frac{1}{q}}. \tag{10}$$

For a vector $a \in \{0,1\}^n$, we define an Affine transformation T_a by

$$T_a(x) \equiv \frac{1}{R+1}x + \frac{R}{R+1}a \quad (x \in \mathbb{R}^n). \tag{11}$$

Let $E_0 \equiv [0,1]^n$. Suppose that we have defined $E_0, E_1, E_2, \dots, E_j$. Define

$$E_{j+1} \equiv \bigcup_{a\in\{0,1\}^n} T_a(E_j), \quad E_{j,0} \equiv [0,(1+R)^{-j}]^n. \tag{12}$$

Proposition 3 *We have*

$$\|\chi_{E_j}\|_{\mathcal{M}^p_q} \sim (1+R)^{-jn/p} = \|\chi_{E_{j,0}}\|_{\mathcal{M}^p_q} = \|\chi_{E_{j,0}}\|_p = \|\chi_{E_j}\|_q, \tag{13}$$

where the implicit constants in $\sim$ *does not depend on* j *but can depend on* p *and* q.

We use the Morrey norm $\|\cdot\|_{\mathcal{M}^p_q}$ is defined by cubes to check this.

Proposition 4 *Let* $1 \le q < p < \infty$. *Then* $\mathcal{M}^p_q \subset L^1 + L^\infty$ *does not hold. That is, there exists an* $\mathcal{M}^p_q$*-function that can not be expressed as the sum of* L^1 *and* L^∞.

Proof Let $n = 1$ for simplicity. Define

$$f = f_p := \sum_{j=100}^{\infty} [\log_2 \log_2 j]^{1/p} \chi_{[j!, j!+[\log_2 \log_2 j]^{-1}]}. \tag{14}$$

See also the proof of Lemma 6. From this we conclude that $\mathcal{M}^p_q$ collects some monsters. □

Remark 5 Note also that

$$\|f_p(\cdot + y) - f_p\|_{\mathcal{M}^p_q} \ge 1.$$

Thus, the set of all elements which are translation-continuous form a proper closed subspace.

Lemma 6 [9] *Let* $1 \le q < p < \infty$ *be fixed. If* $\kappa \gg 1$, *then*

$$F(x) = f(x_1) f(x_2) \cdots f(x_n) \quad (x = (x_1, x_2, \ldots, x_n) \in \mathbb{R}^n),$$

where f *is given by* (14), *belongs to* $\mathcal{M}^p_q(\mathbb{R}^n)$ *and has norm* 1.

Proof Since $\|F\|_{\mathcal{M}^p_q} = (\|f\|_{\mathcal{M}^p_q})^n$, we may assume $n = 1$. Let (a, b) be an interval which intersects the support of f. Testing on the function

$$[\log_2 \log_2 100]^{1/p} \chi_{[100!\kappa, 100!\kappa+[\log_2 \log_2 100]^{-1}]},$$

we see that $\|f\|_{\mathcal{M}^p_q} \ge 1$.

(1) Case 1 : $b - a < 2$. In this case, there exists uniquely $j \in \mathbb{N} \cap [100, \infty)$ such that $[a, b] \cap [j!\kappa, j!\kappa + [\log_2 \log_2 j]^{-1}] \ne \emptyset$. Thus,

$$
\begin{aligned}
&(b-a)^{\frac{1}{p}-\frac{1}{q}}\left(\int_a^b f(t)^q\,\mathrm{d}t\right)^{\frac{1}{q}}\\
&=(b-a)^{\frac{1}{p}-\frac{1}{q}}\left(\int_{\max(a,j!\kappa)}^{\min(b,j!\kappa+[\log_2\log_2 j]^{-1})} f(t)^q\,\mathrm{d}t\right)^{\frac{1}{q}}\\
&\le(\min(b,j!\kappa+[\log_2\log_2 j]^{-1})-\max(a,j!\kappa))^{\frac{1}{p}-\frac{1}{q}}\\
&\quad\times\left(\int_{\max(a,j!\kappa)}^{\min(b,j!\kappa+[\log_2\log_2 j]^{-1})} f(t)^q\,\mathrm{d}t\right)^{\frac{1}{q}}\\
&=[\log_2\log_2 j]^{\frac{1}{p}}(\min(b,j!\kappa+[\log_2\log_2 j]^{-1})-\max(a,j!\kappa))^{\frac{1}{p}}\\
&\le 1.
\end{aligned}
$$

(2) Case 2 : $b-a>2$. Set

$$
m:=\min([a,b]\cap\mathrm{supp}(f)),\quad M:=\max([a,b]\cap\mathrm{supp}(f)).
$$

Choose $j_m, j_M\in\mathbb{N}\cap[100,\infty)$ so that $m\in[j_m!, j_m!+j_m{}^{-1}]$ and $M\in[j_M!, j_M!+j_M{}^{-1}]$. We suppose $j_M>j_m$. Then we have

$$
b-a\ge M-m\ge\kappa(j_M!-j_m!-j_m{}^{-1})\ge\kappa(j_M!-j_m!-1)\ge\frac{\kappa}{50}j_M!.
$$

Thus,

$$
\begin{aligned}
(b-a)^{\frac{1}{p}-\frac{1}{q}}\left(\int_a^b f(t)^q\,\mathrm{d}t\right)^{\frac{1}{q}}&\le(j_M!-j_m!-1)^{\frac{1}{p}-\frac{1}{q}}\left(\int_{j_m!}^{j_M!+1} f(t)^q\,\mathrm{d}t\right)^{\frac{1}{q}}\\
&\le\left(\frac{\kappa}{50}j_M!\right)^{\frac{1}{p}-\frac{1}{q}}\left(\sum_{j=j_m}^{j_M}(\log_2\log_2 j)^{\frac{q-p}{p}}\right)^{\frac{1}{q}}\\
&\le 1
\end{aligned}
$$

as long as $\kappa\gg 1$.

Thus, $f\in\mathcal{M}_q^p$. □

1.6 A Copy of $\ell^\infty(\mathbb{N})$ in the Morrey Space $\mathcal{M}_q^p(\mathbb{R}^n)$ with $1\le q<p<\infty$

We show here that ℓ^∞ is embedded into $\mathcal{M}_q^p(\mathbb{R}^n)$ when $p\ne q$.

Theorem 7 *The Morrey space $\mathcal{M}_q^p(\mathbb{R}^n)$ with $1\le q<p<\infty$ has a copy of $\ell^\infty(\mathbb{N})$*

Proof We define

$$T(\{a_k\}_{k=1}^{\infty}) = \sum_{j=100}^{\infty} a_{j-99}[\log_2 \log_2 j]^{n/p} \chi_{[j!\kappa, j!\kappa+[\log_2 \log_2 j]^{-1}]^n}.$$

Then we have

$$\|T(\{a_k\}_{k=1}^{\infty})\|_{\mathcal{M}_q^p} \le \|\{a_k\}_{k=1}^{\infty}\|_{\ell^\infty} \|F\|_{\mathcal{M}_q^p} = \|\{a_k\}_{k=1}^{\infty}\|_{\ell^\infty}.$$

Meanwhile

$$\begin{aligned}
&\|T(\{a_k\}_{k=1}^{\infty})\|_{\mathcal{M}_q^p} \\
&\ge \|a_j[\log_2 \log_2(j+99)]^{n/p} \chi_{[(j+99)!\kappa,(j+99)!\kappa+[\log_2 \log_2(j+99)]^{-1}]^n}\|_{\mathcal{M}_q^p} \\
&= |a_j|.
\end{aligned}$$

Thus,

$$\|T(\{a_k\}_{k=1}^{\infty})\|_{\mathcal{M}_q^p} \ge \|\{a_k\}_{k=1}^{\infty}\|_{\ell^\infty}.$$

As a result T is an isometry from $\ell^\infty(\mathbb{N})$ to $\mathcal{M}_q^p(\mathbb{R}^n)$. □

Corollary 8 *Let $1 \le q < p < \infty$. Then the isometric copy of c_0 exists in $\mathcal{M}_q^p$. If $1 \le q_1 < q_2 < p < \infty$, then a common isometric copy of c_0 exists in $\mathcal{M}_{q_1}^p$ and $\mathcal{M}_{q_2}^p$.*

Let $1 \le q < p < \infty$. Then $\mathcal{M}_q^p$ is not reflexive. The key idea is to show that there exists a mapping $U : \ell^\infty \to \mathcal{M}_q^p$ such that $\|Ux\|_{\mathcal{M}_q^p} = \|x\|_{\ell^\infty}$.

We define

$$T(\{a_k\}_{k=1}^{\infty}) = \sum_{j=100}^{\infty} a_{j-99}[\log_2 \log_2 j]^{1/p} \chi_{[j!\kappa, j!\kappa+[\log_2 \log_2 j]^{-1}]^n}.$$

2 Sharp Maximal Operators

We define

$$M^\sharp f(x) \equiv \sup_{x \in Q} m_Q(|f - m_Q(f)|) = \sup_{x \in Q} \frac{1}{|Q|} \int_Q |f(y) - m_Q(f)| \, dy.$$

Here m_Q stands for the average.

Theorem 9 *Let $1 < p < \infty$. Then*

$$\|f\|_p \lesssim_p \|M^\sharp f\|_p \tag{15}$$

for every measurable function f *with* $\min(Mf, 1) \in L^p(\mathbb{R}^n)$ [6].

In the inequality

$$\|f\|_p \lesssim_p \|M^\sharp f\|_p,$$

the assumption $\min(Mf, 1) \in L^p(\mathbb{R}^n)$ wants to exclude the case $f \equiv 1$. In view of the definition

$$M^\sharp f(x) = \sup_{x \in Q} \frac{1}{|Q|} \int_Q |f(y) - m_Q(f)| \, dy,$$

this is necessary.

We wanted to exclude this type of assumption and we obtained

$$\|f\|_p \lesssim_p \|M^\sharp f\|_p + \|f\|_{\mathcal{M}_1^p} \quad (1 < p < \infty)$$

for all measurable functions f [34]. Note that neither

$$\|f\|_p \lesssim_p \|f\|_{\mathcal{M}_1^p}$$

nor

$$\|f\|_p \lesssim_p \|M^\sharp f\|_p$$

for all measurable functions f is true.

3 Atomic Decomposition in Morrey Spaces

We have the following decomposition theorem:

Theorem 10 ([11]) *Suppose* $1 < q \le p < \infty, \quad 1 < t \le s < \infty, \quad q < t, \quad p < s.$ *Assume that* $\{Q_j\}_{j=1}^\infty \subset \mathcal{D}(\mathbb{R}^n)$, $\{a_j\}_{j=1}^\infty \subset \mathcal{M}_t^s(\mathbb{R}^n)$ *and* $\{\lambda_j\}_{j=1}^\infty \subset [0, \infty)$ *fulfill*

$$\|a_j\|_{\mathcal{M}_t^s} \le |Q_j|^{1/s}, \quad \mathrm{supp}(a_j) \subset Q_j, \quad \left\| \sum_{j=1}^\infty \lambda_j \chi_{Q_j} \right\|_{\mathcal{M}_q^p} < \infty. \tag{16}$$

Then $f = \sum_{j=1}^\infty \lambda_j a_j$ *converges in* $\mathcal{S}'(\mathbb{R}^n) \cap L^q_{\mathrm{loc}}(\mathbb{R}^n)$ *and satisfies*

$$\|f\|_{\mathcal{M}_q^p} \le C_{p,q,s,t} \left\| \sum_{j=1}^\infty \lambda_j \chi_{Q_j} \right\|_{\mathcal{M}_q^p}. \tag{17}$$

Theorem 11 ([11]) *Suppose* $1 < q \le p < \infty$, $L \in \mathbb{N} \cup \{0\}$. *Let* $f \in \mathcal{M}^p_q(\mathbb{R}^n)$. *Then there exists a triplet* $\{\lambda_j\}_{j=1}^\infty \subset [0, \infty)$, $\{Q_j\}_{j=1}^\infty \subset \mathcal{Q}(\mathbb{R}^n)$ *and* $\{a_j\}_{j=1}^\infty \subset L^\infty(\mathbb{R}^n)$ *such that* $f = \sum_{j=1}^\infty \lambda_j a_j$ *in* $\mathcal{S}'(\mathbb{R}^n)$ *and that, for all* $v > 0$ $|a_j| \le \chi_{Q_j}$,

$$\int_{\mathbb{R}^n} x^\alpha a_j(x)\,\mathrm{d}x = 0, \quad \left\| \left(\sum_{j=1}^\infty (\lambda_j \chi_{Q_j})^v \right)^{1/v} \right\|_{\mathcal{M}^p_q} \le C_v \|f\|_{\mathcal{M}^p_q} \tag{18}$$

for all multi-indices α *with* $|\alpha| \le L$. *Here the constant* $v > 0$ *is independent of* f.

Remark 12 We mentioned that Morrey spaces go beyond $L^1 + L^\infty$. However, according to this decomposition, once the area Q_j (or the cube) is designated, the local behavior of the Morrey function in Q_j is rather tame. Thus, although Morrey spaces can contain "monsters", we have only to keep in mind the above example of the monster. As an application of Theorem 11, we have the following theorem:

Theorem 13 ([11, 30]) *Let* $0 < \alpha < n$, $1 < p \le p_0 < \infty$, $1 < q \le q_0 < \infty$ *and* $1 < r \le r_0 < \infty$. *Define*

$$I_\alpha f(x) = \int_{\mathbb{R}^n} \frac{f(y)}{|x-y|^{n-\alpha}}\,\mathrm{d}y.$$

Suppose that $q > r$, $\dfrac{1}{p_0} > \dfrac{\alpha}{n}$, $\dfrac{1}{q_0} \le \dfrac{\alpha}{n}$, *and that* $\dfrac{1}{r_0} = \dfrac{1}{q_0} + \frac{1}{p_0} - \dfrac{\alpha}{n}$, $\dfrac{r}{r_0} = \dfrac{p}{p_0}$. *Then there exists a constant* C *is independent of* f *and* g *such that*

$$\|g \cdot I_\alpha f\|_{\mathcal{M}^{r_0}_r} \le C \|g\|_{\mathcal{M}^{q_0}_q} \cdot \|f\|_{\mathcal{M}^{p_0}_p}.$$

4 Interpolation

4.1 Interpolation–General Spirit

What is interpolation? Interpolation theory reveals and studies many situations of the following kind. Suppose that X_0, X_1 are Banach spaces both contained continuously in some bigger containing space, and Y_0, Y_1 are Banach spaces both contained continuously in some (other) bigger containing space. Interpolation theory gives us various ways to construct and describe Banach spaces X and Y such that $T : X \to Y$ if $T|X_0 : X_0 \to Y_0$ and $T|X_1 : X_1 \to Y_1$ are bounded. An interpolation theory deals with function spaces X_0 and X_1 as if they are points in the plain; we consider a point which separate "the line segment X_0X_1" into $1-\theta : \theta$.

Let X_0, X_1 be complex quasi-Banach spaces. Then (X_0, X_1) is said to be a compatible couple, if there exists a topological vector space $(X, \mathcal{O}_X)$ into which

$(X_0, \|\star\|_{X_0})$ and $(X_1, \|\star\|_{X_1})$ are continuously embedded. When one needs to specify X, (X_0, X_1) is said to be a compatible couple embedded into a topological vector space X. The space X is sometimes called the *containing space*. The *sum quasi-Banach space* $X_0 + X_1$ is defined to be the algebraic sum of X_0 and X_1 as a linear subspace of X. That is, define

$$X_0 + X_1 \equiv \{x \in X \,:\, x_0 \in X_0,\ x_1 \in X_1,\ x = x_0 + x_1\}. \tag{19}$$

The norm of $X_0 + X_1$ is defined by

$$\|x\|_{X_0+X_1} \equiv \inf\{\|x_0\|_{X_0} + \|x_1\|_{X_1} \,:\, x_0 \in X_0, x_1 \in X_1, x = x_0 + x_1\}. \quad (x \in X_0 + X_1).$$

4.2 Calderón's First and Second Complex Interpolation Space

Definition 14 Let the pair (X_0, X_1) be a compatible couple of Banach spaces.

(1) The space $\mathcal{F}(X_0, X_1)$ is defined as the set of all functions $F \in \mathrm{BC}(\overline{U}, X_0 + X_1) \cap \mathcal{O}(U, X_0 + X_1)$ such that the functions $t \in \mathbb{R} \mapsto F(j + it) \in X_j$ are bounded and continuous on $\mathbb{R}$ for $j = 0, 1$. The space $\mathcal{F}(X_0, X_1)$ is equipped with the norm

$$\|F\|_{\mathcal{F}(X_0,X_1)} := \max\left\{\sup_{t\in\mathbb{R}} \|F(it)\|_{X_0},\ \sup_{t\in\mathbb{R}} \|F(1+it)\|_{X_1}\right\}.$$

(2) Let $\theta \in (0, 1)$. Define the complex interpolation space $[X_0, X_1]_\theta$ with respect to (X_0, X_1) to be the set of all functions $f \in X_0 + X_1$ such that $f = F(\theta)$ for some $F \in \mathcal{F}(X_0, X_1)$. The norm on $[X_0, X_1]_\theta$ is defined by

$$\|f\|_{[X_0,X_1]_\theta} := \inf\{\|F\|_{\mathcal{F}(X_0,X_1)} : f = F(\theta), F \in \mathcal{F}(X_0, X_1)\}.$$

Definition 15 Let the pair (X_0, X_1) be a compatible couple of Banach spaces.

(1) The space $\mathcal{G}(X_0, X_1)$ is defined as the set of all functions $G \in \mathrm{C}(\overline{U}, X_0 + X_1) \cap \mathcal{O}(U, X_0 + X_1)$ with polynomial growth such that $G(\cdot + ih) - G \in \mathcal{F}(X_0, X_1)$ for $j = 0, 1$ and $h \in \mathbb{R}$ and that

$$\|G\|_{\mathcal{G}(X_0,X_1)} := \sup_{h\in\mathbb{R}} |h|^{-1} \|G(\cdot + ih) - G\|_{\mathcal{F}(X_0,X_1)} < \infty.$$

(2) Let $\theta \in (0, 1)$. Define the complex interpolation space $[X_0, X_1]^\theta$ with respect to (X_0, X_1) to be the set of all functions $f \in X_0 + X_1$ such that $f = G'(\theta)$ for some $G \in \mathcal{G}(X_0, X_1)$. The norm on $[X_0, X_1]^\theta$ is defined by

$$\|f\|_{[X_0,X_1]^\theta} := \inf\{\|G\|_{\mathcal{F}(X_0,X_1)} : f = G'(\theta), G \in \mathcal{F}(X_0, X_1)\}.$$

In the case of Lebesgue spaces, we have

$$[L^{p_0}, L^{p_1}]^\theta = [L^{p_0}, L^{p_1}]_\theta = L^p,$$

where $1 \le p_0, p_1, p \le \infty$ and $0 < \theta < 1$ satisfy

$$p_0 \ne p_1, \quad \frac{1}{p} = \frac{1-\theta}{p_0} + \frac{\theta}{p_1}.$$

Reference [3]

Theorem 16 *Suppose that* $p_0, p_1, p, q_0, q_1, q \in (1, \infty)$, *and* $\theta \in (0, 1)$ *satisfies*

$$p_0 > p_1, \quad q_0 \le p_0, \quad q_1 \le p_1, \quad \frac{p_0}{q_0} = \frac{p_1}{q_1}$$

and

$$\frac{1}{p} = \frac{1-\theta}{p_0} + \frac{\theta}{p_1}, \quad \frac{1}{q} = \frac{1-\theta}{q_0} + \frac{\theta}{q_1}.$$

Then we have

(1) $[\mathcal{M}^{p_0}_{q_0}, \mathcal{M}^{p_1}_{q_1}]_\theta = \left\{ f \in \mathcal{M}^p_q \,:\, \lim_{a \to 0^+} \|(1 - \chi_{[a,a^{-1}]}(|f|)) f\|_{\mathcal{M}^p_q} = 0 \right\}$ [8, 17]

(2) $[\mathcal{M}^{p_0}_{q_0}, \mathcal{M}^{p_1}_{q_1}]^\theta = \mathcal{M}^p_q$ [14, 15]

We also note

$$[\tilde{\mathcal{M}}^{p_0}_{q_0}, \tilde{\mathcal{M}}^{p_1}_{q_1}]^\theta = \{f \in \mathcal{M}^p_q : \chi_{(a,b)}(|f|) \in \tilde{\mathcal{M}}^p_q\}.$$

Remark 17 (1) Reference [4] Cobos, Peetre and Persson pointed out that

$$[\mathcal{M}^{p_0}_{q_0}, \mathcal{M}^{p_1}_{q_1}]_\theta \subset \mathcal{M}^p_q$$

as long as $1 \le q_0 \le p_0 < \infty$, $1 \le q_1 \le p_1 < \infty$, and $1 \le q \le p < \infty$ satisfy

$$\frac{1}{p} = \frac{1-\theta}{p_0} + \frac{\theta}{p_1}, \quad \frac{1}{q} = \frac{1-\theta}{q_0} + \frac{\theta}{q_1}.$$

(2) As is shown by Lemarié-Rieusset, when an interpolation functor F satisfies

$$F[\mathcal{M}^{p_0}_{q_0}, \mathcal{M}^{p_1}_{q_1}] = \mathcal{M}^p_q$$

under the condition above then

$$\frac{q_0}{p_0} = \frac{q_1}{p_1}$$

holds [14].

4.3 Real Interpolation Functors

Definition 18 (*Real interpolation functor*) Let (X_0, X_1) be a compatible couple of quasi-Banach spaces, and let θ, p satisfy $0 < \theta < 1, 0 < p \leq \infty$.

(1) For $x \in X_0 + X_1$, define

$$\|x\|_{(X_0,X_1)_{\theta,p}} \equiv \left(\int_0^\infty (t^{-\theta} K(t,x))^p \frac{dt}{t} \right)^{\frac{1}{p}}.$$

(2) The real interpolation quasi-Banach space $((X_0, X_1)_{\theta,p}, \| \star \|_{(X_0,X_1)_{\theta,p}})$ (or shortly $(X_0, X_1)_{\theta,p}$) is the subspace of $X_1 + X_0$ given by

$$(X_0, X_1)_{\theta,p} \equiv \{x \in X_0 + X_1 \, : \|x\|_{(X_0,X_1)_{\theta,p}} < \infty\}.$$

The correspondence $(X_0, X_1) \mapsto (X_0, X_1)_{\theta,p}$ is called the *real interpolation functor* for each θ and p.

In the case of Lebesgue spaces, we have

$$[L^{p_0}, L^{p_1}]_{\theta,p} = L^p,$$

where $1 \leq p_0, p_1, p \leq \infty$ and $0 < \theta < 1$ satisfy

$$p_0 \neq p_1, \quad \frac{1}{p} = \frac{1-\theta}{p_0} + \frac{\theta}{p_1}.$$

Do we have $\mathcal{M}^4_2 = [\mathcal{M}^6_2, \mathcal{M}^3_2]_{\frac{1}{2}}$ or $\mathcal{M}^4_2 = [\mathcal{M}^6_2, \mathcal{M}^3_2]_{\frac{1}{2},q}$ for some q? The answer is no. This is checked by using my fractal example presented in the beginning of this note [14].

5 Weighted Morrey Spaces

It seems that the following two definitions are important.

Definition 19 (1) [12] The Komori–Shirai type weighted Morrey space $\mathcal{M}^p_q(w, w)$ is defined to be:

$$\|f\|_{\mathcal{M}^p_q(w,w)} \equiv \sup_{B \in \mathcal{B}} w(B)^{\frac{1}{p}-\frac{1}{q}} \left(\int_B |f(y)|^q w(y)\,dx \right)^{\frac{1}{q}} \tag{20}$$

for $1 \le q \le p < \infty$.

(2) [29] The Samko type weighted Morrey space $\mathcal{M}^p_q(1, w)$ is defined to be

$$\|f\|_{\mathcal{M}^p_q(1,w)} \equiv \sup_{B \in \mathcal{B}} |B|^{\frac{1}{p}-\frac{1}{q}} \left(\int_B |f(y)|^q w(y)\,dx \right)^{\frac{1}{q}}. \tag{21}$$

We are led to the following questions:

(1) What condition do we need to assume in order that the Hardy-Littlewood maximal operator M is bounded? Here M is given by

$$Mf(x) = \sup_{r>0} \frac{1}{r^n} \int_{B(x,r)} |f(y)|\,dy.$$

(2) What about the other fundamental operators in harmonic analysis? This includes the singular integral operators, the fractional integral operators and the fractional maximal operators.

Here we suppose $1 < q < p < \infty$ since the matters are reduced to weighted Lebesgue spaces when $p = q$.

We are fascinated with considering the following condition:

$$\frac{1}{|Q|} \int_Q |f(y)|\,dy \times \|\chi_Q\|_{\mathcal{M}^p_q(1,w)} \le C\|f\|_{\mathcal{M}^p_q(1,w)}$$

for all measurable functions f, or equivalently,

$$\frac{1}{|Q|} \|\chi_Q\|_{\mathcal{M}^p_q(1,w)'} \times \|\chi_Q\|_{\mathcal{M}^p_q(1,w)} \le C.$$

Here $\mathcal{M}^p_q(1, w)'$ stands for the Köthe dual of $\mathcal{M}^p_q(1, w)$. That is,

$$\mathcal{M}^p_q(1, w)' = \{f \,:\, f \text{ is measurable and } f \cdot g \in L^1 \text{ for all } g \in \mathcal{M}^p_q(1, w)\}.$$

Problem 20 Does this condition suffices to guarantee that the Hardy-Littlewood maximal operator is bounded on $\mathcal{M}^p_q(1, w)$?

(1) This necessary condition can be deduced easily from the necessary condition, since

$$\chi_Q(x) \frac{1}{|Q|} \int_Q f(y)\,dy \le Mf(x)$$

for all cubes Q and non-negative measurable functions f.

(2) For Lebesgue spaces $L^q(w)$, that is, when $p = q$, our condition is equivalent to the A_q condition. So, the answer to our problem is affirmative.

(1) Although we have $\|\chi_Q\|_{\mathcal{M}^p_q(w,w)} = w(Q)^{\frac{1}{p}}$. We do not have any information on $\|\chi_Q\|_{\mathcal{M}^p_q(1,w)}$. However, in [20] Nakamura showed that this quantity can be calculated with the help of sparse family:

$$\|\chi_Q\|_{\mathcal{M}^p_q(1,w)} \sim |Q|^{\frac{1}{p}-\frac{1}{q}} w(Q)^{\frac{1}{q}}.$$

(2) As we have explained, we need to consider the case of $p \neq q$. In this case, there is a gap between "global L^p-property of Morrey spaces" and "the $L^p(w)$-property of Morrey spaces". By this I would like to say that the parameter p in the Morrey space $\mathcal{M}^p_q$ reflect the global property and the parameter q in the Morrey space $\mathcal{M}^p_q$ reflect the local property.

Let us consider weighted Morrey spaces of Samko type.

Note that $w \in A_q$ if and only if

$$[w]_{A_q} = \sup_{Q \in \mathcal{Q}} w(Q) \left(w^{-\frac{1}{q-1}}(Q) \right)^{q-1} |Q|^{-q} < \infty.$$

Theorem 21 ([20]) *Let $1 < q \le p \le \infty$ and w be a weight.*

(1) If M is bounded on $\mathcal{M}^p_q(1, w)$, then w belongs to $\mathcal{B}^p_q$, that is,

$$\|\chi_Q\|_{\mathcal{M}^p_q(1,w)} \sim |Q|^{\frac{1}{p}-\frac{1}{q}} w(Q)^{\frac{1}{q}}.$$

(2) If $w \in \mathcal{B}^p_q \cap A_q$, then M is bounded on $\mathcal{M}^p_q(1, w)$.

What happens for the case of $w_\alpha(x) = |x|^\alpha$?

Theorem 22 ([20, 32])

(1) M is bounded on $\mathcal{M}^p_q(1, w_\alpha)$ if and only if $-n + n\left(1 - \dfrac{q}{p}\right) \le \alpha < n\left(1 - \dfrac{q}{p}\right) + n(q-1)$.

(2) M is bounded on $\mathcal{M}^p_q(w_\alpha, w_\alpha)$ if and only if [24]

$$-n < \alpha < n(p-1), \text{ or equivalently } w_\alpha \in A_p.$$

Reference [20]

Note that the requirement $n\left(1 - \dfrac{q}{p}\right) \le \alpha$ comes from $w \in \mathcal{B}^p_q$.

The following theorem describes a gap between the boundedness of the Hardy-Littlewood maximal operator M and the ith Riesz transform R_i on the weighted

Morrey space $\mathcal{M}^p_q(1, w)$. For a measurable function f, recall that we defined a function Mf by

$$Mf(x) \equiv \sup_{B \in \mathcal{B}} \frac{\chi_B(x)}{|B|} \int_B |f(y)| \mathrm{d}y \quad (x \in \mathbb{R}^n). \tag{22}$$

The mapping $M : f \mapsto Mf$ is called the *Hardy-Littlewood maximal operator*.

The singular integral operators, which are represented by the jth Riesz transform given by,

$$R_j f(x) \equiv \lim_{\varepsilon \downarrow 0} \int_{\mathbb{R}^n \setminus B(x,\varepsilon)} \frac{x_j - y_j}{|x - y|^{n+1}} f(y) \mathrm{d}y, \tag{23}$$

are integral operators with singularity (mainly at the origin).

Theorem 23 ([21]) *Let $1 < q \le p < \infty$. Let w be a weight for which the Hardy-Littlewood maximal operator M is bounded on $\mathcal{M}^p_q(1, w)$. Then the following conditions are equivalent:*

(1) The Riesz transform can be extended to a bounded operator on $\mathcal{M}^p_q(1, w)$.
(2) w satisfies

$$\int_1^\infty \frac{\mathrm{d}t}{t w(B(x, tr))} \lesssim \frac{1}{w(B(x, r))},$$

where the implicit constant does not depend on x and $r > 0$.

The next theorem describes a gap between the boundedness of the fractional maximal operator

$$M_\alpha f(x) = \sup_{r>0} \frac{1}{r^{n-\alpha}} \int_{B(x,r)} |f(y)| \, \mathrm{d}y$$

and the fractional integral operator

$$I_\alpha f(x) = \int_{\mathbb{R}^n} \frac{f(y)}{|x - y|^{n-\alpha}} \mathrm{d}y$$

on the weighted Morrey space $\mathcal{M}^p_q(1, w)$.

Theorem 24 ([22]) *Let $1 < q \le p < \infty$. Let w be a weight for which fractional maximal operator M_α is bounded on $\mathcal{M}^p_q(1, w)$. Then the following conditions are equivalent:*

(1) The operator I_α extends to a bounded operator on $\mathcal{M}^p_q(1, w)$.
(2) w satisfies $\displaystyle\int_1^\infty \frac{\mathrm{d}t}{t w(B(x, tr))} \lesssim \frac{1}{w(B(x, r))}$, where the implicit constant does not depend on x and $r > 0$.

As a corollary, for the singular integral operators, we have the following result:

Theorem 25 ([21]) *The Riesz transforms $R_1, R_2, \ldots, R_n$ are all bounded on $\mathcal{M}^p_q(1, w_\alpha)$ if and only if $n\left(1-\dfrac{q}{p}\right) < \alpha < n\left(1-\dfrac{q}{p}\right) + n(q-1)$.*

We need to rule out the possibility of $n\left(1-\frac{q}{p}\right) = \alpha$ compared to the Hardy-Littlewood maximal operators. In a word, this is because we need to consider the (weighted) integral condition as is observed in the case of generalized Morrey spaces.

The condition $w \in A_q$ is NOT NECESSARY in order that M is bounded on $\mathcal{M}^p_q(1, w)$.

But we should keep in mind that $w \in A_q$ under a mild condition:

Theorem 26 (Sawano and Nakamura) *Let $w \in \mathcal{B}^p_q$. The following are equivalent:*

(1) M is bounded on $\mathcal{M}^p_q(1, w)$ and $w^{-1} \in A_\infty = \bigcup_{u>1} A_u$.
(2) $w \in A_q$.

6 Various Related Spaces

6.1 Weak Morrey Spaces

Consider the norm:

$$\|f\|_{\mathrm{w}\mathcal{M}^p_q} = \sup_{\lambda>0} \lambda \|\chi_{(\lambda,\infty]}(|f|)\|_{\mathcal{M}^p_q}.$$

The Hardy-Littlewood maximal operator is bounded on $\mathcal{M}^p_q$ if $1 < q \le p < \infty$. When $q = 1$, the Hardy-Littlewood maximal operator is bounded on $\mathcal{M}^p_q$ if $1 \le q \le p < \infty$.

Theorem 27 *For all measurable functions f,*

$$\|Mf\|_{\mathrm{w}\mathcal{M}^p_1} \simeq \|f\|_{\mathcal{M}^p_1}.$$

Proof The inequality $\|Mf\|_{\mathrm{w}\mathcal{M}^p_1} \lesssim \|f\|_{\mathcal{M}^p_1}$ is somewhat known. So we omit the proof. We seek to show

$$\|Mf\|_{\mathrm{w}\mathcal{M}^p_1} \gtrsim \|f\|_{\mathcal{M}^p_1}.$$

In fact, for a fixed ball $B = B(x, r)$, we have only to show

$$\|Mf\|_{\mathrm{w}\mathcal{M}^p_1} \gtrsim |B(x,r)|^{\frac{1}{p}-1} \int_{B(x,r)} |f(z)|\,\mathrm{d}z.$$

We remark that

$$\chi_Q(x) \frac{1}{|Q|} \int_Q |f(y)|\,\mathrm{d}y \le Mf(x).$$

Meanwhile, by the definition of the norm,

$$\|Mf\|_{\mathrm{w}\mathcal{M}_1^p} \ge |B(x,r)|^{\frac{1}{p}-1}\lambda|\{z \in B(x,r) \,:\, Mf(z) > \lambda\}|$$

for all $\lambda > 0$. If we choose

$$\lambda = (1-\varepsilon)\frac{1}{|B(x,r)|}\int_{B(x,r)} |f(y)|\,\mathrm{d}y,$$

then we have

$$\|Mf\|_{\mathrm{w}\mathcal{M}_1^p} \ge (1-\varepsilon)|B(x,r)|^{\frac{1}{p}-1}\int_{B(x,r)} |f(y)|\,\mathrm{d}y.$$

Since $\varepsilon > 0$ is arbitrary, we obtain the desired conclusion. □

6.2 Generalized Morrey Spaces

Generalized Morrey spaces play the role of compensating for the failure of the integral operators. It is known that the fractional integral operator I_α maps $\mathcal{M}_q^p$ to $\mathcal{M}_t^s$ when $1 < q \le p < \infty$, $1 < t \le s < \infty$ satisfy

$$\frac{q}{p} = \frac{t}{s}, \quad \frac{1}{s} = \frac{1}{p} - \frac{\alpha}{n}.$$

See [1].

Generalized Morrey spaces are used to compensate for the failure of this boundedness when $s = \infty$. We replace $t^{1/p}$ with a function here. Let $1 \le q < \infty$. One defines

$$\|f\|_{\mathcal{M}_q^\varphi} \equiv \sup_{x\in\mathbb{R}^n, r>0} \varphi(r)\left(\frac{1}{|Q(x,r)|}\int_{Q(x,r)} |f(y)|^q\,\mathrm{d}y\right)^{\frac{1}{q}}. \tag{24}$$

See the paper [30]. Some prefer to define

$$\|f\|_{\mathcal{M}_q^\varphi} \equiv \sup_{x\in\mathbb{R}^n, r>0} \frac{1}{\varphi(r)}\left(\frac{1}{|Q(x,r)|}\int_{Q(x,r)} |f(y)|^q\,\mathrm{d}y\right)^{\frac{1}{q}}. \tag{25}$$

See the original paper by Nakai [23].

We now claim that without the loss of generality we may assume

$$\phi(t) \text{ is nondecreasing and } \phi(t)^p t^{-n} \text{ is nonincreasing.} \tag{26}$$

Indeed, if we let

$$\phi_1(t) = \sup_{t' \in [0,t]} \phi(t')$$

then

$$\|f\|_{p,\phi} \le \|f\|_{p,\phi_1} \le 2^{n/p} \|f\|_{p,\phi}.$$

This holds by using the simple geometric fact that for any cubes $Q \in \mathcal{Q}$ and any positive numbers $t' \le \ell(Q)$

$$\frac{1}{|Q|}\int_Q |f(x)|^p \, dx \le 2^n \sup_{Q' \in \mathcal{Q}:\, Q' \subset Q, \ell(Q')=t'} \frac{1}{|Q'|}\int_{Q'} |f(x)|^p \, dx.$$

Next, if we let

$$\phi_2(t) = t^{n/p} \sup_{t' \ge t} \phi(t') t'^{-n/p}$$

then $\|f\|_{p,\phi} = \|f\|_{p,\phi_2}$.

This observation was made initially by Nakai in 2000 [25].

6.3 Various Orlicz–Morrey Spaces

Let

$$\|f\|_{Q, L \log L} = \inf\left\{\lambda > 0 : \int_Q \frac{|f(x)|}{\lambda} \log\left(3 + \frac{|f(x)|}{\lambda}\right) dx \le |Q|\right\}.$$

The space $\mathcal{M}^p_{L \log L}$ stands for the set of all measurable functions f for which $\|f\|_{\mathcal{M}^p_{L \log L}} = \sup_Q |Q|^{\frac{1}{p}} \|f\|_{Q, L \log L} < \infty$. According to [31], we have $\|Mf\|_{\mathcal{M}^p_1} \simeq \|f\|_{\mathcal{M}^p_{L \log L}}$, when $1 < p < \infty$, a good contrast to $\|Mf\|_{\mathrm{w}\mathcal{M}^p_1} \simeq \|f\|_{\mathcal{M}^p_1}$.

To the best knowledge of the authors, there exist three generalized Orlicz–Morrey spaces.

Definition 28 Let $\Phi : \mathbb{R}^n \times [0, \infty) \to [0, \infty)$ and $\phi : \mathcal{Q} \to (0, \infty)$ be suitable functions. Let f be a measurable function.

(1) For a cube $Q \in \mathcal{Q}$ define the (ϕ, Φ)-*average* over Q of f by:

$$\|f\|_{(\phi,\Phi);Q} \equiv \inf\left\{\lambda > 0 : \frac{\phi(Q)}{|Q|}\int_Q \Phi\left(x, \frac{|f(x)|}{\lambda}\right) dx \le 1\right\}.$$

Define the *generalized Orlicz–Morrey space* $\mathcal{L}_{\phi,\Phi}(\mathbb{R}^n)$ *of the first kind* to be the Banach space equipped with the norm: $\|f\|_{\mathcal{L}_{\phi,\Phi}} \equiv \sup\{\|f\|_{(\phi,\Phi);Q} : Q \in \mathcal{Q}\}$.

(2) For a cube $Q \in \mathcal{Q}$ define the Φ-*average* over Q of f by:

$$\|f\|_{\Phi;Q} \equiv \inf\left\{\lambda > 0 : \frac{1}{|Q|}\int_Q \Phi\left(x, \frac{|f(x)|}{\lambda}\right) dx \le 1\right\}.$$

Define the *generalized Orlicz–Morrey space* $\tilde{\mathcal{M}}_{\phi,\Phi}(\mathbb{R}^n)$ *of the second kind* to be the Banach space equipped with the norm: $\|f\|_{\tilde{\mathcal{M}}_{\phi,\Phi}} \equiv \sup\{\phi(Q)\|f\|_{\Phi;Q} : Q \in \mathcal{Q}\}$.

Let Φ be a Young function. Recall that the *Orlicz norm* $\|f\|_{L^\Phi(E)}$ over a measurable set E in $\mathbb{R}^n$ is defined by:

$$\|f\|_{L^\Phi(E)} \equiv \inf\left\{\lambda > 0 : \int_E \Phi\Big(\frac{|f(x)|}{\lambda}\Big)dx \le 1\right\}. \tag{27}$$

The *generalized Orlicz–Morrey space* $\mathcal{M}_{\phi,\Phi}(\mathbb{R}^n)$ *of the third kind* is defined as the set of all measurable functions f for which the norm

$$\|f\|_{\mathcal{M}_{\phi,\Phi}} \equiv \sup_{Q\in\mathcal{Q}} \frac{1}{\phi(\ell(Q))}\Phi^{-1}\left(\frac{1}{|Q|}\right)\|f\|_{L^\Phi(Q)}$$

is finite.

The spaces $\mathcal{L}_{\phi,\Phi}(\mathbb{R}^n)$, $\tilde{\mathcal{M}}_{\phi,\Phi}(\mathbb{R}^n)$ and $\mathcal{M}_{\phi,\Phi}(\mathbb{R}^n)$ are defined by Nakai in [26] (with Φ independent of x), by Sawano, Sugano and Tanaka in [31] (with Φ independent of x) and by Deringoz, Guliyev and Samko in [5, Definition 2.3], respectively. According to the examples in [7], we can say that the scales $\mathcal{L}$ and $\tilde{\mathcal{M}}$ are different and that $\tilde{\mathcal{M}}$ and $\mathcal{M}$ are different.

6.4 *Morrey Spaces on General Measure on* $\mathbb{R}^n$

Let $1 < q \le p < \infty$, $k > 1$, and let μ be a Radon measure on $\mathbb{R}^n$. Then define

$$\|f\|_{\mathcal{M}^p_q(\mathrm{cube})} \equiv \sup_{x\in\mathbb{R}^n,\, r>0} \mu(Q(x,kr))^{\frac{1}{p}-\frac{1}{q}}\|f\chi_{Q(x,r)}\|_{L^q(\mu)}.$$

According to Sawano and Tanaka a different choice of $k > 1$ yields equivalent norms; see [32]. When $p \ne q$, this type of spaces can be connected to Gaussian metric measure spaces. See the paper by Liu, Sawano and Yang [16].

6.5 *Smoothness Morrey Spaces–Related to Besov Spaces*

Let $\psi \in C^\infty_c$ satisfy $\chi_{B(1)} \le \psi \le \chi_{B(2)}$. Define $\psi_0 \equiv \psi$, $\psi_j \equiv \psi(2^{-j}\cdot) - \psi(2^{-j+1}\cdot)$. Define the Littlewood–Paley decomposition operator S_j by

$$S_j f \equiv \mathcal{F}^{-1}[\psi_j \mathcal{F} f]$$

for $f \in \mathcal{S}'(\mathbb{R}^n)$. We define

$$\|f\|_{\mathcal{N}^s_{pqr}} \equiv \|S_0 f\|_{\mathcal{M}^p_q} + \left(\sum_{j=1}^{\infty} (2^{js} \|S_j f\|_{\mathcal{M}^p_q})^r \right)^{\frac{1}{r}}, \tag{28}$$

$$\|f\|_{\mathcal{E}^s_{pqr}} \equiv \|S_0 f\|_{\mathcal{M}^p_q} + \left\| \left(\sum_{j=1}^{\infty} 2^{jrs} |S_j f|^r \right)^{\frac{1}{r}} \right\|_{\mathcal{M}^p_q} \tag{29}$$

for $f \in \mathcal{S}'(\mathbb{R}^n)$.

See [13, 33, 36].

Taking advantage of $L^p \subset \mathcal{M}^p_q$, we refine the embedding.

Example 29 The embedding $\mathcal{N}^{n/p+\varepsilon}_{pq\infty} \hookrightarrow \mathcal{C}^\varepsilon$ extends and refines the Sobolev embedding [10, 19, 27, 37] which Morrey proved in 1938 [18].

7 Other Open Problems

The Morrey space $\mathcal{M}^{n/\alpha}_q(\mathbb{R}^n)$ with $1 < q < n/\alpha$ contains the extermal function $|x|^{-\alpha}$.

So, I want to propose a problem [2]:

Problem 30 There are many weighted inequalities on Lebesgue spaces such as the one for the Hardy operators. But the constant are not attained by any of them in most cases. I propose to use Morrey spaces to grasp the best constant and send the inequality to the limit $q \to p$. My question is: How many functions to attain the maximum are theres?

Problem 31 Show that the Orlicz–Morrey scales $\mathcal{L}$ and $\mathcal{M}$ are different.

Acknowledgements I thank the organizer for giving me a chance to talk in this conference.

References

1. Adams, D.R.: A note on Riesz potentials. Duke Math. J. **42**, 765–778 (1975)
2. Batbold, T., Sawano, Y.: Sharp bounds for m-linear Hilbert-type operators on the weighted Morrey spaces. Math. Inequalities Appl. **20**(2017). no. 1, 263–283
3. Calderón, A.P.: Intermediate spaces and interpolation, the complex method. Studia Math. **14**, 113–190 (1964). no. 1, 46–56

4. Cobos, F., Peetre, J., Persson, L.E.: On the connection between real and complex interpolation of quasi-Banach spaces. Bull. Sci. Math. **122**, 17–37 (1998)
5. Deringoz, F., Guliyev, V.S., Samko, S.: Boundedness of maximal and singular operators on generalized Orlicz-Morrey spaces. Operator Theory, Operator Algebras and Applications. Operator Theory: Advances and Applications, vol. 242, pp. 139–158. Birkhäuser, Basel (2014)
6. Fefferman, C., Stein, E.: H^p spaces of several variables. Acta Math. **129**, 137–193 (1972)
7. Gala, S., Sawano, Y., Tanaka, H.: A remark on two generalized Orlicz-Morrey spaces. J. Approx. Theory **198**, 1–9 (2015)
8. Hakim, D.I., Sawano, Y.: Interpolation of generalized Morrey spaces. Rev. Math. Complut. **29**(2), 295–340 (2016)
9. Hakim, D.I., Sawano, Y.: Calderón's first and second complex interpolations of closed subspaces of Morrey spaces. J. Fourier Anal. Appl. (online)
10. Haroske, D.D., Skrzypczak, L.: Embeddings of Besov-Morrey spaces on bounded domains. Studia Math. **218**(2), 119–144 (2013)
11. Iida, T., Sawano, Y., Tanaka, H.: Atomic decomposition for Morrey spaces. Z. Anal. Anwend. **33**(2), 149–170 (2014)
12. Komori, Y., Shirai, S.: Weighted Morrey spaces and a singular integral operator. Math. Nachr. **282**(2), 219–231 (2009)
13. Kozono, H., Yamazaki, M.: Semilinear heat equations and the Navier-Stokes equation with distributions in new function spaces as initial data. Commun. PDE **19**, 959–1014 (1994)
14. Lemarie-Rieusset, P.G.: Multipliers and Morrey spaces. Potential Anal. **38**(3), 741–752 (2013)
15. Lemarie-Rieusset, P.G.: Erratum to: Multipliers and Morrey spaces. Potential Anal. **41**(4), 1359–1362 (2014)
16. Liu, L.G., Sawano, Y., Yang, D.: Morrey-type spaces on Gauss measure spaces and boundedness of singular integrals. J. Geom. Anal. **24**(2), 1007–1051 (2014)
17. Lu, Y., Yang, D., Yuan, W.: Interpolation of Morrey spaces on metric measure spaces. Can. Math. Bull. **57**, 598–608 (2014)
18. Morrey Jr., C.B.: On the solutions of quasi-linear elliptic partial differential equations. Trans. Am. Math. Soc. **43**(1), 126–166 (1938)
19. Najafov, A.M.: Some properties of functions from the intersection of Besov-Morrey type spaces with dominant mixed derivatives. Proc. A Razmadze Math. Inst. **139**, 71–82 (2005)
20. Nakamura, S.: Generalized weighted Morrey spaces and classical operators. Math. Nachr. **289**(17–18), 2235–2262 (2016). https://doi.org/10.1002/mana.201500260
21. Nakamura, S., Sawano, Y.: The singular integral operator and its commutator on weighted Morrey spaces. Collect. Math. **68**, 145–174 (2017)
22. Nakamura, S., Sawano, Y., Tanaka, H.: The fractional operators on weighted Morrey spaces. J. Geom. Anal. **28**(2018), no. 2, 1502–1524
23. Nakai, E.: Hardy-Littlewood maximal operator, singular integral operators, and the Riesz potential on generalized Morrey spaces. Math. Nachr. **166**, 95–103 (1994)
24. Nakamura, S., Sawano. Y., Tanaka, H.: Weighted local Morrey spaces. (in preparation)
25. Nakai, E.: A characterization of pointwise multipliers on the Morrey spaces. Sci. Math. **3**, 445–454 (2000)
26. Nakai, E.: The Campanato, Morrey and Hölder spaces on spaces of homogeneous type. Studia Math. **176**(1), 1–19 (2006)
27. Netrusov, Y.V.: Some imbedding theorems for spaces of Besov–Morrey type (Russian) Numerical methods and questions in the organization of calculations, 7. Zap. Nauchn. Sem. Leningrad. Otdel. Mat. Inst. Steklov. (LOMI) **139** (1984), 139–147
28. Peetre, J.: On the theory of $\mathcal{L}_{p,\lambda}$. J. Funct. Anal. **4**, 71–87 (1969)
29. Samko, N.: Weighted Hardy and singular operators in Morrey spaces. J. Math. Anal. Appl. **350**, 56–72 (2009)
30. Sawano, Y., Sugano, S., Tanaka, H.: Generalized fractional integral operators and fractional maximal operators in the framework of Morrey spaces. Trans. Am. Math. Soc. **363**(12), 6481–6503 (2011)

31. Sawano, Y., Sugano, S., Tanaka, H.: Orlicz-Morrey spaces and fractional operators. Potential Anal. **36**(4), 517–556 (2012)
32. Sawano, Y., Tanaka, H.: Morrey spaces for non-doubling measures. Acta Math. Sin. **21**(6), 1535–1544 (2005)
33. Sawano, Y., Tanaka, H.: Decompositions of Besov-Morrey spaces and Triebel-Lizorkin-Morrey spaces. Math. Z. **257**(4), 871–905 (2007)
34. Sawano, Y., Tanaka, H.: Sharp maximal inequalities and commutators on Morrey spaces with non-doubling measures. Taiwan. J. Math. **11**(4), 1091–1112 (2007)
35. Tanaka, H.: Two-weight norm inequalities on Morrey spaces. Annales Academi Scientiarum Fennic Mathematica Volumen **40**, 773–791 (2015)
36. Tang, L., Xu, J.: Some properties of Morrey type Besov-Triebel spaces. Math. Nachr. **278**, 904–917 (2005)
37. Yuan, W., Sickel, W., Yang, D.: Morrey and Campanato Meet Besov, Lizorkin and Triebel. Lecture Notes in Mathematics, vol. 2005, pp. xi+281. Springer, Berlin (2010)

The Grossmann–Royer Transform, Gelfand–Shilov Spaces, and Continuity Properties of Localization Operators on Modulation Spaces

Nenad Teofanov

Abstract This paper offers a review of results concerning localization operators on modulation spaces, and related topics. However, our approach, based on the Grossmann–Royer transform, gives a new insight and (slightly) different proofs. We define the Grossmann–Royer transform as interpretation of the Grossmann–Royer operator in the weak sense. Although such transform is essentially the same as the cross-Wigner distribution, the proofs of several known results are simplified when it is used instead of other time-frequency representations. Due to the importance of their role in applications when dealing with ultrafast decay properties in phase space, we give a detailed account on Gelfand–Shilov spaces and their dual spaces, and extend the Grossmann–Royer transform and its properties in such context. Another family of spaces, modulation spaces, are recognized as appropriate background for time-frequency analysis. In particular, the Gelfand–Shilov spaces are revealed as projective and inductive limits of modulation spaces. For the continuity and compactness properties of localization operators we employ the norms in modulation spaces. We define localization operators in terms of the Grossmann–Royer transform, and show that such definition coincides with the usual definition based on the short-time Fourier transform.

1 Introduction

The Grossmann–Royer operators originate from the problem of physical interpretation of the Wigner distribution, see [44, 62]. It is shown that the Wigner distribution is related to the expectation values of such operators which describe reflections about the phase space point. Apart from this physically plausible interpretation of the Grossmann–Royer operators (sometimes called the parity operators), they are closely related to the Heisenberg operators, the well known objects from quantum

---

N. Teofanov (✉)
Department of Mathematics and Informatics, University of Novi Sad,
Trg Dositeja Obradovica 4, 21000 Novi Sad, Serbia
e-mail: nenad.teofanov@dmi.uns.ac.rs

L. G. Rodino and J. Toft (eds.), *Mathematical Analysis and Applications—Plenary Lectures*, Springer Proceedings in Mathematics & Statistics 262,
https://doi.org/10.1007/978-3-030-00874-1_6

mechanics as "quantized" variants of phase space translations. We refer to [23–25] for such relation, basic properties of those operators, and the definition of the cross-Wigner distribution $W(f, g)$ in terms of the Grossmann–Royer operator Rf.

As it is observed in de Gosson [23] it is a pity that the Grossmann–Royer operators are not universally known. One of the aims of this paper is to promote those operators by defining the Grossmann–Royer transforms $R_g f$ as their weak interpretation. The Grossmann–Royer transform is in fact the cross-Wigner distribution multiplied by 2^d, see Definition 1, Lemma 1 and [25, Definition 12]. Thus all the results valid for $W(f, g)$ can be formulated in terms of $R_g f$. Nevertheless, due to its physical meaning, we treat the Grossmann–Royer transform as one of the time-frequency representations and perform a detailed exposition of its properties. Our calculations confirm the remark from [23, Chapter 8.3] according to which the use of $R_g f$ (instead of $V_g f$, $W(f, g)$ and $A(f, g)$, see Definition 1) "allows one to considerably simplify many proofs".

One aim of this paper is to give the interpretation of localization operators by the means of the Grossmann–Royer transforms, and to reveal its role in time-frequency analysis, but we believe that, due to its intrinsic physical interpretation (as reflections in phase space), the study of the Grossmann–Royer transform is of an interest by itself.

To that end, we extend the definition of the Grossmann–Royer transform from the duality between the Schwartz space $\mathscr{S}(\mathbb{R}^d)$ and its dual space of tempered distributions $\mathscr{S}'(\mathbb{R}^d)$ to the whole range of Gelfand–Shilov spaces and their dual spaces. According to [52], such spaces are "better adapted to the study of the problems of Applied Mathematics", since they describe simultaneous estimates of exponential type for a function and its Fourier transform. Here we give a novel (albeit expected) characterization of Gelfand–Shilov spaces in terms of the Grossmann–Royer transform.

Gelfand–Shilov spaces can be described as projective and inductive limits of modulation spaces, which are recognized as the right spaces for time-frequency analysis [38]. The modulation space norms traditionally measure the joint time-frequency distribution of $f \in \mathscr{S}'$, since their weighted versions are usually equipped with weights of at most polynomial growth at infinity ([29], [38, Ch. 11–13]). The general approach to modulation spaces introduced already in [29] includes the weights of sub-exponential growth (see (19)). However, the study of ultra-distributions requires the use of the weights of exponential or even superexponential growth, cf. [19, 80]. In our investigation of localization operators it is also essential to use the (sharp) convolution estimates for modulation spaces, see Sect. 4.

Localization operators are closely related to the above mentioned cross-Wigner distribution. They were first introduced by Berezin in the study of general Hamiltonians satisfying the so-called Feynman inequality, within a quantization problem in quantum mechanics, [5, 64]. Such operators and their modifications are also called Toeplitz or Berezin–Toeplitz operators, anti-Wick operators and Gabor multipliers, see [34, 77, 78]. We do not intend to discuss the terminology here, and refer to, e.g. [27] for the relation between Toeplitz operators and localization operators.

In signal analysis, different localization techniques are used to describe signals which are as concentrated as possible in general regions of the phase space. In particular, localization operators are related to localization technique developed by Slepian-Polak-Landau, where time and frequency are considered to be two separate spaces, see e.g. the survey article [67]. A different construction is proposed by Daubechies in [22], where time-frequency plane is treated as one geometric whole (phase space). That fundamental contribution, which contains localization in phase space together with basic facts on localization operators and references to applications in optics and signal analysis, initiated further study of the topic. More precisely, Daubechies studied localization operators $A_a^{\varphi_1,\varphi_2}$ with Gaussian windows

$$\varphi_1(x) = \varphi_2(x) = \pi^{-d/4}\exp(-x^2/2), \quad x \in \mathbb{R}^d, \quad \text{and with a radial symbol} \quad a \in L^1(\mathbb{R}^{2d}).$$

Such operators are named Daubechies' operators afterwards. The eigenfunctions of Daubechies' operators are d–dimensional Hermite functions

$$H_k(x) = H_{k_1}(x_1)H_{k_2}(x_2)\dots H_{k_d}(x_d) = \prod_{j=1}^{d} H_{k_j}(x_j), \quad x \in \mathbb{R}^d, \quad k \in \mathbb{N}_0^d, \tag{1}$$

and

$$H_n(t) = (-1)^n \pi^{-1/4}(2^n n!)^{-1/2}\exp(t^2/2)(\exp(-t^2))^{(n)}, \quad t \in \mathbb{R}, \quad n = 0, 1, \dots,$$

and corresponding eigenvalues can be explicitly calculated. Inverse problem for a simply connected localization domain Ω is studied in [1] where it is proved that if one of the eigenfunctions of Daubechies' operator is a Hermite function, then Ω is a disc centered at the origin. Moreover, the Hermite functions belong to test function spaces for ultra-distributions, both in non-quasianalytic and in quasianalytic case, and give rise to important representation theorems, [47]. In our analysis this fact is used in Theorem 2.

The eigenvalues and eigenfunctions in signal analysis are discussed in [60], where the localization operators of the form $\langle L_{\chi_\Omega} f, g\rangle = \iint_\Omega W(f,g)$, were observed. Here Ω is a quite general open and bounded subset in $\mathbb{R}^2$, χ_Ω is its characteristic function, and $W(f,g)$ is the cross-Wigner transform (see Definition 1). In [60] it is proved that the eigenfunctions of L_{χ_Ω} belong to the Gelfand–Shilov space $\mathcal{S}^{(1)}(\mathbb{R}^d)$, see Sect. 3 for the definition.

Apart from applications in signal analysis, as already mentioned, the study of localization operators is motivated (and in fact initiated by Berezin) by the problem of quantization. An exposition of different quantization theories and connection between localization operators and Toeplitz operators is given in e.g. [27, 28]. The quantization problem is in the background of the study of localization operators on $L^p(G)$, $1 \le p \le \infty$, where G is a locally compact group, see [7, 89], where one can find product formulae and Schatten–von Neumann properties of localization operators in that context.

In the context of time-frequency analysis, the groundbreaking contribution is given by Cordero and Gröchenig, [13]. Among other things, their results emphasize the role of modulation spaces in the study of localization operators. Since then, localization operators in the context of modulation spaces and Wiener-amalgam spaces were studied by many authors, cf. [14, 32, 76, 77, 87]. See also the references given there.

In this paper we rewrite the definition of localization operators and Weyl pseudodifferential operators in terms of the Grossmann–Royer transform and recover the main relations between them. Thereafter we give an exposition of results concerning the continuity of localization operators on modulation spaces with polynomial weights, and their compactness properties both for polynomial and exponential weights.

1.1 The Content of the Paper

This paper is mainly a review of the results on the topic of localization operators on modulation spaces with some new insights and (slightly) different proofs. More precisely, we propose a novel approach based on the Grossmann–Royer transform $R_g f$, defined in Sect. 2 as the weak interpretation of the Grossmann–Royer operator. Therefore $R_g f$ is the cross-Wigner distribution in disguise (Definition 1 and [25, Definition 12]), so all the results valid for $W(f, g)$ can be reformulated in terms of $R_g f$. In fact, $R_g f$ is also closely related to other time-frequency representations, $V_g f$ and $A(f, g)$, see Lemma 1. We perform a detailed exposition of the properties of the Grossmann–Royer transform, such as marginal densities and the weak uncertainty principle, Lemma 2 and Proposition 3, respectively. The relation between the Grossmann–Royer transform and the Heisenberg–Weyl (displacement) operators via the symplectic Fourier transform is revealed in Proposition 5.

We start Sect. 3 with a motivation for the study of Gelfand–Shilov spaces, and proceed with a review of their main properties, including their different characterizations (Theorem 1), the kernel theorem (Theorem 2), and a novel (albeit expected) characterization in terms of the Grossmann–Royer transform (Theorem 3).

In Sect. 4 we first discuss weight functions, and then define modulation spaces as subsets of $(\mathscr{S}^{1/2}_{1/2})'(\mathbb{R}^d)$, the dual space of the Gelfand–Shilov space $\mathscr{S}^{1/2}_{1/2}(\mathbb{R}^d)$, see Definition 5. This section is an exposition of known results relevant for the rest of the paper. In particular, we recall that Gelfand–Shilov spaces can be described as the projective and the inductive limits of modulation spaces, Theorem 4 and Remark 6. We also recall the (sharp) convolution estimates for modulation spaces (Proposition 7 and Theorem 5) which will be used in the study of localization operators in Sect. 5.

In Sect. 5 we rewrite the definition of localization operators and Weyl pseudodifferential operators in terms of the Grossmann–Royer transform (Definition 6 and Lemma 4), and recover the main relations between them (Lemmas 5 and 6). Thereafter we give an exposition of known results concerning the continuity of localization operators on modulation spaces (Sect. 5.1), and their compactness properties when the weights are chosen to be polynomial or exponential (Sect. 5.2). As a rule, we only sketch the proofs, and use the Grossmann–Royer transform formalism whenever convenient.

Finally, in Sect. 6 we point out some directions that might be relevant for future investigations. This includes product formulas, multilinear localization operators, continuity over quasianalytic classes, and extension to quasi-Banach modulation spaces.

1.2 Notation

We define $xy = x \cdot y$, the scalar product on $\mathbb{R}^d$. Given a vector $x = (x_1, \ldots, x_d) \in \mathbb{R}^d$, the partial derivative with respect to x_j is denoted by $\partial_j = \frac{\partial}{\partial x_j}$. Given a multi-index $p = (p_1, \ldots, p_d) \geq 0$, i.e., $p \in \mathbb{N}_0^d$ and $p_j \geq 0$, we write $\partial^p = \partial_1^{p_1} \ldots \partial_d^{p_d}$ and $x^p = (x_1, \ldots, x_d)^{(p_1, \ldots, p_d)} = \prod_{i=1}^d x_i^{p_i}$. We write $h \cdot |x|^{1/\alpha} = \sum_{i=1}^d h_i |x_i|^{1/\alpha_i}$. Moreover, for $p \in \mathbb{N}_0^d$ and $\alpha \in \mathbb{R}_+^d$, we set $(p!)^\alpha = (p_1!)^{\alpha_1} \ldots (p_d!)^{\alpha_d}$, while as standard $p! = p_1! \ldots p_d!$. In the sequel, a real number $r \in \mathbb{R}_+$ may play the role of the vector with constant components $r_j = r$, so for $\alpha \in \mathbb{R}_+^d$, by writing $\alpha > r$ we mean $\alpha_j > r$ for all $j = 1, \ldots, d$.

The Fourier transform is normalized to be $\hat{f}(\omega) = \mathscr{F} f(\omega) = \int f(t) e^{-2\pi i t \omega} dt$. We use the brackets $\langle f, g \rangle$ to denote the extension of the inner product $\langle f, g \rangle = \int f(t) \overline{g(t)} dt$ on $L^2(\mathbb{R}^d)$ to the dual pairing between a test function space $\mathscr{A}$ and its dual $\mathscr{A}'$: $\langle \cdot, \cdot \rangle = {}_{\mathscr{A}'}\langle \cdot, \bar{\cdot} \rangle_{\mathscr{A}}$.

By $\check{f}$ we denote the reflection $\check{f}(x) = f(-x)$, and $\langle \cdot \rangle^s$ are polynomial weights

$$\langle (x, \omega) \rangle^s = \langle z \rangle^s = (1 + x^2 + \omega^2)^{s/2}, \quad z = (x, \omega) \in \mathbb{R}^{2d}, \quad s \in \mathbb{R},$$

and $\langle x \rangle = \langle 1 + |x|^2 \rangle^{1/2}$, when $x \in \mathbb{R}^d$.

Translation and modulation operators, T and M are defined by

$$T_x f(\cdot) = f(\cdot - x) \quad \text{and} \quad M_x f(\cdot) = e^{2\pi i x \cdot} f(\cdot), \quad x \in \mathbb{R}^d.$$

The following relations hold

$$M_y T_x = e^{2\pi i x \cdot y} T_x M_y, \quad (T_x f)\hat{} = M_{-x} \hat{f}, \quad (M_x f)\hat{} = T_x \hat{f}, \quad x, y \in \mathbb{R}^d, \ f, g \in L^2(\mathbb{R}^d). \tag{2}$$

The singular values $\{s_k(L)\}_{k=1}^\infty$ of a compact operator $L \in B(L^2(\mathbb{R}^d))$ are the eigenvalues of the positive self-adjoint operator $\sqrt{L^* L}$. For $1 \leq p < \infty$, the Schatten class S_p is the space of all compact operators whose singular values lie in l^p. For consistency, we define $S_\infty := B(L^2(\mathbb{R}^d))$ to be the space of bounded operators on $L^2(\mathbb{R}^d)$. In particular, S_2 is the space of Hilbert-Schmidt operators, and S_1 is the space of trace class operators.

Throughout the paper, we shall use the notation $A \lesssim B$ to indicate $A \leq cB$ for a suitable constant $c > 0$, whereas $A \asymp B$ means that $c^{-1}A \leq B \leq cA$ for some $c \geq 1$. The symbol $B_1 \hookrightarrow B_2$ denotes the continuous and dense embedding of the topological vector space B_1 into B_2.

1.3 Basic Spaces

In general a weight $w(\cdot)$ on $\mathbb{R}^d$ is a non-negative and continuous function. By $L^p_w(\mathbb{R}^d)$, $p \in [1, \infty]$ we denote the weighted Lebesgue space defined by the norm

$$\|f\|_{L^p_w} = \|fw\|_{L^p} = \left(\int |f(x)|^p w(x)^p dx \right)^{1/p},$$

with the usual modification when $p = \infty$.

Similarly, the weighted mixed-norm space $L^{p,q}_w(\mathbb{R}^{2d})$, $p, q \in [1, \infty]$ consists of (Lebesgue) measurable functions on $\mathbb{R}^{2d}$ such that

$$\|F\|_{L^{p,q}_w} = \left(\int_{\mathbb{R}^d} \left(\int_{\mathbb{R}^d} |F(x,\omega)|^p w(x,\omega)^p dx \right)^{q/p} d\omega \right)^{1/q} < \infty.$$

where $w(x, \omega)$ is a weight on $\mathbb{R}^{2d}$.

In particular, when $w(x, \omega) = \langle x\rangle^t \langle \omega\rangle^s$, $s, t \in \mathbb{R}$, we use the notation $L^{p,q}_w(\mathbb{R}^{2d}) = L^{p,q}_{t,s}(\mathbb{R}^{2d})$, and when $w(x) = \langle x\rangle^t$, $t \in \mathbb{R}$, we use the notation $L^p_t(\mathbb{R}^d)$ instead.

The space of smooth functions with compact support on $\mathbb{R}^d$ is denoted by $\mathscr{D}(\mathbb{R}^d)$. The Schwartz class is denoted by $\mathscr{S}(\mathbb{R}^d)$, the space of tempered distributions by $\mathscr{S}'(\mathbb{R}^d)$. Recall, $\mathscr{S}(\mathbb{R}^d)$ is a Fréchet space, the projective limit of spaces $\mathscr{S}_p(\mathbb{R}^d)$, $p \in \mathbb{N}_0$, defined by the norms:

$$\|\varphi\|_{\mathscr{S}_p} = \sup_{|\alpha| \leq p} (1 + |x|^2)^{p/2} |\partial^\alpha \phi(x)| < \infty, \quad p \in \mathbb{N}_0.$$

Note that $\mathscr{D}(\mathbb{R}^d) \hookrightarrow \mathscr{S}(\mathbb{R}^d)$.

The spaces $\mathscr{S}(\mathbb{R}^d)$ and $\mathscr{S}'(\mathbb{R}^d)$ play an important role in various applications since the Fourier transform is a topological isomorphism between $\mathscr{S}(\mathbb{R}^d)$ and $\mathscr{S}(\mathbb{R}^d)$ which extends to a continuous linear transform from $\mathscr{S}'(\mathbb{R}^d)$ onto itself.

In order to deal with particular problems in applications different modifications of the Schwartz type spaces were proposed. An example is given by the Gevrey classes given below. Gelfand–Shilov spaces are another important example, see Sect. 3.

By Ω we denote an open set in $\mathbb{R}^d$, and $K \Subset \Omega$ means that K is compact subset in Ω. For $1 < s < \infty$ we define the Gevrey class $G^s(\Omega)$ by

$$G^s(\Omega) = \{\phi \in C^\infty(\Omega) \mid (\forall K \Subset \Omega)(\exists C > 0)(\exists h > 0) \sup_{x \in K} \left|\partial^\alpha \phi(x)\right| \leq C h^{|\alpha|} |\alpha|!^s\}.$$

We denote by $G_0^s(\Omega)$ the subspace of $G^s(\Omega)$ which consists of compactly supported functions. We have $\mathscr{A}(\Omega) \hookrightarrow \bigcap_{s>1} G^s(\Omega)$ and $\bigcup_{s\geq 1} G^s(\Omega) \hookrightarrow C^\infty(\Omega)$, where $\mathscr{A}(\Omega)$ denotes the space of analytic functions defined by

$$\mathscr{A}(\Omega) = \{\phi \in C^\infty(\Omega) \mid (\forall K \Subset \Omega)(\exists C > 0)(\exists h > 0) \quad \sup_{x\in K} |\partial^\alpha \phi(x)| \leq C h^{|\alpha|} |\alpha|!\}.$$

We end this section with test function spaces for the spaces of ultradistributions due to Komatsu [46]. In fact, both the Gevrey classes and the Gelfand–Shilov spaces can be viewed as particular cases of Komatsu's construction. Let there be given an open set $\Omega \subset \mathbb{R}^d$, and a sequence $(N_q)_{q\in\mathbb{N}_0}$ which satisfies (M.1) and (M.2), see Sect. 3. The function $\phi \in C^\infty(\Omega)$ is called *ultradifferentiable* function of Beurling class (N_q) (respectively of Roumieu class $\{N_q\}$) if, for any $K \subset\subset \Omega$ and for any $h > 0$ (respectively for some $h > 0$),

$$\|\phi\|_{N_q,K,h} = \sup_{x\in K, \alpha\in\mathbb{N}_0^d} \frac{|\partial^\alpha \phi(x)|}{h^{|\alpha|} N_{|\alpha|}} < \infty.$$

We say that $\phi \in \mathscr{E}^{N_q,K,h}(\Omega)$ if $\|\phi\|_{N_q,K,h} < 1$ for given K and $h > 0$, and define the following spaces of ultradifferentiable test functions:

$$\mathscr{E}^{(N_q)}(\Omega) := \text{proj} \lim_{K\subset\subset\Omega} \text{proj} \lim_{h\to 0} \mathscr{E}^{N_q,K,h}(\Omega);$$

$$\mathscr{E}^{\{N_q\}}(\Omega) := \text{proj} \lim_{K\subset\subset\Omega} \text{ind} \lim_{h\to\infty} \mathscr{E}^{N_q,K,h}(\Omega).$$

2 The Grossmann–Royer Transform

In this section we consider the Grossmann–Royer operator $Rf(x,\omega)$ and introduce the corresponding transform. We refer to [23] for the basic properties of $Rf(x,\omega)$ and its relation to the Heiseberg-Weyl operator. In fact, the Grossmann–Royer operator can be viewed as a kind of reflection (therefore we use the letter R to denote it), more precisely the conjugate of a reflection operator by a Heisengerg-Weyl operator, and the product formula ([23, Proposition 150]) reveals that the product of two reflections is a translation. Moreover, it satisfies the symplectic covariance property, [23, Proposition 150].

Those physically plausible interpretations motivate us to define the corresponding transform, which is just the cross-Wigner distribution in disguise. As we shall see, in many situations it is more convenient to choose the Grossmann–Royer transform instead of some of its relatives, the cross-Wigner distribution, the cross-ambiguity function and the short-time Fourier transform.

The Grossmann–Royer operator $R : L^2(\mathbb{R}^d) \to L^2(\mathbb{R}^{2d})$ is given by

$$Rf(x,\omega) = R(f(t))(x,\omega) = e^{4\pi i\omega(t-x)} f(2x-t), \quad f \in L^2(\mathbb{R}^d), \ x, \omega \in \mathbb{R}^d. \tag{3}$$

Definition 1 Let there be given $f, g \in L^2(\mathbb{R}^d)$. The Grossmann–Royer transform of f and g is given by

$$R_g f(x,\omega) = R(f,g)(x,\omega) = \langle Rf, g\rangle = \int e^{4\pi i\omega(t-x)} f(2x-t)\overline{g(t)}dt, \quad x, \omega \in \mathbb{R}^d. \tag{4}$$

The short-time Fourier transform of f with respect to the window g is given by

$$V_g f(x,\omega) = \int e^{-2\pi i t\omega} f(t)\overline{g(t-x)}dt, \quad x, \omega \in \mathbb{R}^d. \tag{5}$$

The cross–Wigner distribution of f and g is

$$W(f,g)(x,\omega) = \int e^{-2\pi i\omega t} f(x+\frac{t}{2})\overline{g(x-\frac{t}{2})}dt, \quad x, \omega \in \mathbb{R}^d, \tag{6}$$

and the cross–ambiguity function of f and g is

$$A(f,g)(x,\omega) = \int e^{-2\pi i\omega t} f(t+\frac{x}{2})\overline{g(t-\frac{x}{2})}dt, \quad x, \omega \in \mathbb{R}^d. \tag{7}$$

Note that $R(f(t))(x,\omega) = e^{-4\pi i\omega x} M_{2\omega}(T_{2x} f)\check{}(t)$ so that

$$R_g f(x,\omega) = e^{-4\pi i\omega x} \langle M_{2\omega}(T_{2x} f)\check{}, \overline{g}\rangle. \tag{8}$$

Lemma 1 *Let $f, g \in L^2(\mathbb{R}^d)$. Then we have:*

$$W(f,g)(x,\omega) = 2^d R_g f(x,\omega),$$
$$V_g f(x,\omega) = e^{-\pi i x\omega} R_{\check{g}} f(\frac{x}{2}, \frac{\omega}{2}),$$
$$A(f,g)(x,\omega) = R_{\check{g}} f(\frac{x}{2}, \frac{\omega}{2}), \quad x, \omega \in \mathbb{R}^d.$$

Proof $W(f,g)(x,\omega) = 2^d R_g f(x,\omega)$ immediately follows from the change of variables $t \mapsto 2(x-t)$ in (6), so the cross–Wigner distribution and the Grossmann–Royer transform are essentially the same.

The change of variables $x - t \mapsto s$ in (5) gives

$$\begin{aligned} V_g f(x,\omega) &= \int e^{-2\pi i(x-s)\omega} f(x-s)\overline{\check{g}(s)}ds \\ &= \int e^{4\pi i\frac{\omega}{2}(s-x)} f(2\frac{x}{2}-s)\overline{\check{g}(s)}ds \end{aligned}$$

$$= e^{-\pi i x\omega} \int e^{4\pi i \frac{\omega}{2}(s-\frac{x}{2})} f(2\frac{x}{2} - s)\overline{\check{g}(s)} ds$$
$$= e^{-\pi i x\omega} R_{\check{g}} f(\frac{x}{2}, \frac{\omega}{2}).$$

Therefore, $R_g f(x, \omega) = e^{\pi i x\omega} V_{\check{g}} f(2x, 2\omega)$. Finally, the change of variables $t - x/2 \mapsto -s$ in (7) gives

$$A(f, g)(x, \omega) = \int e^{2\pi i \omega(s-\frac{x}{2})} f(x - s)\overline{\check{g}(s)} ds$$
$$= \int e^{4\pi i \frac{\omega}{2}(s-\frac{x}{2})} f(2\frac{x}{2} - s)\overline{\check{g}(s)} ds$$
$$= R_{\check{g}} f(\frac{x}{2}, \frac{\omega}{2}).$$

Therefore $R_g f(x, \omega) = A(f, \check{g})(2x, 2\omega)$. □

By Lemma 1 we recapture the well-known formulas [38, 88]:

$$A(f, g)(x, \omega) = e^{\pi i x\omega} V_g f(x, \omega),$$

$$W(f, g)(x, \omega) = 2^d e^{4\pi i x\omega} V_{\check{g}} f(2x, 2\omega),$$

$$W(f, g)(x, \omega) = (\mathscr{F} A(f, g))(x, \omega), \quad x, \omega \in \mathbb{R}^d.$$

The quadratic expressions $Af := A(f, f)$ and $Wf := W(f, f)$ are called the (radar) ambiguity function and the Wigner distribution of f.

We collect the elementary properties of the Grossmann–Royer transform in the next proposition.

Proposition 1 *Let $f, g \in L^2(\mathbb{R}^d)$. The Grossmann–Royer operator is self-adjoint, uniformly continuous on $\mathbb{R}^{2d}$, and the following properties hold:*

1. $\|R_g f\|_\infty \leq \|f\|\|g\|$;
2. $R_g f = \overline{R_f g}$;
3. *The covariance property:*

$$R_{T_x M_\omega g} T_x M_\omega f(p, q) = R_g f(p - x, q - \omega), \quad x, \omega, p, q \in \mathbb{R}^d;$$

4. $R_{\hat{g}} \hat{f}(x, \omega) = R_g f(-\omega, x)$;
5. *For $f_1, f_2, g_1, g_2 \in L^2(\mathbb{R}^d)$, the Moyal identity holds:*

$$\langle R_{g_1} f_1, R_{g_2} f_2 \rangle = \langle f_1, f_2 \rangle \overline{\langle g_1, g_2 \rangle}.$$

6. *$R_g f$ maps $\mathscr{S}(\mathbb{R}^d) \times \mathscr{S}(\mathbb{R}^d)$ into $\mathscr{S}(\mathbb{R}^d \times \mathbb{R}^d)$ and extends to a map from $\mathscr{S}'(\mathbb{R}^d) \times \mathscr{S}'(\mathbb{R}^d)$ into $\mathscr{S}'(\mathbb{R}^d \times \mathbb{R}^d)$.*

Proof We first show that $\langle Rf, g\rangle = \langle f, Rg\rangle$:

$$\langle Rf, g\rangle = \int e^{4\pi i\omega(t-x)} f(2x-t)\overline{g(t)}dt = \int e^{4\pi i\omega(x-s)} f(s)\overline{g(2x-s)}ds$$
$$= \int f(s)\overline{e^{4\pi i\omega(s-x)} g(2x-s)}ds = \langle f, Rg\rangle.$$

The uniform continuity and the estimate in *1.* follow from (8), the continuity of translation, modulation and reflection operators on $L^2(\mathbb{R}^d)$ and the Cauchy–Schwartz inequality.

The property 2. follows from the self-adjointness.

For the covariance we have:

$$\begin{aligned} R_{T_x M_\omega g} T_x M_\omega f(p,q) &= \int e^{4\pi i q(t-p)} T_x M_\omega f(2p-t)\overline{T_x M_\omega g(t)}dt \\ &= \int e^{4\pi i q(t-p)} e^{2\pi i\omega(2p-t)} f(2p-t-x) e^{-2\pi i\omega t}\overline{g}(t-x)dt \\ &= \int e^{4\pi i q(s+x-p)} e^{2\pi i\omega(2p-s-x)} f(2(p-x)-s) e^{-2\pi i\omega(s+x)}\overline{g}(s)ds \\ &= \int e^{4\pi i(q-\omega)(s-(p-x))} f(2(p-x)-s)\overline{g}(s)ds \\ &= R_g f(p-x, q-\omega), \quad x, \omega, p, q \in \mathbb{R}^d. \end{aligned}$$

To prove *4.* we note that the integrals below are absolutely convergent and that the change of order of integration is allowed. Moreover, when suitably interpreted, certain oscillatory integrals are meaningful in $\mathscr{S}'(\mathbb{R}^d)$. We refer to e.g. [65, 88] for such interpretation. In particular, if δ denotes the Dirac distribution, then the Fourier inversion formula in the sence of distributions gives $\int e^{2\pi i x\omega}d\omega = \delta(x)$, wherefrom $\iint \phi(x)e^{2\pi i(x-y)\omega}dxd\omega = \phi(y)$, when $\phi \in \mathscr{S}(\mathbb{R}^d)$.

Therefore we have:

$$\begin{aligned} R_{\hat{g}}\hat{f}(x,\omega) &= \int e^{4\pi i\omega(t-x)} \hat{f}(2x-t)\overline{\hat{g}(t)}dt \\ &= e^{-4\pi i\omega x} \int\int\int e^{4\pi i\omega t} e^{-2\pi i\eta(2x-t)} f(\eta) e^{2\pi i y t}\overline{g(y)}d\eta dy dt \\ &= e^{-4\pi i\omega x} \int\int\int e^{-2\pi i t(-2\omega-\eta-y)} e^{-4\pi i\eta x} f(\eta)\overline{g(y)}d\eta dy dt \\ &= e^{-4\pi i\omega x} \int\int \delta(-2\omega-\eta-y) e^{-4\pi i\eta x} f(\eta)\overline{g(y)}d\eta dy \\ &= e^{-4\pi i\omega x} \int e^{-4\pi i(-2\omega-y)x} f(-2\omega-y)\overline{g(y)}dy \\ &= \int e^{4\pi i(y+\omega)x} f(-2\omega-y)\overline{g(y)}dy \end{aligned}$$

$$= R_g f(-\omega, x), \quad x, \omega \in \mathbb{R}^d.$$

For the Moyal identity, we again use the Fourier transform of Dirac's delta:

$$\begin{aligned}
\langle R_{g_1} f_1, R_{g_2} f_2 \rangle &= \int_{\mathbb{R}^{2d}} \Big(\int e^{4\pi i \omega (t-x)} f_1(2x-t) \overline{g_1(t)} dt \\
&\times \int e^{-4\pi i \omega (s-x)} \overline{f_2}(2x-s) g_2(s) ds \Big) dx d\omega \\
&= \int_{\mathbb{R}^{2d}} \iint e^{-2\pi i \omega (-2(t-s))} f_1(2x-t) \overline{f_2}(2x-s) \overline{g_1(t)} g_2(s) dt ds dx d\omega \\
&= \iiint \delta(-2(t-s)) f_1(2x-t) \overline{f_2}(2x-s) \overline{g_1(t)} g_2(s) dt ds dx \\
&= \langle f_1, f_2 \rangle \overline{\langle g_1, g_2 \rangle}.
\end{aligned}$$

To prove *6.* we note that the Grossman–Royer transform can be written in the form

$$R_g f(x, \omega) = (\mathscr{F}_2 \circ \tau^*) f \otimes \overline{g}(x, \omega), \quad x, \omega \in \mathbb{R}^d, \tag{9}$$

where $\mathscr{F}_2$ denotes the partial Fourier transform with respect to the second variable, and τ^* is the pullback of the operator $\tau : (x, s) \mapsto (2x - s, s)$. The theorem now follows from the invariance of $\mathscr{S}(\mathbb{R}^{2d})$ under τ^* and $\mathscr{F}_2$, and $f \otimes \overline{g} \in \mathscr{S}(\mathbb{R}^{2d})$ when $f, g \in \mathscr{S}(\mathbb{R}^d)$, and the extension to tempered distributions is straightforward. □

The Moyal identity formula implies the inversion formula:

$$f = \frac{1}{\langle g_2, g_1 \rangle} \iint R_{g_1} f(x, \omega) R g_2(x, \omega) dx d\omega,$$

We refer to [38] for details (explained in terms of the short-time Fourier transform).

For the Fourier transform of $R_g f$ we have:

Proposition 2 *Let* $f, g \in L^2(\mathbb{R}^d)$. *Then* $\mathscr{F}(R_g f)(x, \omega) = R_{\check{g}} f(-\frac{\omega}{2}, \frac{x}{2})$, $x, \omega \in \mathbb{R}^d$.

Proof We again use the interpretation of oscillatory integrals in distributional sense.

$$\begin{aligned}
\mathscr{F}(R_g f)(x, \omega) &= \iint e^{-2\pi i (x x' + \omega \omega')} R_g f(x', \omega') dx' d\omega' \\
&= \iiint e^{-2\pi i (x x' + \omega \omega') + 4\pi i \omega' (t - x')} f(2x' - t) \overline{g}(t) dt dx' d\omega' \\
&= \iiint e^{-2\pi i \omega' (\omega - 2t + 2x')} e^{-2\pi i x x'} f(2x' - t) \overline{g}(t) dt dx' d\omega' \\
&= \iint \delta(\omega - 2t + 2x') e^{-2\pi i x x'} f(2x' - t) \overline{g}(t) dt dx'
\end{aligned}$$

$$= \int\int e^{-2\pi i x(t-\omega/2)} f(t-\omega)\overline{g}(t)dt$$

$$= \int\int e^{-4\pi i \frac{x}{2}(t-\omega/2)} f(t-\omega)\overline{g}(t)dt$$

$$= \int\int e^{4\pi i \frac{x}{2}(s-(-\omega/2))} f(2\cdot(-\omega/2)-s)\overline{\check{g}}(s)ds$$

$$= R_{\check{g}} f(-\frac{\omega}{2}, \frac{x}{2}), \quad x, \omega \in \mathbb{R}^d,$$

which proves the proposition. □

The following lemma shows that the Grossmann–Royer transform yields the correct marginal densities. This follows from the well-known marginal densities of the (cross-)Wigner distribution, since the transforms are essentially the same. However, we give an independent proof below. We refer to [23, 38] for the discussion on the probabilistic interpretation of the (cross-)Wigner distribution.

Lemma 2 *Let $f, g \in L^1(\mathbb{R}^d) \cap L^2(\mathbb{R}^d)$. Then*

$$\int_{\mathbb{R}^d} R_g f(x,\omega)d\omega = 2^{-d} f(x)\overline{g}(x),$$

$$\int_{\mathbb{R}^d} R_g f(x,\omega)dx = 2^{-d} \hat{f}(\omega)\overline{\hat{g}}(\omega),$$

and, in particular $2^d \int R_f f(x,\omega)d\omega = |f(x)|^2$, and $2^d \int R_f f(x,\omega)dx = |\hat{f}(\omega)|^2$.

Proof We use the change of variables $s = x - t$ to obtain

$$\begin{aligned}
\int_{\mathbb{R}^d} R_g f(x,\omega)d\omega &= \int\int e^{4\pi i \omega(t-x)} f(2x-t)\overline{g}(t)dtd\omega \\
&= \int\int e^{-2\pi i \omega(2s)} f(x+s)\overline{g}(x-s)dsd\omega \\
&= \int \delta(2s) f(x+s)\overline{g}(x-s)ds \\
&= 2^{-d} f(x)\overline{g}(x), \quad x \in \mathbb{R}^d,
\end{aligned}$$

and similarly

$$\begin{aligned}
\int_{\mathbb{R}^d} R_g f(x,\omega)dx &= \int\int e^{-2\pi i \omega(2x-2t)} f(2x-t)\overline{g}(t)dtdx \\
&= \int\int e^{-2\pi i \omega(2x-t)} f(2x-t)\overline{e^{-2\pi i \omega t} g(t)}dtdx \\
&= 2^{-d} \hat{f}(\omega)\overline{\hat{g}}(\omega), \quad \omega \in \mathbb{R}^d.
\end{aligned}$$

The particular case is obvious. □

Remark 1 If f belongs to the dense subspace of $L^2(\mathbb{R}^d)$ such that $R_f f \in L^1(\mathbb{R}^{2d})$ and the Fubini theorem holds, then Plancherel's theorem follows from Lemma 2:

$$\|\hat{f}\|^2 = 2^d \int\int R_f f(x,\omega)dxd\omega = 2^d \int\int R_f f(x,\omega)d\omega dx = \|f\|^2.$$

Next we show the weak form of the uncertainty principle for the Grossmann–Royer transform.

Proposition 3 *Let there be given $f, g \in L^2(\mathbb{R}^d) \setminus 0$, and let $U \subset \mathbb{R}^{2d}$ and $\varepsilon > 0$ be such that*

$$\int_U |R_g f(x,\omega)|^2 dxd\omega \geq (1-\varepsilon)\|f\|\|g\|.$$

Then $|U| \geq 1-\varepsilon$.

Proof From Proposition 1 *1.* we have

$$(1-\varepsilon)\|f\|\|g\| \leq \int_U |R_g f(x,\omega)|^2 dxd\omega \leq \|R_g f\|_\infty |U| \leq \|f\|\|g\||U|,$$

and the claim follows. □

We end this section with the relation between the Grossmann–Royer operator and the Heisenberg–Weyl operator, also known as displacement operator since it describes translations in phase space. Notice that in [25] the Grossmann–Royer operators are defined in terms of the Heisenberg–Weyl operators, so our definition (3) is formulated as a proposition there.

Definition 2 Let there be given $f, g \in L^2(\mathbb{R}^d)$. The Heisenberg–Weyl operator T: $L^2(\mathbb{R}^d) \to L^2(\mathbb{R}^{2d})$ is given by

$$Tf(x,\omega) = T(f(t))(x,\omega) = e^{2\pi i\omega(t-\frac{x}{2})} f(t-x), \quad x,\omega \in \mathbb{R}^d, \tag{10}$$

and the Heisenberg–Weyl transform is defined to be

$$T_g f(x,\omega) = T(f,g)(x,\omega) = \langle Tf, g\rangle = \int e^{2\pi i\omega(t-\frac{x}{2})} f(t-x)\overline{g(t)}dt, \quad x,\omega \in \mathbb{R}^d. \tag{11}$$

Proposition 4 *Let there be given $f, g \in L^2(\mathbb{R}^d)$. Then we have*

1. $Rf(x,\omega) = T(x,\omega)R(0)Tf(-x,-\omega)$;
2. $R(x,\omega)Rf(p,q) = e^{-4\pi i\sigma((x,\omega),(p,q))} Tf(2(x-p,\omega-q))$, *where σ is the standard symplectic form on phase space $\mathbb{R}^{2d}$:*

$$\sigma((x,\omega),(p,q)) = \omega \cdot p - x \cdot q.$$

Proof 1. Note that $Rf(0) = f(-t)$, wherefrom

$$Tf(-x,-\omega) = e^{-2\pi i\omega(t+\frac{x}{2})} f(t+x) \quad \Rightarrow R(0)Tf(-x,-\omega) = e^{2\pi i\omega(t-\frac{x}{2})} f(x-t).$$

Hence,

$$\begin{aligned} T(x,\omega)R(0)Tf(-x,-\omega) &= T(x,\omega)e^{2\pi i\omega(t-\frac{x}{2})} f(x-t) \\ &= e^{-\pi i\omega x} T(e^{2\pi i\omega t} f(x-t))(x,\omega) \\ &= e^{-\pi i\omega x} e^{2\pi i\omega(t-\frac{x}{2})} e^{2\pi i\omega(t-x)} f(x-(t-x)) \\ &= e^{4\pi i\omega(t-x)} f(2x-t) \\ &= Rf(x,\omega). \end{aligned}$$

To prove 2. we calculate both sides:

$$\begin{aligned} R(x,\omega)Rf(p,q) &= R(e^{4\pi iq(t-p)} f(2p-t))(x,\omega) \\ &= e^{-4\pi iqp} R(e^{4\pi iqt} f(2p-t))(x,\omega) \\ &= e^{-4\pi iqp} e^{4\pi i\omega(t-x)} e^{4\pi iq(2x-t)} f(2p-(2x-t)) \\ &= e^{4\pi i\omega(t-x)} e^{4\pi iq(2x-t-p)} f(t-2(x-p)), \end{aligned}$$

and

$$\begin{aligned} e^{-4\pi i(\omega p-xq)} Tf(2((x,\omega)-(p,q))) &= e^{-4\pi i(\omega p-xq)} e^{2\pi i2(\omega-q)(t-(x-p))} f(t-2(x-p)) \\ &= e^{-4\pi i(\omega p-xq)} e^{4\pi i(\omega-q)t} e^{-4\pi i(\omega-q)(x-p)} f(t-2(x-p)) \\ &= e^{4\pi i\omega(-p+t-x-p)} e^{-4\pi ipq} e^{4\pi iq(x-t+x)} f(t-2(x-p)) \\ &= e^{4\pi i\omega(t-x)} e^{4\pi iq(2x-t-p)} f(t-2(x-p)), \end{aligned}$$

which shows that $R(x,\omega)Rf(p,q) = e^{-4\pi i\sigma((x,\omega),(p,q))} Tf(2(x-p,\omega-q))$. □

Finally, if $\mathscr{F}_\sigma$ denotes the symplectic Fourier transform in $L^2(\mathbb{R}^{2d})$:

$$\begin{aligned} \mathscr{F}_\sigma(F(x,\omega))(p,q) &= \int_{\mathbb{R}^{2d}} e^{-2\pi i\sigma((x,\omega),(p,q))} F(x,\omega)dxd\omega \\ &= \int_{\mathbb{R}^{2d}} e^{-2\pi i(\omega p-xq)} F(x,\omega)dxd\omega, \end{aligned}$$

then the Grossmann–Royer operator and the Heisenberg–Weyl operator are the symplectic Fourier transforms of each other.

Proposition 5 *Let there be given $f, g \in L^2(\mathbb{R}^d)$ and let $\mathscr{F}_\sigma$ be the symplectic Fourier transform. Then*

$$Rf(p,q) = 2^{-d}\mathscr{F}_\sigma(T(f(t))(x,\omega))(-p,-q).$$

Proof

$$\begin{aligned}
\mathscr{F}_\sigma(T(f(t))(x,\omega))(-p,-q) &= \int_{\mathbb{R}^{2d}} e^{2\pi i(xq-\omega p)} e^{2\pi i\omega(t-\frac{x}{2})} f(t-x)dxd\omega \\
&= \int_{\mathbb{R}^d} \left(\int_{\mathbb{R}^d} e^{2\pi i\omega(t-p-\frac{x}{2})} d\omega \right) e^{2\pi iqx)} f(t-x)dx \\
&= \int_{\mathbb{R}^d} \delta(t-p-\frac{x}{2}) e^{2\pi iqx} f(t-x)dx \\
&= 2^d e^{4\pi iq(t-p)} f(t-2(t-p))dx = 2^d Rf(p,q),
\end{aligned}$$

where we again used the Fourier inversion formula in the sense of distributions. □

Since it is easy to see that $\mathscr{F}_\sigma$ is an involution, we also have

$$Tf(p,q) = 2^d \mathscr{F}_\sigma(R(f(t))(x,\omega))(-p,-q).$$

Remark 2 The Weyl operator L_a with the symbol a is introduced in [25, Definition 37] by the means of the symplectic Fourier transform $\mathscr{F}_\sigma a$ of the symbol and the Heisenberg–Weyl operator:

$$L_a f(t) = \int_{\mathbb{R}^{2d}} \mathscr{F}_\sigma a(x,\omega) T(f(t))(x,\omega) dxd\omega.$$

There it is shown that such definition coincides with the usual one, see Sect. 5, and also with the representation given by Lemma 5.

3 Gelfand–Shilov Spaces

Problems of regularity of solutions to partial differential equations (PDEs) play a central role in the modern theory of PDEs. When solutions of certain PDEs are smooth but not analytic, several intermediate spaces of functions are proposed in order to describe its decrease at infinity and also the regularity in $\mathbb{R}^d$. In particular, in the study of properties of solutions of certain parabolic initial-value problems Gelfand and Shilov introduced *the spaces of type S* in [35]. Such spaces provide uniqueness and correctness classes for solutions of Cauchy problems, [36]. We refer to [36] for the fundamental properties of such spaces which are afterwards called Gelfand–Shilov spaces.

More recently, Gelfand–Shilov spaces were used in [9, 10] to describe exponential decay and holomorphic extension of solutions to globally elliptic equations, and in [48] in the regularizing properties of the Boltzmann equation. We refer to [52] for a recent overview and for applications in quantum mechanics and traveling waves, and to [78] for the properties of the Bargmann transform on Gelfand–Shilov spaces. The original definition is generalized already in [36, Ch. IV, App. 1], and more general decay and regularity conditions can be systematically studied by using Komatsu's approach to ultradifferentiable functions developed in [46]. An interesting extension based on the iterates of the harmonic oscillator is studied in [80] under the name Pilipović spaces, see also [54].

In the context of time-frequency analysis, certain Gelfand–Shilov spaces can be described in terms of modulation spaces [38, 43]. The corresponding pseudodifferential calculus is developed in [78, 79]. Gelfand–Shilov spaces are also used in the study of time-frequency localization operators in [18, 19, 70], thus extending the context of the pioneering results of Cordero and Grochenig [13].

In this section we introduce Gelfand–Shilov spaces and list important equivalent characterizations. We also present the kernel theorem which will be used in the study of localization operators.

3.1 Definition

The regularity and decay properties of elements of Gelfand–Shilov spaces are initially measured with respect to sequences of the form $M_p = p^{\alpha p}$, $p \in \mathbb{N}$, $\alpha > 0$ or, equivalently, the Gevrey sequences $M_p = p!^{\alpha}$, $p \in \mathbb{N}$, $\alpha > 0$.

We follow here Komatsu's approach [46] to spaces of ultra-differentiable functions to extend the original definition as follows.

Let $(M_p)_{p\in\mathbb{N}_0}$ be a sequence of positive numbers monotonically increasing to infinity which satisfies:

$(M.1)$ $M_p^2 \le M_{p-1}M_{p+1}, \quad p \in \mathbb{N}$;

$(M.2)$ There exist positive constants A, H such that

$$M_p \le AH^p \min{}_{0\le q\le p} M_{p-q}M_q, \quad p, q \in \mathbb{N}_0,$$

or, equivalently, there exist positive constants A, H such that

$$M_{p+q} \le AH^{p+q} M_p M_q, \quad p, q \in \mathbb{N}_0;$$

We assume $M_0 = 1$, and that $M_p^{1/p}$ is bounded below by a positive constant.

Remark 3 To give an example, we describe (M.1) and (M.2) as follows. Let $(s_p)_{p\in\mathbb{N}_0}$ be a sequence of positive numbers monotonically increasing to infinity ($s_p \nearrow \infty$) so that for every $p, q \in \mathbb{N}_0$ there exist $A, H > 0$ such that

$$\prod_{j=1}^{q} s_{p+j} = s_{p+1} \dots s_{p+q} \le AH^p s_1 \dots s_q = AH^p \prod_{j=1}^{q} s_j. \qquad (12)$$

Then the sequence $(S_p)_{p\in\mathbb{N}_0}$ given by $S_p = \prod_{j=1}^{p} s_j$, $S_0 = 1$, satisfies $(M.1)$ and $(M.2)$.

Conversely, if $(S_p)_{p\in\mathbb{N}_0}$ given by $S_p = \prod_{j=1}^{p} s_j$, $s_j > 0$, $j \in \mathbb{N}$, $S_0 = 1$, satisfies $(M.1)$ then $(s_p)_{p\in\mathbb{N}_0}$ increases to infinity, and if it satisfies $(M.2)$ then (12) holds.

Let $(M_p)_{p\in\mathbb{N}_0}$ and $(N_q)_{q\in\mathbb{N}_0}$ be sequences which satisfy $(M.1)$. We write $M_p \subset N_q$ ($(M_p) \prec (N_q)$, respectively) if there are constants $H, C > 0$ (for any $H > 0$ there is a constant $C > 0$, respectively) such that $M_p \leq C H^p N_p$, $p \in \mathbb{N}_0$. Also, $(M_p)_{p\in\mathbb{N}_0}$ and $(N_q)_{q\in\mathbb{N}_0}$ are said to be equivalent if $M_p \subset N_q$ and $N_q \subset M_p$ hold.

Definition 3 Let there be given sequences of positive numbers $(M_p)_{p\in\mathbb{N}_0}$ and $(N_q)_{q\in\mathbb{N}_0}$ which satisfy $(M.1)$ and $(M.2)$. Let $\mathcal{S}^{N_q,B}_{M_p,A}(\mathbb{R}^d)$ be defined by

$$\mathcal{S}^{N_q,B}_{M_p,A}(\mathbb{R}^d) = \{f \in C^\infty(\mathbb{R}^d) \mid \|x^\alpha \partial^\beta f\|_{L^\infty} \leq C A^\alpha M_{|\alpha|} B^\beta N_{|\beta|},\ \ \forall \alpha, \beta \in \mathbb{N}_0^d\},$$

for some positive constant C, where $A = (A_1, \dots, A_d)$, $B = (B_1, \dots, B_d)$, $A, B > 0$.

Gelfand–Shilov spaces $\Sigma^{N_q}_{M_p}(\mathbb{R}^d)$ and $\mathcal{S}^{N_q}_{M_p}(\mathbb{R}^d)$ are projective and inductive limits of the (Fréchet) spaces $\mathcal{S}^{N_q,B}_{M_p,A}(\mathbb{R}^d)$ with respect to A and B:

$$\Sigma^{N_q}_{M_p}(\mathbb{R}^d) := \text{proj} \lim_{A>0,B>0} \mathcal{S}^{N_q,B}_{M_p,A}(\mathbb{R}^d); \quad \mathcal{S}^{N_q}_{M_p}(\mathbb{R}^d) := \text{ind} \lim_{A>0,B>0} \mathcal{S}^{N_q,B}_{M_p,A}(\mathbb{R}^d).$$

The corresponding dual spaces of $\Sigma^{N_q}_{M_p}(\mathbb{R}^d)$ and $\mathcal{S}^{N_q}_{M_p}(\mathbb{R}^d)$ are the spaces of ultra-distributions of Beurling and Roumier type respectively:

$$(\Sigma^{N_q}_{M_p})'(\mathbb{R}^d) := \text{ind} \lim_{A>0,B>0} (\mathcal{S}^{N_q,B}_{M_p,A})'(\mathbb{R}^d);$$

$$(\mathcal{S}^{N_q}_{M_p})'(\mathbb{R}^d) := \text{proj} \lim_{A>0,B>0} (\mathcal{S}^{N_q,B}_{M_p,A})'(\mathbb{R}^d).$$

Of course, for certain choices of the sequences $(M_p)_{p\in\mathbb{N}_0}$ and $(N_q)_{q\in\mathbb{N}_0}$ the spaces $\Sigma^{N_q}_{M_p}(\mathbb{R}^d)$ and $\mathcal{S}^{N_q}_{M_p}(\mathbb{R}^d)$ are trivial, i.e. they contain only the function $\phi \equiv 0$. Nontrivial Gelfand–Shilov spaces are closed under translation, dilation, multiplication with $x \in \mathbb{R}^d$, and differentiation. Moreover, they are closed under the action of certain differential operators of infinite order (ultradifferentiable operators in the terminology of Komatsu). We refer to [46] for topological properties in a more general context of test function spaces for ultradistributions.

When $(M_p)_{p\in\mathbb{N}_0}$ and $(N_q)_{q\in\mathbb{N}_0}$ are Gevrey sequences: $M_p = p!^r$, $p \in \mathbb{N}_0$ and $N_q = q!^s$, $q \in \mathbb{N}_0$, for some $r, s \geq 0$, then we use the notation

$$\mathcal{S}^{N_q}_{M_p}(\mathbb{R}^d) = \mathcal{S}^s_r(\mathbb{R}^d) \quad \text{and} \quad \Sigma^{N_q}_{M_p}(\mathbb{R}^d) = \Sigma^s_r(\mathbb{R}^d).$$

If, in addition, $s = r$, then we put

$$\mathcal{S}^{\{s\}}(\mathbb{R}^d) = \mathcal{S}^s_s(\mathbb{R}^d) \quad \text{and} \quad \Sigma^{(s)}(\mathbb{R}^d) = \Sigma^s_s(\mathbb{R}^d).$$

The choice of Gevrey sequences (which is the most often used choice in the literature) may serve well as an illuminating example in different contexts. In particular, when discussing the nontriviality we have the following:

1. the space $\mathcal{S}_r^s(\mathbb{R}^d)$ is nontrivial if and only if $s+r>1$, or $s+r=1$ and $sr>0$,
2. if $s+r \geq 1$ and $s<1$, then every $f \in \mathcal{S}_r^s(\mathbb{R}^d)$ can be extended to the complex domain as an entire function,
3. if $s+r \geq 1$ and $s=1$, then every $f \in \mathcal{S}_r^s(\mathbb{R}^d)$ can be extended to the complex domain as a holomorphic function in a strip.
4. the space $\Sigma_r^s(\mathbb{R}^d)$ is nontrivial if and only if $s+r>1$, or, if $s+r=1$ and $sr>0$ and $(s,r) \neq (1/2, 1/2)$.

We refer to [36] or [52] for the proof in the case of $\mathcal{S}_r^s(\mathbb{R}^d)$, and to [54] for the spaces $\Sigma_r^s(\mathbb{R}^d)$, see also [78].

Whenever nontrivial, Gelfand–Shilov spaces contain "enough functions" in the following sense. A test function space Φ is "rich enough" if

$$\int f(x)\varphi(x)dx = 0, \quad \forall \varphi \in \Phi \Rightarrow f(x) \equiv 0 (a.e.).$$

The discussion here above shows that Gelfand–Shilov classes $\mathcal{S}_r^s(\mathbb{R}^d)$ consist of quasi-analytic functions when $s \in (0,1)$. This is in sharp contrast with e.g. Gevrey classes $G^s(\mathbb{R}^d)$, $s>1$, another family of functions commonly used in regularity theory of partial differential equations, whose elements are always non-quasi-analytic. We refer to [61] for microlocal analysis in Gervey classes and note that $G_0^s(\mathbb{R}^d) \hookrightarrow \mathcal{S}_s^s(\mathbb{R}^d) \hookrightarrow G^s(\mathbb{R}^d)$, $s>1$.

When the spaces are nontrivial we have dense and continuous inclusions:

$$\Sigma_r^s(\mathbb{R}^d) \hookrightarrow \mathcal{S}_r^s(\mathbb{R}^d) \hookrightarrow \mathcal{S}(\mathbb{R}^d).$$

In fact, $\mathcal{S}(\mathbb{R}^d)$ can be revealed as a limiting case of $S_r^s(\mathbb{R}^d)$, i.e.

$$\mathcal{S}(\mathbb{R}^d) = \mathcal{S}_\infty^\infty(\mathbb{R}^d) = \lim_{s,r \to \infty} \mathcal{S}_r^s(\mathbb{R}^d),$$

when the passage to the limit $s, r \to \infty$ is interpreted correctly, see [36, p. 169].

We refer to [80] where it is shown how to overcome the minimality condition ($\Sigma_{1/2}^{1/2}(\mathbb{R}^d) = 0$) by transferring the estimates for $\|x^\alpha \partial^\beta f\|_{L^\infty}$ into the estimates of the form $\|H^N f\|_{L^\infty} \lesssim h^N (N!)^{2s}$, for some (for every) $h>0$, where $H = |x|^2 - \Delta$ is the harmonic oscillator.

In what follows, the special role will be played by the Gelfand–Shilov space of analytic functions $\mathcal{S}^{(1)}(\mathbb{R}^d) := \Sigma_1^1(\mathbb{R}^d)$. According to Theorem 1 here below, we have

$$f \in \mathcal{S}^{(1)}(\mathbb{R}^d) \Longleftrightarrow \sup_{x \in \mathbb{R}^d} |f(x) e^{h \cdot |x|}| < \infty \text{ and } \sup_{\omega \in \mathbb{R}^d} |\hat{f}(\omega) e^{h \cdot |\omega|}| < \infty, \quad \forall h > 0.$$

Any $f \in \mathcal{S}^{(1)}(\mathbb{R}^d)$ can be extended to a holomorphic function $f(x+iy)$ in the strip $\{x+iy \in \mathbb{C}^d \,:\, |y| < T\}$ some $T > 0$, [36, 52]. The dual space of $\mathcal{S}^{(1)}(\mathbb{R}^d)$ will be denoted by $\mathcal{S}^{(1)'}(\mathbb{R}^d)$. In fact, $\mathcal{S}^{(1)}(\mathbb{R}^d)$ is isomorphic to the Sato test function space for the space of Fourier hyperfunctions $\mathcal{S}^{(1)'}(\mathbb{R}^d)$, see [20].

3.2 Equivalent Conditions

In this subsection we recall the well known equivalent characterization of Gelfand–Shilov spaces which shows the important behavior of Gelfand–Shilov spaces under the action of the Fourier transform. Already in [36] it is shown that the Fourier transform is a topological isomorphism between $\mathcal{S}^s_r(\mathbb{R}^d)$ and $\mathcal{S}^r_s(\mathbb{R}^d)$ ($\mathcal{F}(\mathcal{S}^s_r) = \mathcal{S}^r_s$), which extends to a continuous linear transform from $(\mathcal{S}^s_r)'(\mathbb{R}^d)$ onto $(\mathcal{S}^r_s)'(\mathbb{R}^d)$. In particular, if $s = r$ and $s \geq 1/2$ then $\mathcal{F}(\mathcal{S}^s_s)(\mathbb{R}^d) = \mathcal{S}^s_s(\mathbb{R}^d)$. Similar assertions hold for $\Sigma^s_r(\mathbb{R}^d)$.

This invariance properties easily follow from the following theorem which also enlightens fundamental properties of Gelfand–Shilov spaces implicitly contained in their definition. Among other things, it states that the decay and regularity estimates of $f \in \mathcal{S}^{N_q}_{M_p}(\mathbb{R}^d)$ can be studied separately.

Before we state the theorem, we introduce another notion. The *associated function* for a given sequence (M_p) is defined by

$$M(\rho) = \sup_{p\in\mathbb{N}_0} \ln \frac{\rho^p M_0}{M_p}, \quad 0 < \rho < \infty.$$

For example, the associated function for the Gevrey sequence $M_p = p!^r$, $p \in \mathbb{N}_0$ behaves at infinity as $|\cdot|^{1/r}$, cf. [53]. In fact, the interplay between the defining sequence and its associated function plays an important role in the theory of ultra-distributions.

Theorem 1 *Let there be given sequences of positive numbers $(M_p)_{p\in\mathbb{N}_0}$ and $(N_q)_{q\in\mathbb{N}_0}$ which satisfy $(M.1)$ and $(M.2)$ and $p! \subset M_p N_p$ ($p! \prec M_p N_p$, respectively). Then the following conditions are equivalent:*

1. $f \in \mathcal{S}^{N_q}_{M_p}(\mathbb{R}^d)$ ($f \in \Sigma^{N_q}_{M_p}(\mathbb{R}^d)$, *respectively*).
2. *There exist constants $A, B \in \mathbb{R}^d$, $A, B > 0$ (for every $A, B \in \mathbb{R}^d$, $A, B > 0$ respectively), and there exist $C > 0$ such that*

$$\|e^{M(|Ax|)}\partial^q f(x)\|_{L^\infty} \leq C B^q N_{|q|}, \quad \forall p, q \in \mathbb{N}^d_0.$$

3. *There exist constants $A, B \in \mathbb{R}^d$, $A, B > 0$ (for every $A, B \in \mathbb{R}^d$, $A, B > 0$, respectively), and there exist $C > 0$ such that*

$$\|x^p f(x)\|_{L^\infty} \leq C A^p M_{|p|} \quad \text{and} \quad \|\partial^q f(x)\|_{L^\infty} \leq C B^q N_{|q|}, \quad \forall p, q \in \mathbb{N}^d_0.$$

4. *There exist constants* $A, B \in \mathbb{R}^d, A, B > 0$ *(for every* $A, B \in \mathbb{R}^d$, $A, B > 0$, *respectively), and there exist* $C > 0$ *such that*

$$\|x^p f(x)\|_{L^\infty} \le C A^p M_{|p|} \quad \text{and} \quad \|\omega^q \hat{f}(\omega)\|_{L^\infty} \le C B^q N_{|q|}, \quad \forall p, q \in \mathbb{N}_0^d.$$

5. *There exist constants* $A, B \in \mathbb{R}^d$, $A, B > 0$ *(for every* $A, B \in \mathbb{R}^d$, $A, B > 0$, *respectively), such that*

$$\| f(x)\, e^{M(|Ax|)} \|_{L^\infty} < \infty \quad \text{and} \quad \| \hat{f}(\omega)\, e^{N(|B\omega|)} \|_{L^\infty} < \infty,$$

where $M(\cdot)$ *and* $N(\cdot)$ *are the associated functions for the sequences* $(M_p)_{p \in \mathbb{N}_0}$ *and* $(N_q)_{q \in \mathbb{N}_0}$ *respectively.*

Proof Theorem 1 is for the first time proved in [21] and reinvented many times afterwards, see e.g. [19, 43, 45, 52, 55]. As an illustration, and to give a flavor of the technique, we show *1.* $\Leftrightarrow$ *2.* For the simplicity, we observe the Gevrey sequences $M_p = p!^r$ and $N_q = q!^s$, $p \in \mathbb{N}_0, r, s > 0$.

Recall, $f \in \mathscr{S}_{M_p}^{N_q}(\mathbb{R}^d) = \mathscr{S}_r^s(\mathbb{R}^d)$ if and only if there exist constants $h, C > 0$ such that

$$\sup_{x \in \mathbb{R}^d} \left| x^p \partial^q f(x) \right| \le C h^{|p|+|q|} p!^r q!^s \quad \forall p, q \in \mathbb{N}_0^d.$$

To avoid the use of inequalities related to multi-indices we consider $d = 1$. Put $F_q(x) = \partial^q f(x) / (h^{|q|} q!^s)$. We have

$$\sup_x h^{-|p|} p!^{-r} |x|^p |F_q(x)| \le C,$$

so that

$$\sup_x h^{-|p|/r} p!^{-1} |x|^{p/r} |F_q(x)|^{1/r} \le C^{1/r}$$

uniformly in p. Therefore

$$\sup_x \sum_{p \in \mathbb{N}_0} \left(\frac{(|x| h^{-1})^{1/r}}{2} \right)^{|p|} p!^{-1} |F_q(x)|^{1/r} \le C^{1/r} \sum_{p \in \mathbb{N}_0} \frac{1}{2^{|p|}}.$$

Put $A = h^{-1/r} 2^{-1}$, and conclude that there exist constants $A, B, C > 0$ such that

$$\left| \partial^q f(x) \right| \le C B^{|q|} q!^s e^{-A|x|^{1/r}}, \quad \forall x \in \mathbb{R}, \quad \forall q \in \mathbb{N}_0,$$

which gives 2.

Assume now that 2. holds. Put $F_q(x) = \partial^q f(x) / (C^{|q|} q!^s)$. Then, $\left| \partial^q f(x) \right| \le C^{1+|q|} q!^s e^{-A|x|^{1/r}}$ for all $x \in \mathbb{R}$ implies the following chain of inclusions.

$$
\begin{aligned}
\left|F_q(x)\right|^{1/r} e^{\frac{A}{r}|x|^{1/r}} < \infty &\Rightarrow \sup_x \sum_{p\in\mathbb{N}_0} \frac{1}{p!}(\frac{A}{r})^p |x|^{p/r} \left|F_q(x)\right|^{1/r} < \infty \\
&\Rightarrow \sup_x \frac{1}{p!}(\frac{A}{r})^p |x|^{p/r} \left|F_q(x)\right|^{1/r} < \infty \\
&\Rightarrow \sup_x \frac{1}{p!^r}(\frac{A}{r})^{rp} |x|^{p} \left|F_q(x)\right| < \infty \\
&\Rightarrow \left|x^p \partial^q f(x)\right| \le \tilde{C}\left((\frac{r}{A})^r\right)^{|p|} C^{|q|} p!^r q!^s \\
&\Rightarrow \left|x^p \partial^q f(x)\right| \le \tilde{C} h^{|p|+|q|} p!^r q!^s,
\end{aligned}
$$

so that $f \in \mathscr{S}^{N_q}_{M_p}(\mathbb{R})$.

The proof for $\Sigma^{N_q}_{M_p}(\mathbb{R}^d)$ is (almost) the same. □

By the above characterization $\mathscr{F}\mathscr{S}^{N_q}_{M_p}(\mathbb{R}^d) = \mathscr{S}^{M_p}_{N_q}(\mathbb{R}^d)$. Observe that when M_p and N_q are chosen to be Gevrey sequences, then $\mathscr{S}^{1/2}_{1/2}(\mathbb{R}^d)$ is the smallest non-empty Gelfand–Shilov space invariant under the Fourier transform, see also Remark 4 below. Theorem 1 implies that $f \in \mathscr{S}^{1/2}_{1/2}(\mathbb{R}^d)$ if and only if $f \in \mathscr{C}^{\infty}(\mathbb{R}^d)$ and there exist constants $h > 0, k > 0$ such that

$$
\| f e^{h|\cdot|^2} \|_{L^\infty} < \infty \quad \text{and} \quad \| \hat{f} e^{k|\cdot|^2} \|_{L^\infty} < \infty. \tag{13}
$$

Therefore the Hermite functions given by (1) belong to $\mathscr{S}^{1/2}_{1/2}(\mathbb{R}^d)$. This is an important fact when dealing with Gelfand–Shilov spaces, cf. [47, 54]. We refer to [80] for the situation below the "critical exponent" 1/2.

Remark 4 Note that $\Sigma^{1/2}_{1/2}(\mathbb{R}^d) = \{0\}$ and $\Sigma^{s}_{s}(\mathbb{R}^d)$ is dense in the Schwartz space whenever $s > 1/2$. One may consider a "fine tuning", that is the spaces $\Sigma^{N_q}_{M_p}(\mathbb{R}^d)$ such that

$$
\{0\} = \Sigma^{1/2}_{1/2}(\mathbb{R}^d) \hookrightarrow \Sigma^{N_q}_{M_p}(\mathbb{R}^d) \hookrightarrow \mathscr{S}^{N_q}_{M_p}(\mathbb{R}^d) \hookrightarrow \Sigma^{s}_{s}(\mathbb{R}^d), \quad s > 1/2.
$$

For that reason, we define sequences $(M_p)_{p\in\mathbb{N}_0}$ and $(N_q)_{q\in\mathbb{N}_0}$ by

$$
M_p := p!^{\frac{1}{2}} \prod_{k=0}^{p} l_k = p!^{\frac{1}{2}} L_p, \quad p \in \mathbb{N}_0, \quad N_q := q!^{\frac{1}{2}} \prod_{k=0}^{q} r_k = q!^{\frac{1}{2}} R_q, \quad q \in \mathbb{N}_0 \tag{14}
$$

where $(r_p)_{p\in\mathbb{N}_0}$ and $(l_p)_{p\in\mathbb{N}_0}$ are sequences of positive numbers monotonically increasing to infinity such that (12) holds with the letter s replaced by r and l respectively and which satisfy: For every $\alpha \in (0, 1]$ and every $k > 1$ so that $kp \in \mathbb{N}$, $p \in \mathbb{N}$,

$$
\max\{(\frac{r_{kp}}{r_p})^2, (\frac{l_{kp}}{l_p})^2\} \le k^\alpha, \quad p \in \mathbb{N}. \tag{15}
$$

Then $p! \prec M_p N_p$ and the sequences $(R_p)_{p\in\mathbb{N}_0}$ and $(L_p)_{p\in\mathbb{N}_0}$ ($R_p = r_1 \dots r_p$, $L_p = l_1 \dots l_p$, $p \in \mathbb{N}$ $R_0 = 1$, and $L_0 = 1$) satisfy conditions $(M.1)$ and $(M.2)$. Moreover,

$$\max\{R_p, L_p\} \leq p!^{\alpha/2}, p \in \mathbb{N},$$

for every $\alpha \in (0, 1]$. (For $p, q, k \in \mathbb{N}_0^d$ we have $L_{|p|} = \prod_{|k|\leq|p|} l_{|k|}$, and $R_{|q|} = \prod_{|k|\leq|q|} r_{|q|}$.) Such sequences are used in the study of localization operators in the context of quasianalytic spaces in [18].

3.3 Kernel Theorem

For the study the action of linear operators it is convenient to use their kernels. In particular, when dealing with multilinear extensions of localization operators, the kernel theorem for Glefand-Shilov spaces appears to be a crucial tool, cf. [73]. Such theorems extend the famous Schwartz kernel theorem (see [63, 86]) to the spaces of ultradistributions. We refer to [58] for the proof in the case of non-quasianalytic Gelfand–Shilov spaces, and here we give a sketch of the proof for a general case from [70]. The only difference is that in quasianalytic case, the density arguments from [58] can not be used. Instead, we use arguments based on Hermite expansions in Gelfand–Shilov spaces, see [47, 50].

We need additional conditions for a sequence of positive numbers $(M_p)_{p\in\mathbb{N}_0}$:
$\{N.1\}$ There exist positive constants A, H such that

$$p!^{1/2} \leq AH^p M_p, \quad p \in \mathbb{N}_0,$$

and
$(N.1)$ For every $H > 0$ there exists $A > 0$ such that

$$p!^{1/2} \leq AH^p M_p, \quad p \in \mathbb{N}_0.$$

The conditions $\{N.1\}$ and $(N.1)$ are taken from [49] where they are called *nontriviality conditions* for the spaces $\mathcal{S}^{M_p}_{M_p}(\mathbb{R}^d)$ and $\Sigma^{M_p}_{M_p}(\mathbb{R}^d)$ respectively. In fact, the following lemma is proved in [47].

Lemma 3 *Let there be given a sequence of positive numbers $(M_p)_{p\in\mathbb{N}_0}$ which satisfies $(M.1)$ and*

$(M.2)'$ There exist positive constants A, H such that $M_{p+1} \leq AH^p M_p$, $p \in \mathbb{N}_0$. Then the following are equivalent:

1. *The Hermite functions are contained in $\mathcal{S}^{M_p}_{M_p}(\mathbb{R}^d)$ (in $\Sigma^{M_p}_{M_p}(\mathbb{R}^d)$, respectively).*
2. *$(M_p)_{p\in\mathbb{N}_0}$ satisfies $\{N.1\}$ ($(M_p)_{p\in\mathbb{N}_0}$ satisfies $(N.1)$, respectively).*
3. *There are positive constants A, B and H such that*

$$p!^{1/2} M_q \leq AB^{p+q} H^p M_{p+q}, \quad p, q \in \mathbb{N}_0.$$

There is $B > 0$ such that for every $H > 0$ there exists $A > 0$ such that

$$p!^{1/2} M_q \leq A B^{p+q} H^p M_{p+q}, \quad p, q \in \mathbb{N}_0.$$

We note that the condition $(M.2)'$ is weaker than the condition $(M.2)$, and refer to [47, Remark 3.3] for the proof of Lemma 3.

Theorem 2 *Let there be given a sequence of positive numbers $(M_p)_{p\in\mathbb{N}_0}$ which satisfies $(M.1)$, $(M.2)$ and $\{N.1\}$. Then the following isomorphisms hold:*

1. $\mathscr{S}^{M_p}_{M_p}(\mathbb{R}^{d_1+d_2}) \cong \mathscr{S}^{M_p}_{M_p}(\mathbb{R}^{d_1})\hat{\otimes}\mathscr{S}^{M_p}_{M_p}(\mathbb{R}^{d_2})$

$$\cong \mathscr{L}_b((\mathscr{S}^{M_p}_{M_p})'(\mathbb{R}^{d_1}), \mathscr{S}^{M_p}_{M_p}(\mathbb{R}^{d_2})),$$

2. $(\mathscr{S}^{M_p}_{M_p})'(\mathbb{R}^{d_1+d_2}) \cong (\mathscr{S}^{M_p}_{M_p})'(\mathbb{R}^{d_1})\hat{\otimes}(\mathscr{S}^{M_p}_{M_p})'(\mathbb{R}^{d_2})$

$$\cong \mathscr{L}_b(\mathscr{S}^{M_p}_{M_p}(\mathbb{R}^{d_1}), (\mathscr{S}^{M_p}_{M_p})'(\mathbb{R}^{d_2})).$$

If the sequence $(M_p)_{p\in\mathbb{N}_0}$ satisfies $(M.1)$, $(M.2)$ and $(N.1)$ instead, then the following isomorphisms hold:

3. $\Sigma^{M_p}_{M_p}(\mathbb{R}^{d_1+d_2}) \cong \Sigma^{M_p}_{M_p}(\mathbb{R}^{d_1})\hat{\otimes}\Sigma^{M_p}_{M_p}(\mathbb{R}^{d_2})$

$$\cong \mathscr{L}_b((\Sigma^{M_p}_{M_p})'(\mathbb{R}^{d_1}), \Sigma^{M_p}_{M_p}(\mathbb{R}^{d_2})),$$

4. $(\Sigma^{M_p}_{M_p})'(\mathbb{R}^{d_1+d_2}) \cong (\Sigma^{M_p}_{M_p})'(\mathbb{R}^{d_1})\hat{\otimes}(\Sigma^{M_p}_{M_p})'(\mathbb{R}^{d_2})$

$$\cong \mathscr{L}_b(\Sigma^{M_p}_{M_p}(\mathbb{R}^{d_1}), (\Sigma^{M_p}_{M_p})'(\mathbb{R}^{d_2})).$$

Proof By [47, Remark 3.3] it follows that $\{N.1\}$ is equivalent to

$$H_k(x) \in \mathscr{S}^{M_p}_{M_p}(\mathbb{R}^{d_1}), \quad x \in \mathbb{R}^{d_1}, k \in \mathbb{N}_0^{d_1}, \quad \text{and} \quad H_l(y) \in \mathscr{S}^{M_p}_{M_p}(\mathbb{R}^{d_2}), \quad y \in \mathbb{R}^{d_2}, l \in \mathbb{N}_0^{d_2},$$

where $H_k(x)$ and $H_l(y)$ are the Hermite functions given by (1). Now, by representation theorems from [47, 49] and the fact that $H_{(k,l)}(x, y) \in \mathscr{S}^{M_p}_{M_p}(\mathbb{R}^{d_1+d_2})$, $(x, y) \in \mathbb{R}^{d_1+d_2}$, $(k, l) \in \mathbb{N}_0^{d_1+d_2}$, it follows that $\mathscr{S}^{M_p}_{M_p}(\mathbb{R}^{d_1})\otimes\mathscr{S}^{M_p}_{M_p}(\mathbb{R}^{d_2})$ is dense in $\mathscr{S}^{M_p}_{M_p}(\mathbb{R}^{d_1+d_2})$.

In order to obtain the isomorphism $\mathscr{S}^{M_p}_{M_p}(\mathbb{R}^{d_1+d_2}) \cong \mathscr{S}^{M_p}_{M_p}(\mathbb{R}^{d_1})\hat{\otimes}\mathscr{S}^{M_p}_{M_p}(\mathbb{R}^{d_2})$ it is sufficient to prove that $\mathscr{S}^{M_p}_{M_p}(\mathbb{R}^{d_1+d_2})$ induces the $\pi = \varepsilon$ topology on the product space $\mathscr{S}^{M_p}_{M_p}(\mathbb{R}^{d_1})\hat{\otimes}\mathscr{S}^{M_p}_{M_p}(\mathbb{R}^{d_2})$. The topologies $\mathscr{S}^{M_p}_{M_p}(\mathbb{R}^{d_1})\otimes_\pi\mathscr{S}^{M_p}_{M_p}(\mathbb{R}^{d_2})$ and $\mathscr{S}^{M_p}_{M_p}(\mathbb{R}^{d_1})\otimes_\varepsilon \mathscr{S}^{M_p}_{M_p}(\mathbb{R}^{d_2})$ coincide since $\mathscr{S}^{M_p}_{M_p}(\mathbb{R}^{d})$ is a nuclear space (see e.g. [49]). We refer to [86, Chapter 43] for the definition and basic facts on the π and ε topologies.

The idea of the proof is to show that the topology π on $\mathscr{S}^{M_p}_{M_p}(\mathbb{R}^{d_1})\otimes\mathscr{S}^{M_p}_{M_p}(\mathbb{R}^{d_2})$ is stronger than the one induced from $\mathscr{S}^{M_p}_{M_p}(\mathbb{R}^{d_1+d_2})$, and that the ε topology on $\mathscr{S}^{M_p}_{M_p}(\mathbb{R}^{d_1})\otimes\mathscr{S}^{M_p}_{M_p}(\mathbb{R}^{d_2})$ is weaker than the induced one. This will imply the isomorphism

$$\mathscr{S}^{M_p}_{M_p}(\mathbb{R}^{d_1+d_2}) \cong \mathscr{S}^{M_p}_{M_p}(\mathbb{R}^{d_1})\hat{\otimes}\mathscr{S}^{M_p}_{M_p}(\mathbb{R}^{d_2}).$$

That proof is quite technical, and we give here only a sketch, cf. [70] for details. For the π topology, we use a convenient separately continuous bilinear mapping which implies the continuity of the inclusion $\mathscr{S}^{M_p}_{M_p}(\mathbb{R}^{d_1})\otimes_\pi\mathscr{S}^{M_p}_{M_p}(\mathbb{R}^{d_2}) \to \mathscr{S}^{M_p}_{M_p}(\mathbb{R}^{d_1+d_2})$.

For the ε topology, the proof is more technically involved. Namely, for a given equicontinuous subsets $A' \subset \mathscr{S}^{M_p}_{M_p}(\mathbb{R}^{d_1})$ and $B' \subset \mathscr{S}^{M_p}_{M_p}(\mathbb{R}^{d_2})$, we use a particularly chosen family of norms which defines a topology equivalent to the one given by Definition 3 to estimate $|\langle F_x \otimes \tilde{F}_y, \Phi(x, y)\rangle|$, $F_x \in A'$ and $\tilde{F}_y \in B'$.

Next, we use the fact that $(\mathscr{S}^{M_p}_{M_p})'(\mathbb{R}^{d_1})$ and $\mathscr{S}^{M_p}_{M_p}(\mathbb{R}^{d_2})$ are complete and that $(\mathscr{S}^{M_p}_{M_p})'(\mathbb{R}^{d_1})$ is barreled. Moreover, $\mathscr{S}^{M_p}_{M_p}(\mathbb{R}^{d_1})$ is nuclear and complete, so that [86, Proposition 50.5] implies that $\mathscr{L}_b((\mathscr{S}^{M_p}_{M_p})'(\mathbb{R}^{d_1}), \mathscr{S}^{M_p}_{M_p}(\mathbb{R}^{d_2}))$ is complete and that

$$\mathscr{S}^{M_p}_{M_p}(\mathbb{R}^{d_1})\hat{\otimes}\mathscr{S}^{M_p}_{M_p}(\mathbb{R}^{d_2}) \cong \mathscr{L}_b((\mathscr{S}^{M_p}_{M_p})'(\mathbb{R}^{d_1}), \mathscr{S}^{M_p}_{M_p}(\mathbb{R}^{d_2})).$$

This proves (1) and we leave the other claims to the reader, see also [58, 59]. □

The isomorphisms in Theorem 2 (2) tells us that for a given kernel-distribution $k(x, y)$ on $\mathbb{R}^{d_1+d_2}$ we may associate a continuous linear mapping k of $\mathscr{S}^{M_p}_{M_p}(\mathbb{R}^{d_2})$ into $(\mathscr{S}^{M_p}_{M_p})'(\mathbb{R}^{d_1})$ as follows:

$$\langle k_\varphi, \phi\rangle = \langle k(x, y), \phi(x)\varphi(y)\rangle, \quad \phi \in \mathscr{S}^{M_p}_{M_p}(\mathbb{R}^{d_1}),$$

which is commonly written as $k_\varphi(\cdot) = \int k(\cdot, y)\varphi(y)dy$. By Theorem 2 (b) it follows that the correspondence between $k(x, y)$ and k is an isomorphism. Note also that the transpose ${}^tk$ of the mapping k is given by ${}^tk_\phi(\cdot) = \displaystyle\int k(x, \cdot)\phi(x)dx$.

By the above isomorphisms we conclude that for any continuous and linear mapping between $\mathscr{S}^{M_p}_{M_p}(\mathbb{R}^{2d})$ to $(\mathscr{S}^{M_p}_{M_p})'(\mathbb{R}^{2d})$ one can assign a uniquely determined kernel with the above mentioned properties. We will use this fact in the proof of Theorem 6, and refer to [86, Chapter 52] for applications of kernel theorems in linear partial differential equations.

Remark 5 The choice of the Fourier transform invariant spaces of the form $\mathscr{S}^{M_p}_{M_p}(\mathbb{R}^d)$ in Theorem 2 is not accidental. We refer to [37] where it is proved that if the Hermite expansion $\sum_{k\in\mathbb{N}^d} a_k H_k(x)$ converges to f (a_k are the Hermite coefficients of f) in the sense of $\mathscr{S}^s_r(\mathbb{R}^d)$ ($\Sigma^s_r(\mathbb{R}^d)$, respectively), $r < s$, then it belongs to $\mathscr{S}^r_r(\mathbb{R}^d)$ ($\Sigma^r_r(\mathbb{R}^d)$, respectively).

3.4 Time-Frequency Analysis of Gelfand–Shilov Spaces

In this section we extend the action of the time-frequency representations from Sect. 2 to the Gelfand–Shilov spaces and their dual spaces. To that end we observe the following modification of Definition 3.

Definition 4 Let there be given sequences of positive numbers $(M_p)_{p\in\mathbb{N}_0}$, $(N_q)_{q\in\mathbb{N}_0}$, $(\tilde{M}_p)_{p\in\mathbb{N}_0}$, $(\tilde{N}_q)_{q\in\mathbb{N}_0}$ which satisfy $(M.1)$ and $(M.2)$. We define $\mathscr{S}^{N_q,\tilde{N}_q,B}_{M_p,\tilde{M}_p,A}(\mathbb{R}^{2d})$ to be the set of smooth functions $f \in C^\infty(\mathbb{R}^{2d})$ such that

$$\|x^{\alpha_1}\omega^{\alpha_2}\partial_x^{\beta_1}\partial_\omega^{\beta_2} f\|_{L^\infty} \le C A^{|\alpha_1+\alpha_2|} M_{|\alpha_1|}\tilde{M}_{|\alpha_2|} B^{|\beta_1+\beta_2|} N_{|\beta_1|}\tilde{N}_{|\beta_2|},$$
$$\forall \alpha_1, \alpha_2, \beta_1, \beta_2 \in \mathbb{N}_0^d\},$$

and for some $A, B, C > 0$. *Gelfand–Shilov spaces* are projective and inductive limits of $\mathscr{S}^{N_q,\tilde{N}_q,B}_{M_p,\tilde{M}_p,A}(\mathbb{R}^{2d})$:

$$\Sigma^{N_q,\tilde{N}_q}_{M_p,\tilde{M}_p}(\mathbb{R}^{2d}) := \text{proj} \lim_{A>0,B>0} \mathscr{S}^{N_q,\tilde{N}_q,B}_{M_p,\tilde{M}_p,A}(\mathbb{R}^{2d});$$

$$\mathscr{S}^{N_q,\tilde{N}_q}_{M_p,\tilde{M}_p}(\mathbb{R}^{2d}) := \text{ind} \lim_{A>0,B>0} \mathscr{S}^{N_q,\tilde{N}_q,B}_{M_p,\tilde{M}_p,A}(\mathbb{R}^{2d}).$$

Clearly, the corresponding dual spaces are given by

$$(\Sigma^{N_q,\tilde{N}_q}_{M_p,\tilde{M}_p})'(\mathbb{R}^{2d}) := \text{ind} \lim_{A>0,B>0} (\mathscr{S}^{N_q,\tilde{N}_q,B}_{M_p,\tilde{M}_p,A})'(\mathbb{R}^{2d});$$

$$(\mathscr{S}^{N_q,\tilde{N}_q}_{M_p,\tilde{M}_p})'(\mathbb{R}^{2d}) := \text{proj} \lim_{A>0,B>0} (\mathscr{S}^{N_q,\tilde{N}_q,B}_{M_p,\tilde{M}_p,A})'(\mathbb{R}^{2d}).$$

By Theorem 1, the Fourier transform is a homeomorphism from $\Sigma^{N_q,\tilde{N}_q}_{M_p,\tilde{M}_p}(\mathbb{R}^{2d})$ to $\Sigma^{M_p,\tilde{M}_p}_{N_q,\tilde{N}_q}(\mathbb{R}^{2d})$ and, if $\mathscr{F}_1 f$ denotes the partial Fourier transform of $f(x,\omega)$ with respect to the x variable, and if $\mathscr{F}_2 f$ denotes the partial Fourier transform of $f(x,\omega)$ with respect to the ω variable, then $\mathscr{F}_1$ and $\mathscr{F}_2$ are homeomorphisms from $\Sigma^{N_q,\tilde{N}_q}_{M_p,\tilde{M}_p}(\mathbb{R}^{2d})$ to $\Sigma^{N_q,\tilde{M}_p}_{M_p,\tilde{N}_q}(\mathbb{R}^{2d})$ and $\Sigma^{N_q,\tilde{M}_p}_{M_p,\tilde{N}_q}(\mathbb{R}^{2d})$, respectively. Similar facts hold when $\Sigma^{N_q,\tilde{N}_q}_{M_p,\tilde{M}_p}(\mathbb{R}^{2d})$ is replaced by $\mathscr{S}^{N_q,\tilde{N}_q}_{M_p,\tilde{M}_p}(\mathbb{R}^{2d})$, $(\Sigma^{N_q,\tilde{N}_q}_{M_p,\tilde{M}_p})'(\mathbb{R}^{2d})$ or $(\mathscr{S}^{N_q,\tilde{N}_q}_{M_p,\tilde{M}_p})'(\mathbb{R}^{2d})$.

When $M_p = \tilde{M}_p$ and $N_q = \tilde{N}_q$ we use usual abbreviated notation: $\mathscr{S}^{N_q}_{M_p}(\mathbb{R}^{2d}) = \mathscr{S}^{N_q,\tilde{N}_q}_{M_p,\tilde{M}_p}(\mathbb{R}^{2d})$ and similarly for other spaces.

Let $(M_p)_{p\in\mathbb{N}_0}$ satisfy $(M.1)$, $(M.2)$ and $\{N.1\}$ ($(N.1)$, respectively). For any given $f, g \in \mathscr{S}^{M_p}_{M_p}(\mathbb{R}^d)$ ($f, g \in \Sigma^{M_p}_{M_p}(\mathbb{R}^d)$, respectively) the Grossmann–Royer transform of f and g is given by (4), i.e.

$$R_g f(x, \omega) = \int e^{4\pi i\omega(t-x)} f(2x - t)\overline{g(t)}dt, \quad x, \omega \in \mathbb{R}^d,$$

and the definition can be extended to $f \in (\mathcal{S}_{M_p}^{M_p})'(\mathbb{R}^d)$ ($f \in (\Sigma_{M_p}^{M_p})'(\mathbb{R}^d)$, respectively) by duality.

Similarly, another time-frequency representations, the short-time Fourier transform V_g, the cross-Wigner distribution $W(f, g)$, and the cross-ambiguity function $A(f, g)$ given by (5), (6) and (7), can be extended to $f \in (\mathcal{S}_{M_p}^{M_p})'(\mathbb{R}^d)$ ($f \in (\Sigma_{M_p}^{M_p})'(\mathbb{R}^d)$, respectively) when $g \in \mathcal{S}_{M_p}^{M_p}(\mathbb{R}^d)$ ($g \in \Sigma_{M_p}^{M_p}(\mathbb{R}^d)$, respectively).

The following theorem and its variations is a folklore, in particular in the framework of the duality between $\mathcal{S}(\mathbb{R}^{2d})$ and $\mathcal{S}'(\mathbb{R}^{2d})$. For Gelfand–Shilov spaces we refer to e.g. [43, 68, 70, 78].

Theorem 3 *Let there be given sequences $(M_p)_{p\in\mathbb{N}_0}$ and $(N_q)_{q\in\mathbb{N}_0}$ which satisfy (M.1), (M.2) and $\{N.1\}$, and let $TFR(f, g) \in \{R_g f, V_g f, W(f, g), A\}$. If $f, g \in \mathcal{S}_{M_p}^{N_q}(\mathbb{R}^d)$, then $TFR(f, g) \in \mathcal{S}_{M_p,N_q}^{N_q,M_p}(\mathbb{R}^{2d})$ and extends uniquely to a continuous map from $(\mathcal{S}_{M_p}^{N_q})'(\mathbb{R}^d) \times (\mathcal{S}_{N_q}^{M_p})'(\mathbb{R}^d)$ into $(\mathcal{S}_{M_p,N_q}^{N_q,M_p})'(\mathbb{R}^{2d})$.*

Conversely, if $TFR(f, g) \in \mathcal{S}_{M_p,N_q}^{N_q,M_p}(\mathbb{R}^{2d})$ then $f, g \in \mathcal{S}_{M_p}^{N_q}(\mathbb{R}^d)$.

Let the sequences $(M_p)_{p\in\mathbb{N}_0}$ and $(N_q)_{q\in\mathbb{N}_0}$ satisfy (M.1), (M.2) and $(N.1)$ instead. If $f, g \in \Sigma_{M_p}^{N_q}(\mathbb{R}^d)$, then $TFR(f, g) \in \Sigma_{M_p,N_q}^{N_q,M_p}(\mathbb{R}^{2d})$ and extends uniquely to a continuous map from $(\Sigma_{M_p}^{N_q})'(\mathbb{R}^d) \times (\Sigma_{N_q}^{M_p})'(\mathbb{R}^d)$ into $(\Sigma_{M_p,N_q}^{N_q,M_p})'(\mathbb{R}^{2d})$.

Conversely, if $TFR(f, g) \in \Sigma_{M_p,N_q}^{N_q,M_p}(\mathbb{R}^{2d})$ then $f, g \in \Sigma_{M_p}^{N_q}(\mathbb{R}^d)$.

Proof Since Gelfand–Shilov spaces are closed under reflections, dilations and modulations, by Lemma 1 it is enough to give the proof for the Grossmann–Royer transform, and the same conclusion holds for other time-frequency representations. But the proof is essentially the same as the proof of Proposition 1 *6*. We recall (9):

$$R_g f(x, \omega) = (\mathcal{F}_2 \circ \tau^*) f \otimes \overline{g}(x, \omega), \quad x, \omega \in \mathbb{R}^d,$$

Since the pullback operator τ^* is a continuous bijection on $\mathcal{S}_{M_p,M_p}^{N_q,N_q}(\mathbb{R}^{2d})$, and $\mathcal{F}_2$ is a continuous bijection between $\mathcal{S}_{M_p,M_p}^{N_q,N_q}(\mathbb{R}^{2d})$ and $\mathcal{S}_{M_p,N_q}^{N_q,M_p}(\mathbb{R}^{2d})$, we obtain

$$R_g f \in \mathcal{S}_{M_p,N_q}^{N_q,M_p}(\mathbb{R}^{2d}) \Leftrightarrow f \otimes \overline{g} \in \mathcal{S}_{M_p,M_p}^{N_q,N_q}(\mathbb{R}^{2d}) \Leftrightarrow f, g \in \mathcal{S}_{M_p}^{N_q}(\mathbb{R}^d).$$

Moreover, $R_g f$ can be extended to a map from $(\mathcal{S}_{M_p}^{N_q})'(\mathbb{R}^d) \times (\mathcal{S}_{N_q}^{M_p})'(\mathbb{R}^d)$ into $(\mathcal{S}_{M_p,N_q}^{N_q,M_p})'(\mathbb{R}^{2d})$ by duality.

To prove that $R_g f \in \mathcal{S}_{M_p,N_q}^{N_q,M_p}(\mathbb{R}^{2d})$ when $f, g \in \mathcal{S}_{M_p}^{N_q}(\mathbb{R}^d)$, we could also perform direct calculations based on the following observations.

Assume that $g \in \mathscr{S}^{\tilde{N}_q}_{\tilde{M}_p}(\mathbb{R}^d)$ where $(\tilde{M}_p)_{p\in\mathbb{N}_0}$ and $(\tilde{N}_q)_{q\in\mathbb{N}_0}$ satisfy (M.1), (M.2), $\tilde{M}_p \subset M_p$ and $\tilde{N}_q \subset N_q$, which is a slightly more general situation. Than $f(x) \otimes g(t) \in \mathscr{S}^{N_q,\tilde{N}_q}_{M_p,\tilde{M}_p}(\mathbb{R}^d \times \mathbb{R}^d)$.

Put $\varphi(x,t) := f(2x-t)g(t)$. If we show

$$\sup_{x,t\in\mathbb{R}^d} |x^\alpha t^\beta \varphi(x,t)| \leq C h^{|\alpha|+|\beta|} M_{|\alpha|} M_{|\beta|}, \tag{16}$$

and

$$\sup_{x,t\in\mathbb{R}^d} |\partial_x^\alpha \partial_t^\beta \varphi(x,t)| \leq C k^{|\alpha|+|\beta|} N_{|\alpha|} N_{|\beta|} \tag{17}$$

for some $h, k > 0$, then by Theorem 1 it follows that $\varphi \in \mathscr{S}^{N_q,N_q}_{M_p,M_p}(\mathbb{R}^{2d})$.

The first inequality easily follows from assumptions on f and g and a change of variables:

$$\sup_{x,t\in\mathbb{R}^d} |x^\alpha t^\beta f(2x-t)g(t)| \leq 2^{-|\alpha|} \sup_{y,t\in\mathbb{R}^d} |(y+t)^\alpha t^\beta f(y)g(t)|,$$

and (16) follows from the assumptions on f, g and $\tilde{M}_p \subset M_p$. To prove (17), we use the Leibniz formula which gives

$$\begin{aligned}
|\partial_x^\alpha \partial_t^\beta \varphi(x,t)| &= |\sum_{\gamma\leq\beta} \binom{\beta}{\gamma} \frac{1}{2^{|\alpha|+|\beta|}} \partial_x^\alpha \partial_t^\gamma f(2x-t) \partial_t^{\beta-\gamma} g(t)| \\
&\leq C_{\alpha,\beta} \sup_{x,t\in\mathbb{R}^d} |\partial_x^\alpha \partial_t^\gamma f(2x-t) \partial_t^{\beta-\gamma} g(t)|.
\end{aligned}$$

Next we use $\tilde{N}_q \subset N_q$ and conditions (M.1) and (M.2) applied to the sequence (N_q) to obtain (17). Therefore, $\varphi \in \mathscr{S}^{N_q,N_q}_{M_p,M_p}(\mathbb{R}^{2d})$.

Now, the partial inverse Fourier transform of φ with respect to the second variable is continuous bijection between $\mathscr{S}^{N_q,N_q}_{M_p,M_p}(\mathbb{R}^{2d})$ and $\mathscr{S}^{N_q,M_p}_{M_p,N_q}(\mathbb{R}^{2d})$, and those spaces are closed under dilations, so that

$$R_g f(x,\omega) = e^{-4\pi i\omega x} \int e^{2\pi i\omega(2t)} \varphi(x,t)dt \in \mathscr{S}^{N_q,M_p}_{M_p N_q}(\mathbb{R}^{2d})$$

if and only if $\varphi \in \mathscr{S}^{N_q,N_q}_{M_p,M_p}(\mathbb{R}^{2d})$. The extension to a map from $(\mathscr{S}^{N_q}_{M_p})'(\mathbb{R}^d) \times (\mathscr{S}^{M_p}_{N_q})'(\mathbb{R}^d)$ into $(\mathscr{S}^{N_q,M_p}_{M_p,N_q})'(\mathbb{R}^{2d})$ is straightforward. □

4 Modulation Spaces

The modulation space norms traditionally measure the joint time-frequency distribution of $f \in \mathcal{S}'$, we refer, for instance, to [29], [38, Ch. 11–13] and the original literature quoted there for various properties and applications. It is usually sufficient to observe modulation spaces with weights which admit at most polynomial growth at infinity. However the study of ultra-distributions requires a more general approach that includes the weights of exponential or even superexponential growth, cf. [19, 80]. Note that the general approach introduced already in [29] includes the weights of sub-exponential growth (see (19)). We refer to [30, 31] for related but even more general constructions, based on the general theory of coorbit spaces.

Weight Functions. In the sequel v will always be a continuous, positive, even, submultiplicative function (submultiplicative weight), i.e., $v(0) = 1$, $v(z) = v(-z)$, and $v(z_1 + z_2) \leq v(z_1)v(z_2)$, for all $z, z_1, z_2 \in \mathbb{R}^{2d}$. Moreover, v is assumed to be even in each group of coordinates, that is, $v(x, \omega) = v(-\omega, x) = v(-x, \omega)$, for any $(x, \omega) \in \mathbb{R}^{2d}$. Submultipliciativity implies that $v(z)$ is *dominated* by an exponential function, i.e.

$$\exists\, C, k > 0 \quad \text{such that} \quad v(z) \leq C e^{k\|z\|}, \quad z \in \mathbb{R}^{2d}, \tag{18}$$

and $\|z\|$ is the Euclidean norm of $z \in \mathbb{R}^{2d}$. For example, every weight of the form

$$v(z) = e^{s\|z\|^b} (1 + \|z\|)^a \log^r(e + \|z\|) \tag{19}$$

for parameters $a, r, s \geq 0$, $0 \leq b \leq 1$ satisfies the above conditions.

For our investigation of localization operators we will mostly use the exponential weights defined by

$$w_s(z) = w_s(x, \omega) = e^{s\|(x,\omega)\|}, \quad z = (x, \omega) \in \mathbb{R}^{2d}, \tag{20}$$

$$\tau_s(z) = \tau_s(x, \omega) = e^{s\|\omega\|}. \tag{21}$$

Notice that arguing on $\mathbb{R}^{4d}$ we may read

$$\tau_s(z, \zeta) = w_s(\zeta) \qquad z, \zeta \in \mathbb{R}^{2d}, \tag{22}$$

which will be used in the sequel.

Associated to every submultiplicative weight we consider the class of so-called v-*moderate* weights $\mathcal{M}_v$. A positive, even weight function m on $\mathbb{R}^{2d}$ belongs to $\mathcal{M}_v$ if it satisfies the condition

$$m(z_1 + z_2) \leq C v(z_1) m(z_2) \quad \forall z_1, z_2 \in \mathbb{R}^{2d}.$$

We note that this definition implies that $\frac{1}{v} \lesssim m \lesssim v$, $m \neq 0$ everywhere, and that $1/m \in \mathcal{M}_v$.

Depending on the growth of the weight function m, different Gelfand–Shilov classes may be chosen as fitting test function spaces for modulation spaces, see [19, 68, 80]. The widest class of weights allowing to define modulation spaces is the weight class $\mathcal{N}$. A weight function m on $\mathbb{R}^{2d}$ belongs to $\mathcal{N}$ if it is a continuous, positive function such that

$$m(z) = o(e^{cz^2}), \quad \text{for } |z| \to \infty, \quad \forall c > 0, \tag{23}$$

with $z \in \mathbb{R}^{2d}$. For instance, every function $m(z) = e^{s|z|^b}$, with $s > 0$ and $0 \leq b < 2$, is in $\mathcal{N}$. Thus, the weight m may grow faster than exponentially at infinity. For example, the choice $m \in \mathcal{N} \setminus \mathcal{M}_v$ is related to the spaces of quasianalytic functions, [18]. We notice that there is a limit in enlarging the weight class for modulation spaces, imposed by Hardy's theorem: if $m(z) \geq Ce^{cz^2}$, for some $c > \pi/2$, then the corresponding modulation spaces are trivial [42]. We refer to [39] for a survey on the most important types of weights commonly used in time-frequency analysis.

Definition 5 Let $m \in \mathcal{N}$, and g a non-zero *window* function in $\mathcal{S}_{1/2}^{1/2}(\mathbb{R}^d)$. For $1 \leq p, q \leq \infty$ the *modulation space* $M_m^{p,q}(\mathbb{R}^d)$ consists of all $f \in (\mathcal{S}_{1/2}^{1/2})'(\mathbb{R}^d)$ such that $V_g f \in L_m^{p,q}(\mathbb{R}^{2d})$ (weighted mixed-norm spaces). The norm on $M_m^{p,q}$ is

$$\|f\|_{M_m^{p,q}} = \|V_g f\|_{L_m^{p,q}} = \left(\int_{\mathbb{R}^d} \left(\int_{\mathbb{R}^d} |V_g f(x,\omega)|^p m(x,\omega)^p \, dx \right)^{q/p} d\omega \right)^{1/q}$$

(with obvious changes if either $p = \infty$ or $q = \infty$). If $p, q < \infty$, the modulation space $M_m^{p,q}$ is the norm completion of $\mathcal{S}_{1/2}^{1/2}$ in the $M_m^{p,q}$-norm. If $p = \infty$ or $q = \infty$, then $M_m^{p,q}$ is the completion of $\mathcal{S}_{1/2}^{1/2}$ in the weak* topology.

In this paper we restrict ourselves to v-moderate weights $\mathcal{M}_v$. Then, for $f, g \in \mathcal{S}^{(1)}(\mathbb{R}^d) = \Sigma_1^1(\mathbb{R}^d)$ the above integral is convergent thanks to Theorem 3. Namely, in view of (18), for a given $m \in \mathcal{M}_v$ there exist $l > 0$ such that $m(x,\omega) \leq Ce^{l\|(x,\omega)\|}$ and therefore

$$\left| \int_{\mathbb{R}^d} \left(\int_{\mathbb{R}^d} |V_g f(x,\omega)|^p m(x,\omega)^p \, dx \right)^{q/p} d\omega \right|$$
$$\leq C \left| \int_{\mathbb{R}^d} \left(\int_{\mathbb{R}^d} |V_g f(x,\omega)|^p e^{lp\|(x,\omega)\|} \, dx \right)^{q/p} d\omega \right| < \infty$$

since by Theorems 1 and 3 we have $|V_g f(x,\omega)| < Ce^{-s\|(x,\omega)\|}$ for every $s > 0$. This implies $\mathcal{S}^{(1)} \subset M_m^{p,q}$.

In particular, when m is a polynomial weight of the form $m(x,\omega) = \langle x \rangle^t \langle \omega \rangle^s$ we will use the notation $M_{s,t}^{p,q}(\mathbb{R}^d)$ for the modulation spaces which consists of all $f \in \mathcal{S}'(\mathbb{R}^d)$ such that

$$\|f\|_{M^{p,q}_{s,t}} \equiv \left(\int_{\mathbb{R}^d} \left(\int_{\mathbb{R}^d} |V_\phi f(x,\omega)\langle x\rangle^t \langle\omega\rangle^s|^p \, dx \right)^{q/p} d\omega \right)^{1/q} < \infty$$

(with obvious interpretation of the integrals when $p = \infty$ or $q = \infty$).

If $p = q$, we write M^p_m instead of $M^{p,p}_m$, and if $m(z) \equiv 1$ on $\mathbb{R}^{2d}$, then we write $M^{p,q}$ and M^p for $M^{p,q}_m$ and $M^{p,p}_m$, and so on.

In the next proposition we show that $M^{p,q}_m(\mathbb{R}^d)$ are Banach spaces whose definition is independent of the choice of the window $g \in M^1_v \setminus \{0\}$. In order to do so, we need the adjoint of the short-time Fourier transform.

For given window $g \in \mathscr{S}^{(1)}$ and a function $F(x, \xi) \in L^{p,q}_m(\mathbb{R}^{2d})$ we define $V^*_g F$ by

$$\langle V^*_g F, f\rangle := \langle F, V_g f\rangle,$$

whenever the duality is well defined.

In our context, [38, Proposition 11.3.2] can be rewritten as follows.

Proposition 6 *Fix $m \in \mathscr{M}_v$ and $g, \psi \in \mathscr{S}^{(1)}$, with $\langle g, \psi\rangle \neq 0$. Then*

1. *$V^*_g : L^{p,q}_m(\mathbb{R}^{2d}) \to M^{p,q}_m(\mathbb{R}^d)$, and*

$$\|V^*_g F\|_{M^{p,q}_m} \leq C \|V_\psi g\|_{L^1_v} \|F\|_{L^{p,q}_m}. \tag{24}$$

2. *The inversion formula holds: $I_{M^{p,q}_m} = \langle g, \psi\rangle^{-1} V^*_g V_\psi$, where $I_{M^{p,q}_m}$ stands for the identity operator.*
3. *$M^{p,q}_m(\mathbb{R}^d)$ are Banach spaces whose definition is independent on the choice of $g \in \mathscr{S}^{(1)} \setminus \{0\}$.*
4. *The space of admissible windows can be extended from $\mathscr{S}^{(1)}$ to M^1_v.*

Proof We refer to [19] for the proof which is based on the proof of [38, Proposition 11.3.2.]. Note that for *4.* we need the density of $\mathscr{S}^{(1)}$ in $M^{p,q}_m$. This fact is not obvious, we refer to [11] for the proof. Then *4.* follows by using standard arguments of [38, Theorem 11.3.7].

Note that this result actually implies that Definition 5 coincides with the usual definition of modulation spaces with weights of polynomial and sub-exponential growth (see, for example [13, 29, 38, 56]). □

The following theorem lists some basic properties of modulation spaces. We refer to [29, 38, 43, 55, 69, 78] for the proof.

Theorem 4 *Let $p, q, p_j, q_j \in [1, \infty]$ and $s, t, s_j, t_j \in \mathbb{R}$, $j = 1, 2$. Then:*

1. *$M^{p,q}_{s,t}(\mathbb{R}^d)$ are Banach spaces, independent of the choice of $\phi \in \mathscr{S}(\mathbb{R}^d) \setminus 0$;*
2. *if $p_1 \leq p_2$, $q_1 \leq q_2$, $s_2 \leq s_1$ and $t_2 \leq t_1$, then*

$$\mathscr{S}(\mathbb{R}^d) \subseteq M^{p_1,q_1}_{s_1,t_1}(\mathbb{R}^d) \subseteq M^{p_2,q_2}_{s_2,t_2}(\mathbb{R}^d) \subseteq \mathscr{S}'(\mathbb{R}^d);$$

3. *$\cap_{s,t} M^{p,q}_{s,t}(\mathbb{R}^d) = \mathscr{S}(\mathbb{R}^d)$, $\cup_{s,t} M^{p,q}_{s,t}(\mathbb{R}^d) = \mathscr{S}'(\mathbb{R}^d)$;*

4. *Let* $1 \leq p, q \leq \infty$, *and let* w_s *be given by* (20). *Then*

$$\Sigma_1^1(\mathbb{R}^d) = \mathscr{S}^{(1)}(\mathbb{R}^d) = \bigcap_{s\geq 0} M^{p,q}_{w_s}(\mathbb{R}^d), \quad (\Sigma_1^1)'(\mathbb{R}^d) = \bigcup_{s\geq 0} M^{p,q}_{1/w_s}(\mathbb{R}^d),$$

$$\mathscr{S}_1^1(\mathbb{R}^d) = \mathscr{S}^{\{1\}}(\mathbb{R}^d) = \bigcup_{s>0} M^{p,q}_{w_s}(\mathbb{R}^d), \quad (\mathscr{S}_1^1)'(\mathbb{R}^d) = \bigcap_{s>0} M^{p,q}_{1/w_s}(\mathbb{R}^d).$$

5. *For* $p, q \in [1, \infty)$, *the dual of* $M^{p,q}_{s,t}(\mathbb{R}^d)$ *is* $M^{p',q'}_{-s,-t}(\mathbb{R}^d)$, *where* $\frac{1}{p} + \frac{1}{p'} = \frac{1}{q} + \frac{1}{q'} = 1$.

Remark 6 Alternatively, $\mathscr{S}_1^1(\mathbb{R}^d)$ can also be viewed as a projective limit (and its dual space as an inductive limit) of modulation spaces as follows:

$$\mathscr{S}_1^1(\mathbb{R}^d) = \bigcap_{m\in\cap\mathscr{M}_{ws}} M^{p,q}_{m}(\mathbb{R}^d), \quad (\mathscr{S}_1^1)'(\mathbb{R}^d) = \bigcup_{m\in\cap\mathscr{M}_{ws}} M^{p,q}_{1/m}(\mathbb{R}^d),$$

where w_s is given by (20), see [81].

In the context of quasianalytic Gelfand–Shilov spaces, we recall (a special case of) [78, Theorem 3.9]: Let $s, t > 1/2$ and set

$$w_h(x, \omega) \equiv e^{h(|x|^{1/t}+|\omega|^{1/s})}, \quad h > 0, \; x, \omega \in \mathbb{R}^d.$$

Then

$$\Sigma_t^s(\mathbb{R}^d) = \bigcap_{h>0} M^{p,q}_{w_h}(\mathbb{R}^d), \quad (\Sigma_t^s)'(\mathbb{R}^d) = \bigcup_{h>0} M^{p,q}_{1/w_h}(\mathbb{R}^d),$$

$$\mathscr{S}_t^s(\mathbb{R}^d) = \bigcup_{h>0} M^{p,q}_{w_h}(\mathbb{R}^d), \quad (\mathscr{S}_t^s)'(\mathbb{R}^d) = \bigcap_{h>0} M^{p,q}_{1/w_h}(\mathbb{R}^d).$$

Modulation spaces include the following well-know function spaces:

1. $M^2(\mathbb{R}^d) = L^2(\mathbb{R}^d)$, and $M^2_{t,0}(\mathbb{R}^d) = L^2_t(\mathbb{R}^d)$;
2. The Feichtinger algebra: $M^1(\mathbb{R}^d) = S_0(\mathbb{R}^d)$;
3. Sobolev spaces: $M^2_{0,s}(\mathbb{R}^d) = H^2_s(\mathbb{R}^d) = \{f \mid \hat{f}(\omega)\langle\omega\rangle^s \in L^2(\mathbb{R}^d)\}$;
4. Shubin spaces: $M^2_s(\mathbb{R}^d) = L^2_s(\mathbb{R}^d) \cap H^2_s(\mathbb{R}^d) = Q_s(\mathbb{R}^d)$, cf. [65].

4.1 Convolution Estimates for Modulation Spaces

Different theorems concerning the convolution relation between modulation spaces can be found in the literature. We recall the convolution estimates given in [13, Proposition 2.4] and in [84], which is sufficient for our purposes.

Proposition 7 *Let $m \in \mathscr{M}_v$ defined on $\mathbb{R}^{2d}$ and let $m_1(x) = m(x,0)$ and $m_2(\omega) = m(0,\omega)$, the restrictions to $\mathbb{R}^d \times \{0\}$ and $\{0\} \times \mathbb{R}^d$, and likewise for v. Let $\nu(\omega) > 0$ be an arbitrary weight function on $\mathbb{R}^d$ and*
$1 \le p, q, r, s, t \le \infty$. If

$$\frac{1}{p} + \frac{1}{q} - 1 = \frac{1}{r}, \quad \textit{and} \quad \frac{1}{t} + \frac{1}{t'} = 1,$$

then

$$M^{p,st}_{m_1 \otimes \nu}(\mathbb{R}^d) * M^{q,st'}_{v_1 \otimes v_2 \nu^{-1}}(\mathbb{R}^d) \hookrightarrow M^{r,s}_m(\mathbb{R}^d) \tag{25}$$

with norm inequality $\|f * h\|_{M^{r,s}_m} \lesssim \|f\|_{M^{p,st}_{m_1\otimes\nu}} \|h\|_{M^{q,st'}_{v_1\otimes v_2 \nu^{-1}}}$.

When the weights in Proposition 7 are chosen to be of the form $\langle x\rangle^t \langle\omega\rangle^s$, sharper continuity properties can be proved. For such results on multiplication and convolution in modulation spaces and in weighted Lebesgue spaces we observe the *Young functional*:

$$\mathsf{R}(\mathrm{p}) = \mathsf{R}(p_0, p_1, p_2) \equiv 2 - \frac{1}{p_0} - \frac{1}{p_1} - \frac{1}{p_2}, \quad \mathrm{p} = (p_0, p_1, p_2) \in [1,\infty]^3. \tag{26}$$

When $\mathsf{R}(\mathrm{p}) = 0$, the Young inequality for convolution reads as

$$\|f_1 * f_2\|_{L^{p_0'}} \le \|f_1\|_{L^{p_1}} \|f_2\|_{L^{p_2}}, \quad f_j \in L^{p_j}(\mathbb{R}^d), \quad j = 1, 2.$$

The following theorem is an extension of the Young inequality to the case of weighted Lebesgue spaces and modulation spaces when $0 \le \mathsf{R}(\mathrm{p}) \le 1/2$.

Theorem 5 *Let $s_j, t_j \in \mathbb{R}$, $p_j, q_j \in [1,\infty]$, $j = 0, 1, 2$. Assume that $0 \le \mathsf{R}(\mathrm{p}) \le 1/2$, $\mathsf{R}(\mathrm{q}) \le 1$,*

$$0 \le t_j + t_k, \quad j, k = 0, 1, 2, \quad j \ne k, \tag{27}$$
$$0 \le t_0 + t_1 + t_2 - d \cdot \mathsf{R}(\mathrm{p}), \ \textit{and} \tag{28}$$
$$0 \le s_0 + s_1 + s_2, \tag{29}$$

with strict inequality in (28) when $\mathsf{R}(\mathrm{p}) > 0$ and $t_j = d \cdot \mathsf{R}(\mathrm{p})$ for some $j = 0, 1, 2$.
Then $(f_1, f_2) \mapsto f_1 * f_2$ on $C_0^\infty(\mathbb{R}^d)$ extends uniquely to a continuous map from

1. *$L^{p_1}_{t_1}(\mathbb{R}^d) \times L^{p_2}_{t_2}(\mathbb{R}^d)$ to $L^{p_0'}_{-t_0}(\mathbb{R}^d)$;*
2. *$M^{p_1,q_1}_{s_1,t_1}(\mathbb{R}^d) \times M^{p_2,q_2}_{s_2,t_2}(\mathbb{R}^d)$ to $M^{p_0',q_0'}_{-s_0,-t_0}(\mathbb{R}^d)$.*

For the proof we refer to [84]. It is based on the detailed study of an auxiliary three-linear map over carefully chosen regions in $\mathbb{R}^d$ (see Sects. 3.1 and 3.2 in [84]). This result extends multiplication and convolution properties obtained in [57]. Moreover, the sufficient conditions from Theorem 5 are also necessary in the following sense.

Theorem 6 *Let $p_j, q_j \in [1,\infty]$ and $s_j, t_j \in \mathbb{R}$, $j = 0, 1, 2$. Assume that at least one of the following statements hold true:*

1. *the map $(f_1, f_2) \mapsto f_1 * f_2$ on $C_0^\infty(\mathbb{R}^d)$ is continuously extendable to a map from $L^{p_1}_{t_1}(\mathbb{R}^d) \times L^{p_2}_{t_2}(\mathbb{R}^d)$ to $L^{p_0'}_{-t_0}(\mathbb{R}^d)$;*
2. *the map $(f_1, f_2) \mapsto f_1 * f_2$ on $C_0^\infty(\mathbb{R}^d)$ is continuously extendable to a map from $M^{p_1,q_1}_{s_1,t_1}(\mathbb{R}^d) \times M^{p_2,q_2}_{s_2,t_2}(\mathbb{R}^d)$ to $M^{p_0',q_0'}_{-s_0,-t_0}(\mathbb{R}^d)$;*

Then (27) *and* (28) *hold true.*

5 Localization Operators

We refer to [13, 18, 19, 71] for the continuity properties of localization operators on modulation spaces, and here we give a reformulation of such results by using the Grossmann–Royer transform instead of the cross-Wigner distribution. Furthermore, we use the Grossmann–Royer transform to define localization operators and to show that such operators are Weyl pseudodifferential operators. For our purposes the duality between $\mathscr{S}^{(1)}(\mathbb{R}^d)$ and $(\mathscr{S}^{(1)})'(\mathbb{R}^d)$ will suffice, and we use it here for the simplicity and for the clarity of exposition.

Definition 6 Let $f \in \mathscr{S}^{(1)}(\mathbb{R}^d)$. The *localization operator* $A_a^{\varphi_1,\varphi_2}$ with *symbol* $a \in \mathscr{S}^{(1)'}(\mathbb{R}^{2d})$ and *windows* $\varphi_1, \varphi_2 \in \mathscr{S}^{(1)}(\mathbb{R}^d)$ is given by

$$A_a^{\varphi_1,\varphi_2} f(t) = \int_{\mathbb{R}^{2d}} a(x,\omega) R_{\check{\varphi}_1} f(\frac{x}{2}, \frac{\omega}{2}) R(\check{\varphi}_2(t))(\frac{x}{2}, \frac{\omega}{2})\, dx d\omega. \tag{30}$$

In the weak sense,

$$\langle A_a^{\varphi_1,\varphi_2} f, g\rangle = \langle a(x,\omega) R_{\check{\varphi}_1} f(\frac{x}{2}, \frac{\omega}{2}), R_{\check{\varphi}_2} g(\frac{x}{2}, \frac{\omega}{2})\rangle \tag{31}$$

$$= \langle a(x,\omega), R_{\check{\varphi}_1} f(\frac{x}{2}, \frac{\omega}{2}), R_{\check{\varphi}_2} g(\frac{x}{2}, \frac{\omega}{2})\rangle, \quad f, g \in \mathscr{S}^{(1)}(\mathbb{R}^d), \tag{32}$$

so that $A_a^{\varphi_1,\varphi_2}$ is well-defined continuous operator from $\mathscr{S}^{(1)}(\mathbb{R}^d)$ to $(\mathscr{S}^{(1)})'(\mathbb{R}^d)$, cf. Proposition 1.

Lemma 4 *Let there be given $f \varphi_1, \varphi_2 \in \mathscr{S}^{(1)}(\mathbb{R}^d)$ and $a \in \mathscr{S}^{(1)'}(\mathbb{R}^{2d})$. Then $A_a^{\varphi_1,\varphi_2}$ given by* (30) *coincides with the usual localization operator $\tilde{A}_a^{\varphi_1,\varphi_2}$ defined by the short-time Fourier transform:*

$$\tilde{A}_a^{\varphi_1,\varphi_2} f(t) = \int_{\mathbb{R}^{2d}} a(x,\omega) V_{\varphi_1} f(x,\omega) M_\omega T_x \varphi_2(t)\, dx d\omega.$$

Proof From Lemma 1 it follows that

$$\begin{aligned}\tilde{A}_a^{\varphi_1,\varphi_2} f(t) &= \int_{\mathbb{R}^{2d}} a(x,\omega) e^{-\pi i x\omega} R_{\check{\varphi}_1} f(\frac{x}{2},\frac{\omega}{2}) M_\omega T_x \varphi_2(t)\, dx d\omega \\ &= \int_{\mathbb{R}^{2d}} a(x,\omega) e^{-\pi i x\omega} R_{\check{\varphi}_1} f(\frac{x}{2},\frac{\omega}{2}) e^{\pi i x\omega} R(\check{\varphi}_2(t))(\frac{x}{2},\frac{\omega}{2})\, dx d\omega \\ &= A_a^{\varphi_1,\varphi_2} f(t),\end{aligned}$$

since

$$\begin{aligned}R(\check{\varphi}_2(t))(\frac{x}{2},\frac{\omega}{2}) &= e^{4\pi i \frac{\omega}{2}(t-\frac{x}{2})} \check{\varphi}_2(2\frac{x}{2}-t) \\ &= e^{-\pi i \omega x} e^{2\pi i \omega t} \varphi_2(t-x) \\ &= e^{-\pi i \omega x} M_\omega T_x \varphi_2(t),\end{aligned}$$

and the lemma is proved. □

Next we show that localization operators can be represented as Weyl pseudodifferential operators.

Recall, if $\sigma \in \mathscr{S}^{(1)}(\mathbb{R}^{2d})$ then the Weyl pseudodifferential operator L_σ is defined as the oscillatory integral:

$$L_\sigma f(x) = \iint \sigma(\frac{x+y}{2},\omega) f(y) e^{2\pi(x-y)\cdot\omega} dy d\omega, \quad f \in \mathscr{S}^{(1)}(\mathbb{R}^{2d}).$$

It extends to each $\sigma \in \mathscr{S}^{(1)'}(\mathbb{R}^{2d})$, and then L_σ is continuous from $\mathscr{S}^{(1)}(\mathbb{R}^{2d})$ to $\mathscr{S}^{(1)'}(\mathbb{R}^{2d})$, and σ is called *the Weyl symbol* of the pseudodifferential operator L_σ.

Lemma 5 *Let L_σ be the Weyl pseudodifferential operator with the Weyl symbol $\sigma \in \mathscr{S}^{(1)'}(\mathbb{R}^{2d})$. Then we have*

$$L_\sigma f(t) = 2^d \int_{\mathbb{R}^{2d}} \sigma(x,\omega)(Rf(t))(x,\omega) dx d\omega, \quad t \in \mathbb{R}^d,$$

or, in the weak sense

$$\langle L_\sigma f, g\rangle = 2^d \langle \sigma, R_f g\rangle, \qquad f, g \in \mathscr{S}^{(1)}(\mathbb{R}^d). \tag{33}$$

Proof The lemma is the same as [25, Proposition 40]. We give here a different proof. In fact we only use Fubini's theorem and the change of variables $2x - y \mapsto t$:

$$\begin{aligned}2^d \langle \sigma, R_f g\rangle &= 2^d \int\int\int \sigma(x,\omega) e^{-4\pi i \omega(y-x)} \overline{g}(2x-y) f(y) dy dx d\omega \\ &= 2^d \int\int \left(\int \sigma(x,\omega) e^{4\pi i \omega x} \overline{g}(2x-y) dx\right) e^{-4\pi i \omega y} f(y) dy d\omega \\ &= \int\int\int \sigma(\frac{t+y}{2},\omega) e^{2\pi i \omega(t+y)} \overline{g}(t) dt e^{-4\pi i \omega y} f(y) dy d\omega\end{aligned}$$

$$= \int\int\int \sigma(\frac{t+y}{2}, \omega) e^{2\pi i \omega (t-y)} f(y)\overline{g}(t) dt dy d\omega,$$

therefore $2^d \langle \sigma, R_f g\rangle = \langle L_\sigma f, g\rangle$. □

Lemmas 1 and 5 imply the well-known formula: $\langle L_\sigma f, g\rangle = \langle \sigma, W(g, f)\rangle$, cf. [34, 65, 88].

Next we establish the so called Weyl connection, which shows that the set of localization operators is a subclass of the set of Weyl operators. Although the same result can be found elsewhere ([6, 34, 71]), it is given here in order to be self-contained. The proof is based on kernel theorem for Gelfand–Shilov spaces, and direct calculation.

Lemma 6 *If $a \in \mathscr{S}^{(1)'}(\mathbb{R}^{2d})$ and $\varphi_1, \varphi_2 \in \mathscr{S}^{(1)}(\mathbb{R}^d)$, then the localization operator $A_a^{\varphi_1,\varphi_2}$ is Weyl pseudodifferential operator with the Weyl symbol $\sigma = 2^{-d} a * R_{\varphi_1}\varphi_2$, in other words,*

$$A_a^{\varphi_1,\varphi_2} = 2^{-d} L_{a*R_{\varphi_1}\varphi_2}. \tag{34}$$

Proof By the kernel Theorem 2 it follows that for any linear and continuous operator T from $\mathscr{S}^{(1)}(\mathbb{R}^{2d})$ to $\mathscr{S}^{(1)'}(\mathbb{R}^{2d})$, there exists a uniquely determined $k \in \mathscr{S}^{(1)'}(\mathbb{R}^{2d})$ such that

$$\langle Tf, g\rangle = \langle k, g \otimes \overline{f}\rangle, \quad f, g \in \mathscr{S}^{(1)}(\mathbb{R}^{2d}),$$

see also [50, 71, 85].

We first calculate the kernel of $A_a^{\varphi_1,\varphi_2}$, and than show that it coincides to the kernel of L_σ when $\sigma = 2^{-d} a * R_{\varphi_1}\varphi_2$.

From (31) and Proposition 1 2. it follows:

$$\begin{aligned}
&\langle A_a^{\varphi,\phi} f, g\rangle = \langle a(x,\omega) R_{\check{\varphi}_1} f(\tfrac{x}{2}, \tfrac{\omega}{2}), R_{\check{\varphi}_2} g(\tfrac{x}{2}, \tfrac{\omega}{2})\rangle \\
&= \iint_{\mathbb{R}^{2d}} a(x,\omega) \left(\int_{\mathbb{R}^d} f(y)\overline{R(\check{\varphi}_1(y))}(\tfrac{x}{2}, \tfrac{\omega}{2}) dy\right) \left(\int_{\mathbb{R}^d} \overline{g}(t) R(\check{\varphi}_2(t))(\tfrac{x}{2}, \tfrac{\omega}{2}) dt\right) dx d\omega \\
&= \int_{\mathbb{R}^d}\int_{\mathbb{R}^d} f(y)\overline{g}(t) \left(\iint_{\mathbb{R}^{2d}} a(x,\omega)\overline{R(\check{\varphi}_1(y))}(\tfrac{x}{2}, \tfrac{\omega}{2}) R(\check{\varphi}_2(t))(\tfrac{x}{2}, \tfrac{\omega}{2}) dx d\omega\right) dt dy \\
&= \langle k, g \otimes \overline{f}\rangle,
\end{aligned}$$

where

$$k(t,y) = \int_{\mathbb{R}^{2d}} a(x,\omega)\overline{R(\check{\varphi}_1(y))}(\frac{x}{2}, \frac{\omega}{2}) R(\check{\varphi}_2(t))(\frac{x}{2}, \frac{\omega}{2}) dx d\omega. \tag{35}$$

Next, we calculate the kernel of $L_{a*R_{\varphi_1}\varphi_2}$. We use the covariance property of the Grossmann–Royer transform, Proposition 1 *3.*, to obtain

$$
\begin{aligned}
a * R_{\varphi_1}\varphi_2(p,q) \\
&= \iint_{\mathbb{R}^{2d}} a(x,\omega) R_{\varphi_1}\varphi_2(p-x, q-\omega) dx d\omega \\
&= \iint_{\mathbb{R}^{2d}} a(x,\omega) R_{T_x M_\omega \varphi_1} T_x M_\omega \varphi_2(p,q) dx d\omega \\
&= \iint_{\mathbb{R}^{2d}} a(x,\omega) \left(\int_{\mathbb{R}^d} e^{4\pi i q(t-p)} T_x M_\omega \varphi_2(2p-t) \overline{T_x M_\omega \varphi_1}(t) dt \right) dx d\omega.
\end{aligned}
$$

Now,

$$
\begin{aligned}
\langle a * R_{\varphi_1}\varphi_2, R_f g\rangle \\
&= \iint_{\mathbb{R}^{4d}} a(x,\omega) \left(\int_{\mathbb{R}^d} e^{4\pi i q(t-p)} T_x M_\omega \varphi_2(2p-t) \overline{T_x M_\omega \varphi_1}(t) dt \right) dx d\omega \\
&\quad \times \left(\int_{\mathbb{R}^d} e^{-4\pi i q(s-p)} \overline{g}(2p-s) f(s) ds \right) dp dq \\
&= \iint_{\mathbb{R}^{3d}} a(x,\omega) \left(\int_{\mathbb{R}^d} e^{4\pi i q(t-s)} dq \int_{\mathbb{R}^d} T_x M_\omega \varphi_2(2p-t) \overline{T_x M_\omega \varphi_1}(t) dt \right) \\
&\quad \times \left(\int_{\mathbb{R}^d} \overline{g}(2p-s) f(s) ds \right) dp dx d\omega
\end{aligned}
$$

$$
\begin{aligned}
&= \iint_{\mathbb{R}^{3d}} a(x,\omega) \delta(t-s) \int_{\mathbb{R}^d} T_x M_\omega \varphi_2(2p-t) \overline{T_x M_\omega \varphi_1}(t) dt \\
&\quad \times \left(\int_{\mathbb{R}^d} \overline{g}(2p-s) f(s) ds \right) dp dx d\omega \\
&= \iint_{\mathbb{R}^{4d}} a(x,\omega) T_x M_\omega \varphi_2(2p-t) \overline{T_x M_\omega \varphi_1}(t) \overline{g}(2p-t) f(t) dt dp dx d\omega
\end{aligned}
$$

$$
\begin{aligned}
&= \iint_{\mathbb{R}^{4d}} a(x,\omega) T_x M_\omega \varphi_2(2p-t) \overline{T_x M_\omega \varphi_1}(t) \overline{g}(2p-t) f(t) dt dp dx d\omega \\
&= 2^{-d} \iint_{\mathbb{R}^{4d}} a(x,\omega) T_x M_\omega \varphi_2(s) \overline{T_x M_\omega \varphi_1}(t) \overline{g}(s) f(t) dt ds dx d\omega \\
&= 2^{-d} \iint_{\mathbb{R}^{4d}} a(x,\omega) R(\check{\varphi}_2(s))(\frac{x}{2}, \frac{\omega}{2}) \overline{R((\check{\varphi}_1(s))}(\frac{x}{2}, \frac{\omega}{2}) \overline{g}(s) f(t) dt ds dx d\omega,
\end{aligned}
$$

where we deal with the oscillatory integral as we did before, and use the change of variables $2p - t \mapsto t$.

Therefore

$$
\langle L_{a*R_{\varphi_1}\varphi_2} f, g\rangle = \langle k, g \otimes \overline{f}\rangle,
$$

where the k is given by (35). By the uniqueness of the kernel we conclude that $A_a^{\varphi,\phi} = L_{a*R_{\varphi_1}\varphi_2}$, and the proof is finished. □

Note that Lemma 6 can be proved in quasianalytic case by the same arguments. However, in this paper we do not need such an extension. Notice also that in the literature the symbol a in Lemma 6 is called the *anti-Wick symbol* of the Weyl pseudodifferential operator L_σ.

Lemma 6 describes localization operators in terms of the convolution which is a smoothing operator. This implies different boundedness results of localization operators even if a is an ultradistribution. In what follows we review some of these results.

5.1 Continuity Properties

In this subsection we recall the continuity properties from [71] obtained by using the relation between the Weyl pseudodifferential operators and localization operators, Lemma 6, and convolution results for modulation spaces from Theorem 5.

We also use sharp continuity results from [15]. There it is shown that the sufficient conditions for the continuity of the cross-Wigner distribution on modulation spaces are also necessary (in the un-weighted case). Related results can be found elsewhere, e.g. in [71, 74, 75]. In many situations such results overlap. For example, Proposition 10 in [72] coincides with certain sufficient conditions from [15, Theorem 1.1] when restricted to $\mathsf{R}(\mathsf{p}) = 0$, $t_0 = -t_1$, and $t_2 = |t_0|$. For our purposes it is convenient to rewrite [15, Theorem 1.1] in terms of the Grossmann–Royer transform.

Theorem 7 *Let there be given $s \in \mathbb{R}$ and $p_i, q_i, p, q \in [1, \infty]$, such that*

$$p \leq p_i, q_i \leq q, \quad i = 1, 2 \tag{36}$$

and

$$\min\left\{\frac{1}{p_1} + \frac{1}{p_2}, \frac{1}{q_1} + \frac{1}{q_2}\right\} \geq \frac{1}{p} + \frac{1}{q}. \tag{37}$$

If $f, g \in \mathscr{S}(\mathbb{R}^d)$, then the map $(f, g) \mapsto R_g f$ extends to sesquilinear continuous map from $M^{p_1,q_1}_{|s|}(\mathbb{R}^d) \times M^{p_2,q_2}_{s}(\mathbb{R}^d)$ to $M^{p,q}_{s,0}(\mathbb{R}^{2d})$ and

$$\|R_g f\|_{M^{p,q}_{s,0}} \lesssim \|f\|_{M^{p_1,q_1}_{|s|}} \|g\|_{M^{p_2,q_2}_{s}}. \tag{38}$$

Viceversa, if there exists a constant $C > 0$ such that

$$\|R_g f\|_{M^{p,q}} \lesssim \|f\|_{M^{p_1,q_1}} \|g\|_{M^{p_2,q_2}}.$$

then (36) *and* (37) *must hold.*

Proof We refer to [15, Section 3] for the proof. It is given there in terms of the cross-Wigner distribution, which is the same as the Grossmann–Royer transform, up to the constant factor 2^d. □

Let σ be the Weyl symbol of L_σ. By [38, Theorem 14.5.2] if $\sigma \in M^{\infty,1}(\mathbb{R}^{2d})$ then L_σ is bounded on $M^{p,q}(\mathbb{R}^d)$, $1 \le p, q \le \infty$. This result has a long history starting with the Calderon–Vaillancourt theorem on boundedness of pseudodifferential operators with smooth and bounded symbols on $L^2(\mathbb{R}^d)$, [8]. It is extended by Sjöstrand in [66] where $M^{\infty,1}$ is used as appropriate symbol class. Sjöstrand's results were thereafter extended in [38, 40, 41, 74–76].

Theorem 8 *Let the assumptions of Theorem 5 hold. If* $\varphi_j \in M^{p_j}_{t_j}(\mathbb{R}^d)$, $j = 1, 2$, *and* $a \in M^{\infty,r}_{u,v}(\mathbb{R}^{2d})$ *where* $1 \le r \le p_0$, $u \ge t_0$ *and* $v \ge d\mathsf{R}(\mathrm{p})$ *with* $v > d\mathsf{R}(\mathrm{p})$ *when* $\mathsf{R}(\mathrm{p}) > 0$, *then* $A_a^{\varphi_1,\varphi_2}$ *is bounded on* $M^{p,q}(\mathbb{R}^d)$, *for all* $1 \le p, q \le \infty$ *and the operator norm satisfies the uniform estimate*

$$\|A_a^{\varphi_1,\varphi_2}\|_{op} \lesssim \|a\|_{M^{\infty,r}_{u,v}} \|\varphi_1\|_{M^{p_1}_{t_1}} \|\varphi_2\|_{M^{p_2}_{t_2}}.$$

Proof Let $\varphi_j \in M^{p_j}_{t_j}(\mathbb{R}^d)$, $j = 1, 2$. Then, by Theorem 7 it follows that $R_{\varphi_1}\varphi_2 \in M^{1,p_0}_{-t_0,0}(\mathbb{R}^{2d})$. This fact, together with Theorem 5 *(2)* implies that

$$a * R_{\varphi_1}\varphi_2 \in M^{\tilde{p},1}(\mathbb{R}^{2d}), \quad \tilde{p} \ge 2,$$

if the involved parameters fulfill the conditions of the theorem. Concerning the Lebesgue parameters it is easy to see that $\tilde{p} \ge 2$ is equivalent to $\mathsf{R}(\mathrm{p}) = \mathsf{R}(p, \infty, 1) \in [0, 1/2]$, and that $1 \le r \le p_0$ is equivalent to $\mathsf{R}(\mathrm{q}) = \mathsf{R}(\infty, r, p_0') \le 1$. It is also straightforward to check that the choice of the weight parameters u and v implies that $a * R_{\varphi_1}\varphi_2 \in M^{\tilde{p},1}(\mathbb{R}^{2d})$, $\tilde{p} \ge 2$.

In particular, if $\tilde{p} = \infty$ then $a * R_{\varphi_1}\varphi_2 \in M^{\infty,1}(\mathbb{R}^{2d})$. From [38, Theorem 14.5.2] (and Lemma 6) it follows that $A_a^{\varphi_1,\varphi_2}$ is bounded on $M^{p,q}(\mathbb{R}^d)$, $1 \le p, q \le \infty$.

The operator norm estimate also follows from [38, Theorem 14.5.2]. □

Remark 7 When $p_1 = p_2 = 1$, $r = p_0 = \infty$ and $t_1 = t_2 = -t_0 = s \ge 0$, $u = -s$, $v = 0$ we recover the celebrated Cordero–Gröchenig Theorem, [13, Theorem 3.2], in the case of polynomial weights, with the uniform estimate

$$\|A_a^{\varphi_1,\varphi_2}\|_{op} \lesssim \|a\|_{M^{\infty}_{-s,0}} \|\varphi_1\|_{M^1_s} \|\varphi_2\|_{M^1_s}$$

in our notation.

Another version of Theorem 8 with symbols from weighted modulation spaces can be obtained by using [38, Theorem 14.5.6] instead. We leave this to the reader as an exercise.

5.2 Schatten–von Neumann Properties

In this subsection we recall the compactness properties from [71].

References to the proof of the following well known theorem can be found in [13].

Theorem 9 *Let σ be the Weyl symbol of L_σ.*

1. *If $\sigma \in M^1(\mathbb{R}^{2d})$ then $\|L_\sigma\|_{S_1} \lesssim \|\sigma\|_{M^1}$.*
2. *If $\sigma \in M^p(\mathbb{R}^{2d})$, $1 \leq p \leq 2$, then $\|L_\sigma\|_{S_p} \lesssim \|\sigma\|_{M^p}$.*
3. *If $\sigma \in M^{p,p'}(\mathbb{R}^{2d})$, $2 \leq p \leq \infty$, then $\|L_\sigma\|_{S_p} \lesssim \|\sigma\|_{M^{p,p'}}$.*

The Schatten–von Neumann properties in the following Theorem are formulated in the spirit of [13], see also [74, 75]. Note that more general weights are considered in [76, 77], leading to different type of results. We also refer to [33] for compactness properties obtained by a different approach.

Theorem 10 *Let $\mathsf{R}(\mathrm{p})$ be the Young functional given by* (26), *$s \geq 0$ and $t \geq d\mathsf{R}(\mathrm{p})$ with $t > d\mathsf{R}(\mathrm{p})$ when $\mathsf{R}(\mathrm{p}) > 0$.*

1. *If $1 \leq p \leq 2$ and $p \leq r \leq 2p/(2-p)$ then the mapping $(a, \varphi_1, \varphi_2) \mapsto A_a^{\varphi_1,\varphi_2}$ is bounded from $M^{r,q}_{-s,t} \times M^1_s \times M^p_s$, into S_p, that is*

$$\|A_a^{\varphi_1,\varphi_2}\|_{S_p} \lesssim \|a\|_{M^{r,q}_{-s,t}} \|\varphi_1\|_{M^1_s} \|\varphi_2\|_{M^p_s}.$$

2. *If $2 \leq p \leq \infty$ and $p \leq r$ then the mapping $(a, \varphi_1, \varphi_2) \mapsto A_a^{\varphi_1,\varphi_2}$ is bounded from $M^{r,q}_{-s,t} \times M^1_s \times M^{p'}_s$, into S_p, that is*

$$\|A_a^{\varphi_1,\varphi_2}\|_{S_p} \lesssim \|a\|_{M^{r,q}_{-s,t}} \|\varphi_1\|_{M^1_s} \|\varphi_2\|_{M^{p'}_s}.$$

Proof 1. By Theorem 7 it follows that $R_{\varphi_1}\varphi_2 \in M^{1,p_w}_{-t_0,0}(\mathbb{R}^{2d})$, with $t_0 \geq -s$ and $p_w \in [2p/(p+2), p]$. Therefore $R_{\varphi_1}\varphi_2 \in M^{1,p}_{s,0}(\mathbb{R}^{2d})$.

To apply Theorem 5 we notice that for $\mathsf{R}(\mathrm{p}) = (p', r, 1)$ the condition $\frac{1}{2} \geq \mathsf{R}(\mathrm{p}) \geq 0$ is equivalent to $p \leq r \leq 2p/(2-p)$, and that for $\mathsf{R}(\mathrm{p}) = (p', q, p)$ the condition $\mathsf{R}(\mathrm{q}) \leq 1$ is equivalent to $1 \leq q$. Now, Theorem 5 implies that $a * R_{\varphi_1}\varphi_2 \in M^p(\mathbb{R}^{2d})$, and the result follows from Theorem 9 2.

2. By Theorem 7 it follows that $R_{\varphi_1}\varphi_2 \in M^{1,p_w}_{-t_0,0}(\mathbb{R}^{2d})$, with $t_0 \geq -s$ and $p_w \in [p', 2p'/(p'+2), p]$. Therefore $R_{\varphi_1}\varphi_2 \in M^{1,p'}_{s,0}(\mathbb{R}^{2d})$.

Now, for $2 \leq p \leq \infty$, $\frac{1}{2} \geq \mathsf{R}(\mathrm{p}) \geq 0$ is equivalent to $p \leq r$, and for $\mathsf{R}(\mathrm{q}) = (p, q, p')$ the condition $\mathsf{R}(\mathrm{q}) \leq 1$ is equivalent to $1 \leq q$. The statement follows from Theorems 5 *2.* and 9 *3.* □

We remark that we corrected a typo from the formulation of [71, Theorem 3.9] also note that a particular choice: $r = p$, $q = \infty$ and $t = 0$ recovers [13, Theorem 3.4].

In the next theorem we give some necessary conditions. The proof follows from the proofs of Theorems 4.3 and 4.4 in [13] and is therefore omitted.

Theorem 11 *Let the assumptions of Theorem 5 hold and let $a \in \mathscr{S}'(\mathbb{R}^{2d})$.*

1. *If there exists a constant $C = C(a) > 0$ depending only on the symbol a such that*
$$\|A_a^{\varphi_1,\varphi_2}\|_{S_\infty} \leq C\|\varphi_1\|_{M_{t_1}^{p_1}}\|\varphi_2\|_{M_{t_2}^{p_2}},$$
for all $\varphi_1, \varphi_2 \in \mathscr{S}(\mathbb{R}^d)$, then $a \in M_{u,v}^{\infty,r}(\mathbb{R}^{2d})$ where $1 \leq r \leq p_0$, $u \geq t_0$ and $v \geq d\mathsf{R}(\mathrm{p})$ with $v > d\mathsf{R}(\mathrm{p})$ when $\mathsf{R}(\mathrm{p}) > 0$.
2. *If there exists a constant $C = C(a) > 0$ depending only on the symbol a such that*
$$\|A_a^{\varphi_1,\varphi_2}\|_{S_2} \leq C\|\varphi_1\|_{M^1}\|\varphi_2\|_{M^1}$$
for all $\varphi_1, \varphi_2 \in \mathscr{S}(\mathbb{R}^d)$, then $a \in M^{2,\infty}(\mathbb{R}^{2d})$.

We finish the section with results from [19] related to the weights which may have exponential growth. To that end we use the boundedness result Theorem 7 formulated in terms of such weights. The proof of the next lemma is a modified version of the proof of [13, Proposition 2.5] and therefore omitted.

Lemma 7 *Let w_s, τ_s be the weights defined in (20) and (21). If $1 \leq p \leq \infty$, $s \geq 0$, $\varphi_1 \in M_{w_s}^1(\mathbb{R}^d)$ and $\varphi_2 \in M_{w_s}^p(\mathbb{R}^d)$, then $R_{\varphi_1}\varphi_2 \in M_{\tau_s}^{1,p}(\mathbb{R}^{2d})$, and*

$$\|R_{\varphi_1}\varphi_2\|_{M_{\tau_s}^{1,p}} \lesssim \|\varphi_1\|_{M_{w_s}^1}\|\varphi_2\|_{M_{w_s}^p}. \tag{39}$$

Theorem 12 *Let $A_a^{\varphi_1,\varphi_2}$ be the localization operator with the symbol a and windows φ_1 and φ_2.*

1. *If $s \geq 0$, $a \in M_{1/\tau_s}^{\infty}(\mathbb{R}^{2d})$, and $\varphi_1, \varphi_2 \in M_{w_s}^1(\mathbb{R}^d)$, then $A_a^{\varphi_1,\varphi_2}$ is bounded on $M^{p,q}(\mathbb{R}^d)$ for all $1 \leq p, q \leq \infty$, and the operator norm satisfies the uniform estimate*
$$\|A_a^{\varphi_1,\varphi_2}\|_{op} \lesssim \|a\|_{M_{1/\tau_s}^{\infty}}\|\varphi_1\|_{M_{w_s}^1}\|\varphi_2\|_{M_{w_s}^1}.$$
2. *If $1 \leq p \leq 2$, then the mapping $(a, \varphi_1, \varphi_2) \mapsto A_a^{\varphi_1,\varphi_2}$ is bounded from $M_{1/\tau_s}^{p,\infty}(\mathbb{R}^{2d}) \times M_{w_s}^1(\mathbb{R}^d) \times M_{w_s}^p(\mathbb{R}^d)$ into S_p, in other words,*
$$\|A_a^{\varphi_1,\varphi_2}\|_{S_p} \lesssim \|a\|_{M_{1/\tau_s}^{p,\infty}}\|\varphi_1\|_{M_{w_s}^1}\|\varphi_2\|_{M_{w_s}^p}.$$
3. *If $2 \leq p \leq \infty$, then the mapping $(a, \varphi_1, \varphi_2) \mapsto A_a^{\varphi_1,\varphi_2}$ is bounded from $M_{1/\tau_s}^{p,\infty} \times M_{w_s}^1 \times M_{w_s}^{p'}$ into S_p, and*
$$\|A_a^{\varphi_1,\varphi_2}\|_{S_p} \lesssim \|a\|_{M_{1/\tau_s}^{p,\infty}}\|\varphi_1\|_{M_{w_s}^1}\|\varphi_2\|_{M_{w_s}^{p'}}.$$

Proof 1. We use the convolution relation (25) to show that the Weyl symbol $a * R_{\varphi_1}\varphi_2$ of $A_a^{\varphi_1,\varphi_2}$ is in $M^{\infty,1}$. If $\varphi_1, \varphi_2 \in M^1_{w_s}(\mathbb{R}^d)$, then by (39), we have $R_{\varphi_1}\varphi_2 \in M^1_{\tau_s}(\mathbb{R}^{2d})$. Applying Proposition 7 in the form $M^\infty_{1/\tau_s} * M^1_{\tau_s} \subseteq M^{\infty,1}$, we obtain $\sigma = a * R_{\varphi_1}\varphi_2 \in M^{\infty,1}$. The result now follows from Theorem 9 *1.*

Similarly, the proof of *2.* and *3.* is based on results of Proposition 7 and Theorem 9 *2.* and *3.*

For the sake of completeness, we state the necessary boundedness result, which follows by straightforward modifications of [13, Theorem 4.3].

Theorem 13 *Let $a \in \mathcal{S}^{(1)'}(\mathbb{R}^{2d})$ and $s \geq 0$. If there exists a constant $C = C(a) > 0$ depending only on a such that*

$$\|A_a^{\varphi_1,\varphi_2}\|_{S_\infty} \leq C\, \|\varphi_1\|_{M^1_{w_s}} \|\varphi_2\|_{M^1_{w_s}} \tag{40}$$

for all $\varphi_1, \varphi_2 \in \mathcal{S}^{(1)}(\mathbb{R}^d)$, then $a \in M^\infty_{1/\tau_s}$.

Note that a trace-class result for certain quasianalytic distributions (based on the heat kernel and parametrix techniques) is given in [19].

6 Further Extensions

The results of previous sections can be extended in several directions. We mention here some references to product formulas, Shubin type pseudodifferential operators, multilinear localization operators, and generalizations to quasi-Banach spaces and quasianalytic spaces of test functions and their dual spaces (of Fourier ultra-hyperfunctions). The list of topics and references is certainly incomplete and we leave it to the reader to accomplish it according to his/hers preferences.

The product of two localization operators can be written as a localization operator in very few cases. Exact formulas hold only in special cases, e.g. when the windows are Gaussians, [89]. Therefore, a symbolic calculus is developed in [14] where such product is written as a sum of localization operators plus a remainder term expressed in a Weyl operator form. This is done in linear case in the framework of the symbols which may have subexponential growth.

The multilinear localization operators were first introduced in [17] and their continuity properties are formulated in terms of modulation spaces. The key point is the interpretation of these operators as multilinear Kohn–Nirenberg pseudodifferential operators. The multilinear pseudodifferential operators were already studied in the context of modulation spaces in [2–4], see also a more recent contribution [51] where such approach is strengthened and applied to the bilinear and trilinear Hilbert transforms.

To deal with multilinear localization operators, instead of standard modulation spaces $M^{p,q}$ observed in [17], continuity properties in [72, 73] are formulated in terms of a modified version of modulation spaces denoted by $\mathscr{M}^{p,q}_{s,t}$, and given as follows.

By a slight abuse of the notation (as it is done in e.g. [51]), $\mathbf{f}$ denotes both the vector $\mathbf{f} = (f_1, f_2, \dots, f_n)$ and the tensor product $\mathbf{f} = f_1 \otimes f_2 \otimes \dots \otimes f_n$. This will not cause confusion, since the meaning of $\mathbf{f}$ will be clear from the context.

For example, if $t = (t_1, t_2, \dots, t_n)$, and $F_j = F_j(t_j)$, $t_j \in \mathbb{R}^d$, $j = 1, 2, \dots, n$, then

$$\prod_{j=1}^{n} F_j(t_j) = F_1(t_1) \cdot F_2(t_2) \cdot \dots \cdot F_n(t_n) = F_1(t_1) \otimes F_2(t_2) \otimes \dots \otimes F_n(t_n) = \mathbf{F}(t). \tag{41}$$

To give an interpretation of multilinear operators in the weak sense we note that, when $\mathbf{f} = (f_1, f_2, \dots, f_n)$ and $\boldsymbol{\varphi} = (\varphi_1, \varphi_2, \dots, \varphi_n)$, $f_j, \varphi_j \in \mathscr{S}^{(1)}(\mathbb{R}^d)$, $j = 1, 2, \dots, n$, we put

$$R_{\boldsymbol{\varphi}}\mathbf{f}(x, \omega) = \int_{\mathbb{R}^{nd}} \mathbf{f}(2x - t) \prod_{j=1}^{n} e^{4\pi i \omega_j (t_j - x_j)} \varphi_j(t_j) dt, \tag{42}$$

see also (41) for the notation.

According to (42), $\mathscr{M}^{p,q}_{s,t}(\mathbb{R}^{nd})$ denotes the the set of $\mathbf{f} = (f_1, f_2, \dots, f_n)$, $f_j \in \mathscr{S}'(\mathbb{R}^d)$, $j = 1, 2, \dots, n$, such that

$$\|\mathbf{f}\|_{\mathscr{M}^{p,q}_{s,t}} \equiv \left(\int_{\mathbb{R}^{nd}} \left(\int_{\mathbb{R}^{nd}} |V_{\boldsymbol{\varphi}}\mathbf{f}(x, \omega) \langle x \rangle^t \langle \omega \rangle^s|^p \, dx \right)^{q/p} d\omega \right)^{1/q} < \infty,$$

where $\boldsymbol{\varphi} = (\varphi_1, \varphi_2, \dots, \varphi_n)$, $\varphi_j \in \mathscr{S}(\mathbb{R}^d) \setminus 0$, $j = 1, 2, \dots, n$, is a given window function.

The kernel theorem for $\mathscr{S}(\mathbb{R}^d)$ and $\mathscr{S}'(\mathbb{R}^d)$ (see [86]) implies that there is an isomorphism between $\mathscr{M}^{p,q}_{s,t}(\mathbb{R}^{nd})$ and $M^{p,q}_{s,t}(\mathbb{R}^{nd})$ (which commutes with the operators from (2)). This allows us to identify $\mathbf{f} \in \mathscr{M}^{p,q}_{s,t}(\mathbb{R}^{nd})$ with (its isomorphic image) $F \in M^{p,q}_{s,t}(\mathbb{R}^{nd})$ (and vice versa).

Next we give multilinear version of Definition 6.

Definition 7 Let $f_j \in \mathscr{S}(\mathbb{R}^d)$, $j = 1, 2, \dots, n$, and $\mathbf{f} = (f_1, f_2, \dots, f_n)$. The *multilinear localization operator* $A_a^{\varphi,\phi}$ with *symbol* $a \in \mathscr{S}(\mathbb{R}^{2nd})$ and *windows*

$$\varphi = (\varphi_1, \varphi_2, \dots, \varphi_n) \text{ and } \phi = (\phi_1, \phi_2, \dots, \phi_n), \quad \varphi_j, \phi_j \in \mathscr{S}(\mathbb{R}^d), \quad j = 1, 2, \dots, n,$$

is given by

$$A_a^{\varphi,\phi}\mathbf{f}(t) = \int_{\mathbb{R}^{2nd}} a(x, \omega) \prod_{j=1}^{n} \left(R_{\check{\varphi}_j} f_j(\frac{x_j}{2}, \frac{\omega_j}{2}) R(\check{\phi}_j(t))(\frac{x_j}{2}, \frac{\omega_j}{2}) \right) dx d\omega,$$

where $x_j, \omega_j, t_j \in \mathbb{R}^d$, $j = 1, 2, \ldots, n$, and $x = (x_1, x_2, \ldots, x_n)$, $\omega = (\omega_1, \omega_2 \ldots, \omega_n)$, $t = (t_1, t_2 \ldots, t_n)$.

Let $\mathscr{R}$ denote the trace mapping that assigns to each function F defined on $\mathbb{R}^{nd}$ a function defined on $\mathbb{R}^d$ by the formula

$$\mathscr{R} : F \mapsto F\,\big|_{t_1=t_2=\cdots=t_n}, \quad t_j \in \mathbb{R}^d,\ j = 1, 2, \ldots, n.$$

Then $\mathscr{R} A_a^{\varphi,\phi}$ is the multilinear operator given in [17, Definition 2.2].

The approach to multilinear localization operators related to Weyl pseudodifferential operators is given in [72, 73]. Both Weyl and Kohn–Nirenberg correspondences are particular cases of the so-called $\tau-$pseudodifferential operators, $\tau \in [0, 1]$ ($\tau = 1/2$ gives Weyl operators, and $\tau = 0$ we reveals Kohn–Nirenberg operators). We refer to [65] for such operators, and e.g. [12, 16, 26, 83] for recent contributions in that context (see also the references given there).

The properties of such multilinear operators and their extension to Gelfand–Shilov spaces and their dual spaces will be the subject of a separate contribution.

Furthermore, to obtain continuity properties in the framework of quasianalytic Gelfand–Shilov spaces and corresponding dual spaces analogous to those given in Sect. 5, another techniques should be used, cf. [18].

Finally, we mention possible extension to quasi-Banach modulation spaces, when the Lebesgue parameters p and q are allowed to take the values in $(0, 1)$ as well. For such spaces and broad classes of pseudodifferential operators acting on them we refer to [82, 83].

Acknowledgements This research is supported by MPNTR of Serbia, projects no. 174024, and DS 028 (TIFMOFUS).

References

1. Abreu, L.D., Dörfler, M.: An inverse problem for localization operators. Inverse Probl. **28**(11), 115001, 16 (2012)
2. Bényi, Á., Gröchenig, K.H., Heil, C., Okoudjou, K.A.: Modulation spaces and a class of bounded multilinear pseudodifferential operators. J. Oper. Theory **54**(2), 387–399 (2005)
3. Bényi, Á., Okoudjou, K.A.: Bilinear pseudodifferential operators on modulation spaces. J. Fourier Anal. Appl. **10**(3), 301–313 (2004)
4. Bényi, Á., Okoudjou, K.A.: Modulation space estimates for multilinear pseudodifferential operators. Studia Math. **172**(2), 169–180 (2006)
5. Berezin, F.A.: Wick and anti-Wick symbols of operators. Mat. Sb. (N.S.) **86**(128), 578–610 (1971)
6. Boggiatto, P., Cordero, E., Gröchenig, K.: Generalized anti-Wick operators with symbols in distributional Sobolev spaces. Integral Equ. Oper. Theory **48**, 427–442 (2004)
7. Boggiatto, P., Oliaro, A., Wong, M.W.: L^p boundedness and compactness of localization operators. J. Math. Anal. Appl. **322**(1), 193–206 (2006)
8. Calderón, A.-P., Vaillancourt, R.: On the boundedness of pseudo-differential operators. J. Math. Soc. Japn. **23**, 374–378 (1971)

9. Cappiello, M., Gramchev, T., Rodino, L.: Sub-exponential decay and uniform holomorphic extensions for semilinear pseudodifferential equations. Commun. Partial Differ. Equ. **35**, 846–877 (2010)
10. Cappiello, M., Gramchev, T., Rodino, L.: Entire extensions and exponential decay for semilinear elliptic equations. Journal d'Analyse Mathématique **111**, 339–367 (2010)
11. Cordero, E.: Gelfand-Shilov window classes for weighted modulation spaces. Integral Transforms Spec. Funct. **18**(11–12), 829–837 (2007)
12. Cordero, E., D'Elia, L., Trapasso, S.I.: Norm estimates for τ-pseudodifferential operators in Wiener amalgam and modulation spaces. arXiv:1803.07865
13. Cordero, E., Gröchenig, K.: Time-frequency analysis of localization operators. J. Funct. Anal. **205**(1), 107–131 (2003)
14. Cordero, E., Gröchenig, K.: Symbolic calculus and Fredholm property for localization operators. J. Fourier Anal. Appl. **12**(4), 371–392 (2006)
15. Cordero, E., Nicola, F.: Sharp integral bounds for Wigner distributions. Int. Math. Res. Not. **2016**(00), 1–29 (2016)
16. Cordero, E., Nicola, F., Trapasso, S.I.: Almost diagonalization of τ-pseudodifferential operators with symbols in Wiener amalgam and modulation spaces. arXiv:1802.10314
17. Cordero, E., Okoudjou, K.A.: Multilinear localization operators. J. Math. Anal. Appl. **325**(2), 1103–1116 (2007)
18. Cordero, E., Pilipović, S., Rodino, L., Teofanov, N.: Localization operators and exponential weights for modulation spaces. Mediterr. J. Math. **2**(4), 381–394 (2005)
19. Cordero, E., Pilipović, S., Rodino, L., Teofanov, N.: Quasianalytic Gelfand-Shilov spaces with application to localization operators. Rocky Mt. J. Math. **40**(4), 1123–1147 (2010)
20. Chung, J., Chung, S.-Y., Kim, D.: A characterization for Fourier hyperfunctions. Publ. Res. Inst. Math. Sci. **30**(2), 203–208 (1994)
21. Chung, J., Chung, S.-Y., Kim, D.: Characterization of the Gelfand-Shilov spaces via Fourier transforms. Proc. Am. Math. Soc. **124**(7), 2101–2108 (1996)
22. Daubechies, I.: Time-frequency localization operators: a geometric phase space approach. IEEE Trans. Inf. Theory **34**(4), 605–612 (1988)
23. de Gosson, M.: Symplectic Methods in Harmonic Analysis and in Mathematical Physics. Birkhäuser, Basel (2011)
24. de Gosson, M.: Born-Jordan Quantization, Theory and Applications. Springer, Berlin (2016)
25. de Gosson, M.: The Wigner Transform. World Scientific, London (2017)
26. Delgado, J., Ruzhansky, M., Wang, B.: Approximation property and nuclearity on mixed-norm L^p, modulation and Wiener amalgam spaces. J. Lond. Math. Soc. **94**(2), 391–408 (2016)
27. Engliš, M.: Toeplitz operators and localization operators. Trans. Am. Math. Soc. **361**(2), 1039–1052 (2009)
28. Engliš, M.: An excursion into Berezin-Toeplitz quantization and related topics. In: Bahns, D., Bauer, W., Witt, I. (eds.) Operator Theory Advances and Applications, vol. 251, pp. 69–115. Birkhäuser, Basel (2016)
29. Feichtinger, H.G., Modulation spaces on locally compact Abelian groups, Technical report, University Vienna (1983), also in Krishna, M., Radha, R., Thangavelu S. (eds.): Wavelets and Their Applications, pp. 99–140. Allied Publishers, Chennai (2003)
30. Feichtinger, H.G., Gröchenig, K.: Banach spaces related to integrable group representations and their atomic decompositions I. J. Funct. Anal. **86**(2), 307–340 (1989)
31. Feichtinger, H.G., Gröchenig, K.: Banach spaces related to integrable group representations and their atomic decompositions II. Monatsh. f. Math. **108**, 129–148 (1989)
32. Feichtinger, H.G., Nowak, K.: A first survey of Gabor multipliers. In: Feichtinger, H., Strohmer, T. (eds.) Advances in Gabor Analysis, pp. 99–128. Birkhäuser, Basel (2003)
33. Fernández, C., Galbis, A.: Compactness of time-frequency localization operators on $L^2(\mathbb{R}^d)$. J. Funct. Anal. **233**(2), 335–350 (2006)
34. Folland, G.B.: Harmonic Analysis in Phase Space. Princeton University Press, Princeton (1989)
35. Gelfand, I.M., Shilov, G.E.: On a new method in uniqueness theorems for solution of Cauchy's problem for systems of linear partial differential equations. (Russian) Dokl. Akad. Nauk SSSR (N.S.) **102**, 1065–1068 (1955)

36. Gelfand, I.M., Shilov, G.E.: Generalized Functions, vol. II, III, Academic Press, New York (1968), reprinted by the AMS (2016)
37. Gramchev, T., Lecke, A., Pilipović, S., Rodino, L.: Gelfand-Shilov type spaces through Hermite expansions. In: Pilipović, S., Toft, J. (eds.) Pseudo-Differential Operators and Generalized Functions. Operator Theory: Advances and Applications, vol. 245. Birkhäuser, Basel (2015)
38. Gröchenig, K.: Foundations of Time-frequency analysis. Birkhäuser, Boston (2001)
39. Gröchenig, K.H.: Weight functions in time-frequency analysis. In: Rodino, L., Schulze, B.-W., Wong, M.W. (eds.) Pseudodifferential Operators: Partial Differential Equations and Time-Frequency Analysis. Fields Institute Communication, vol. 52, pp. 343–366 (2007)
40. Gröchenig, K.H., Heil, C.: Modulation spaces and pseudo-differential operators. Integral Equ. Oper. Theory **34**, 439–457 (1999)
41. Gröchenig, K.H., Heil, C.: Modulation spaces as symbol classes for pseudodifferential operators. In: Krishna, M., Radha, R., Thangavelu, S. (eds.) Wavelets and Their Applications, pp. 151–170. Allied Publishers, Chennai (2003)
42. Gröchenig, K., Zimmermann, G.: Hardy's theorem and the short-time Fourier transform of Schwartz functions. J. Lond. Math. Soc. **63**, 205–214 (2001)
43. Gröchenig, K., Zimmermann, G.: Spaces of test functions via the STFT. J. Funct. Spaces Appl. **2**, 25–53 (2004)
44. Grossmann, A.: Parity operator and quantization of d-functions. Commun. Math. Phys. **48**(3), 191–194 (1976)
45. Kamiński, A., Perišić, D., Pilipović, S.: On various integral transformations of tempered ultradistributions. Demonstr. Math. **33**(3), 641–655 (2000)
46. Komatsu, H.: Ultradistributions I, structure theorems and a characterization. J. Fac. Sci. Sect. I A Math. **20**, 25–105 (1973)
47. Langenbruch, M.: Hermite functions and weighted spaces of generalized functions. Manuscripta Math. **119**, 269–285 (2006)
48. Lerner, N., Morimoto, Y., Pravda-Starov, K., Xu, C.-J.: Gelfand-Shilov smoothing properties of the radially symmetric spatially homogeneous Boltzmann equation without angular cutoff. J. Differ. Equ. **256**(2), 797–831 (2014)
49. Lozanov-Crvenković, Z., Perišić, D.: Hermite expansions of elements of Gelfand-Shilov spaces in quasianalytic and non quasianalytic case. Novi Sad J. Math. **37**(2), 129–147 (2007)
50. Lozanov–Crvenković, Z., Perišić, D., Tasković, M., Gelfand-Shilov spaces, structural and kernel theorems. arXiv:0706.2268v2
51. Molahajloo, S., Okoudjou, K.A., Pfander, G.E.: Boundedness of multilinear pseudodifferential operators on modulation spaces. J. Fourier Anal. Appl. **22**(6), 1381–1415 (2016)
52. Nicola, F., Rodino, L.: Global Pseudo-Differential Calculus on Euclidean Spaces. Pseudo-Differential Operators. Theory and Applications, vol. 4. Birkhäuser, Basel (2010)
53. Petche, H.-J.: Generalized functions and the boundary values of holomorphic functions. J. Fac. Sci. Univ. Tokyo, Sec. IA **31**(2), 391–431 (1984)
54. Pilipović, S.: Tempered ultradistributions. Bollettino della Unione Matematica Italiana **7**(2-B), 235–251 (1988)
55. Pilipović, S., Teofanov, N.: Wilson bases and ultra-modulation spaces. Math. Nachr. **242**, 179–196 (2002)
56. Pilipović, S., Teofanov, N.: Pseudodifferential operators on ultra-modulation spaces. J. Funct. Anal. **208**, 194–228 (2004)
57. Pilipović, S., Teofanov, N., Toft, J.: Micro-local analysis in Fourier Lebesgue and modulation spaces, II. J. Pseudo-Differ. Oper. Appl. **1**, 341–376 (2010)
58. Prangoski, B.: Pseudodifferential operators of infinite order in spaces of tempered ultradistributions. J. Pseudo Differ. Oper. Appl. **4**(4), 495–549 (2013)
59. Prangoski, B.: Laplace transform in spaces of ultradistributions. Filomat **27**(5), 747–760 (2013)
60. Ramanathan, J., Topiwala, P.: Time-frequency localization via the Weyl correspondence. SIAM J. Math. Anal. **24**(5), 1378–1393 (1993)
61. Rodino, L.: Linear Partial Differential Operators in Gevrey Spaces. World Scientific, Singapore (1993)

62. Royer, A.: Wigner function as the expectation value of a parity operator. Phys. Rev. A **15**(2), 449–450 (1977)
63. Schwartz, L.: Théorie des distributions. Hermann, Paris (1950–1951)
64. Shifman, M. (ed.): Felix Berezin: Life and Death of the Mastermind of Supermathematics. World Scientific, Singapore (2007)
65. Shubin, M.A.: Pseudodifferential Operators and Spectral Theory, 2nd edn. Springer, Berlin (2001)
66. Sjöstrand, J.: An algebra of pseudodifferential operators. Math. Res. Lett. **1**, 185–192 (1994)
67. Slepian, D.: Some comments on Fourier analysis, uncertainty and modeling. SIAM Rev. **25**(3), 379–393 (1983)
68. Teofanov, N.: Ultradistributions and time-frequency analysis. In: Boggiatto, P., Rodino, L., Toft, J., Wong, M.W. (eds.)Operator Theory: Advances and Applications, vol. 164, pp. 173–191. Birkhäuser, Basel (2006)
69. Teofanov, N.: Modulation spaces, Gelfand-Shilov spaces and pseudodifferential operators. Sampl. Theory Signal Image Process **5**(2), 225–242 (2006)
70. Teofanov, N.: Gelfand-Shilov spaces and localization operators. Funct. Anal. Approx. Comput. **7**(2), 135–158 (2015)
71. Teofanov, N.: Continuity and Schatten-von Neumann properties for localization operators on modulation spaces. Mediterr. J. Math. **13**(2), 745–758 (2016)
72. Teofanov, N.: Bilinear localization operators on modulation spaces. J. Funct. Spaces (2018). https://doi.org/10.1155/2018/7560870
73. Teofanov, N.: Continuity properties of multilinear localization operators on modulation spaces. In: Boggiatto, P., Cordero, E., Feichtinger, H., de Gosson, M., Nicola, F., Oliaro, A., Tabacco, A. (eds.) Landscapes of Time-frequency Analysis, in preparation
74. Toft, J.: Continuity properties for modulation spaces with applications to pseudo-differential calculus, I. J. Funct. Anal. **207**, 399–429 (2004)
75. Toft, J.: Continuity properties for modulation spaces with applications to pseudo-differential calculus, II. Ann. Global Anal. Geom. **26**, 73–106 (2004)
76. Toft, J.: Continuity and Schatten properties for pseudo-differential operators on modulation spaces. In: Toft, J., Wong, M.W., Zhu, H. (eds.) Operator Theory Advances and Applications, vol. 172, pp. 173–206. Birkhäuser, Basel (2007)
77. Toft, J.: Continuity and Schatten properties for Toeplitz operators on modulation spaces. In: Toft, J., Wong, M.W., Zhu, H. (eds.) Operator Theory Advances and Applications, vol. 172, pp. 313–328. Birkhäuser, Basel (2007)
78. Toft, J.: The Bargmann transform on modulation and Gelfand-Shilov spaces, with applications to Toeplitz and pseudo-differential operators. J. Pseudo Differ. Oper. Appl. **3**(2), 145–227 (2012)
79. Toft, J.: Multiplication properties in Gelfand-Shilov pseudo-differential calculus. In: Molahajloo, S., Pilipovi'c, S., Toft, J., Wong, M.W. (eds.) Pseudo-Differential Operators, Generalized Functions and Asymptotics. Operator Theory: Advances and Applications, vol. 231, pp. 117–172. Birkhäuser, Basel (2013)
80. Toft, J.: Images of function and distribution spaces under the Bargmann transform. J. Pseudo Differ. Oper. Appl. **8**(1), 83–139 (2017)
81. Toft, J.: Matrix parameterized pseudo-differential calculi on modulation spaces. In: Oberguggenberger, M., Toft, J., Vindas, J., Wahlberg, P. (eds.) Generalized Functions and Fourier Analysis. Operator Theory: Advances and Applications, vol. 260. Birkhäuser, Basel (2017)
82. Toft, J.: Continuity and compactness for pseudo-differential operators with symbols in quasi-Banach spaces or Hörmander classes. Anal. Appl. (Singap.) **15**(3), 353–389 (2017)
83. Toft, J.: Schatten properties, nuclearity and minimality of phase shift invariant spaces. Applied Comput. Harmon. Anal. https://doi.org/10.1016/j.acha.2017.04.003
84. Toft, J., Johansson, K., Pilipović, S., Teofanov, N.: Sharp convolution and multiplication estimates in weighted spaces. Anal. Appl. **13**(5), 457–480 (2015)
85. Toft, J., Khrennikov, A., Nilsson, B., Nordebo, S.: Decompositions of Gelfand-Shilov kernels into kernels of similar class. J. Math. Anal. Appl. **396**(1), 315–322 (2012)

86. Treves, F.: Topological Vector Spaces, Distributions and Kernels. Academic Press, New York (1967)
87. Weisz, F.: Multiplier theorems for the short-time Fourier transform. Integral Equ. Oper. Theory **60**(1), 133–149 (2008)
88. Wong, M.W.: Weyl Transforms. Springer, Berlin (1998)
89. Wong, M.W.: Wavelet Transforms and Localization Operators. Operator Theory: Advances and Applications, vol. 136. Birkhäuser Verlag, Basel (2002)

Printed in the United States
By Bookmasters